LAYER TOOLBAR

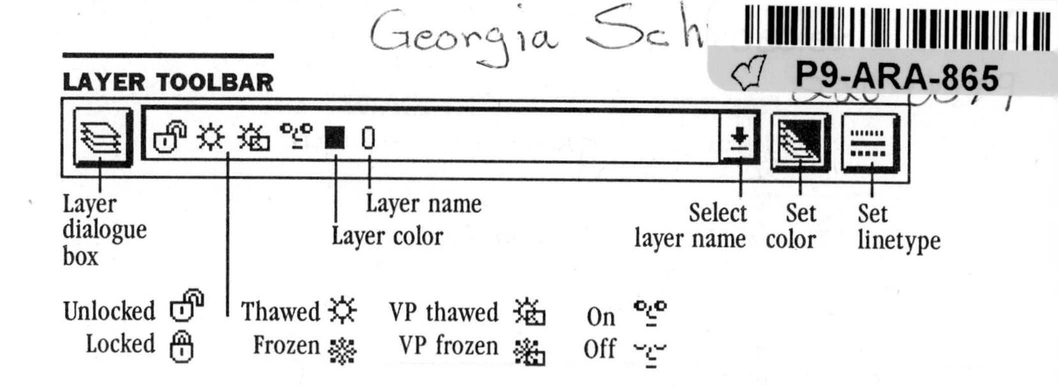

LINETYPE TOOLBAR

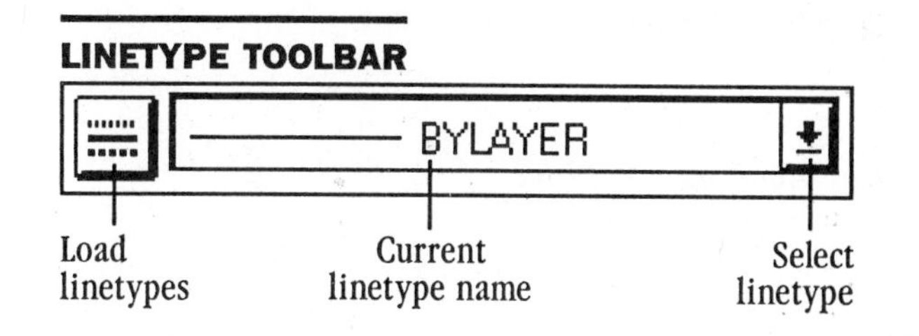

FILTER MODES

Logical operators:

*	Equal to any value.
=	Equal.
!=	Not equal.
<	Less than.
>	Greater than.
<=	Less than or equal.
>=	Greater than or equal.

Grouping operators:

****BEGIN**	Begin group.
AND	Intersection.
OR	Union.
NOT	Exclude.
XOR	Exclusive or.
****END**	End of group.

FLOATING TOOLBOX

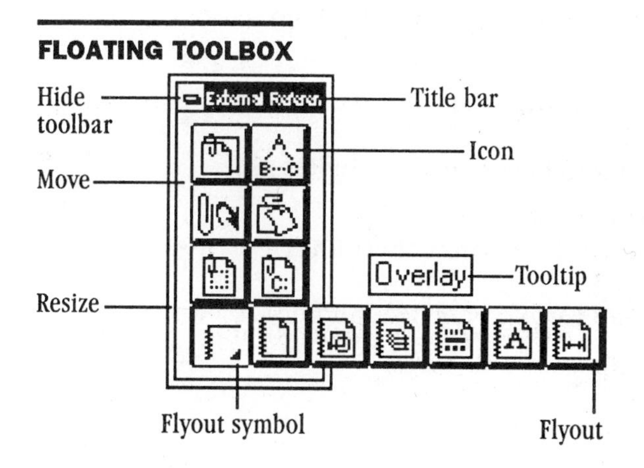

TEXT ALIGNMENT MODES

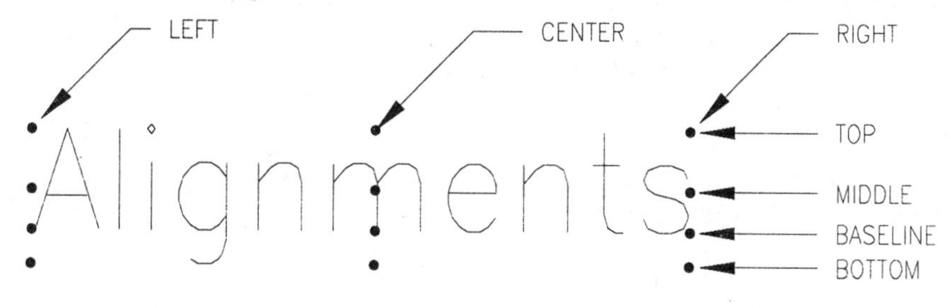

The Illustrated AutoCAD Quick Reference

Release 13 for Windows

Ralph Grabowski

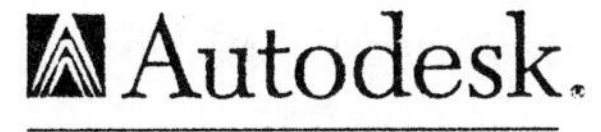

Press

I(T)P™ An International Thomson Publishing Company

Albany • Bonn • Boston • Cincinnati • Detroit • London • Madrid
Melbourne • Mexico City • New York • Pacific Grove • Paris • San Francisco
Singapore • Tokyo • Toronto • Washington

NOTICE TO THE READER

Cover Design: **Michael Speke**
Book Design: **Ralph H Grabowski** *(ralphg@haven.uniserve.com)*
Typesetting: **XYZ Publishing, Ltd.** *(CompuServe 72700,3205)*

DELMAR STAFF

Publisher: **Michael McDermott**
Acquisitions Editor: **Mary Beth Ray** *(CompuServe 73234,3664)*
Project Developmental Editor: **Jenna Daniels** *(CompuServe 76433,1677)*
Production Coordinator: **Andrew Crouth** *(CompuServe 74507,250)*
Art and Design Coordinator: **Lisa Bower**
Publishing Assistant: **Karianne Simone** *(CompuServe 76433,1702)*

Printed in the United States of America

2 3 4 5 6 7 8 9 10 XXX 01 00 99 98 97 96

Library of Congress Cataloging-in-Publication Data

Grabowski, Ralph.
The illustrated AutoCAD quick reference for Windows Release 13/ Ralph Grabowski.
p. cm.
ISBN 0-8273-7149-7
1. Computer graphics. 2. AutoCAD for Windows I. Title.
T385.G69245 1995
620'.0042'02855369-dc20 94-43748
CIP

Table of Contents

00 *Command is new to AutoCAD Release 13; or, the command has been renamed from Release 12.*

Command is specific to the Windows version of AutoCAD.

Dimension

E

F

G

H

I

L

M

N

O

P

Q

R

S

T

U

FOR MORE INFORMATION, CONTACT:

Delmar Publishers
3 Columbia Circle, Box 15015
Albany, New York 12212-5015

International Thomson Publishing Europe
Berkshire House 168-173
High Holborn
London, WC1V 7AA
England

Thomas Nelson Australia
102 Dodds Street
South Melbourne, 3205
Victoria
Australia

Nelson Canada
1120 Birchmont Road
Scarborough, Ontario
M1K 5G4
Canada

International Thomson Editores
Campos Eliseos 385, Piso 7
Col Polanco
11560 Mexico D F Mexico

International Thomson Publishing, GmbH
Konigswinterer Strasse 418
D-53227 Bonn
Germany

International Thomson Publishing Asia
221 Henderson Road
#05-10 Henderson Building
Singapore 0315

International Thomson Publishing – Japan
Hirakawacho Kyowa Building, 3F
2-2-1 Hirakawacho
Chiyoda-ku, Tokyo 102
Japan

How to Use This Book

The *Illustrated AutoCAD Quick Reference (Windows edition)* presents concise facts about all commands found in AutoCAD Release 13 for Windows. The clear format of this reference book demonstrates each command starting on its own page, plus these exclusive features:

- The 30 commands that are undocumented or underdocumented by Autodesk in the Release 13 manuals.
- Ten "Quick Start" mini-tutorials that help you get started quicker.
- Over 100 definitions of acronyms and hard-to-understand terms.
- More than 600 context-sensitive tips.
- All system variables, including those not listed by the **SetVar** command, in Appendix A.
- Obsolete commands and features that no longer work in Release 13, in Appendix B.

Each command includes the following information:

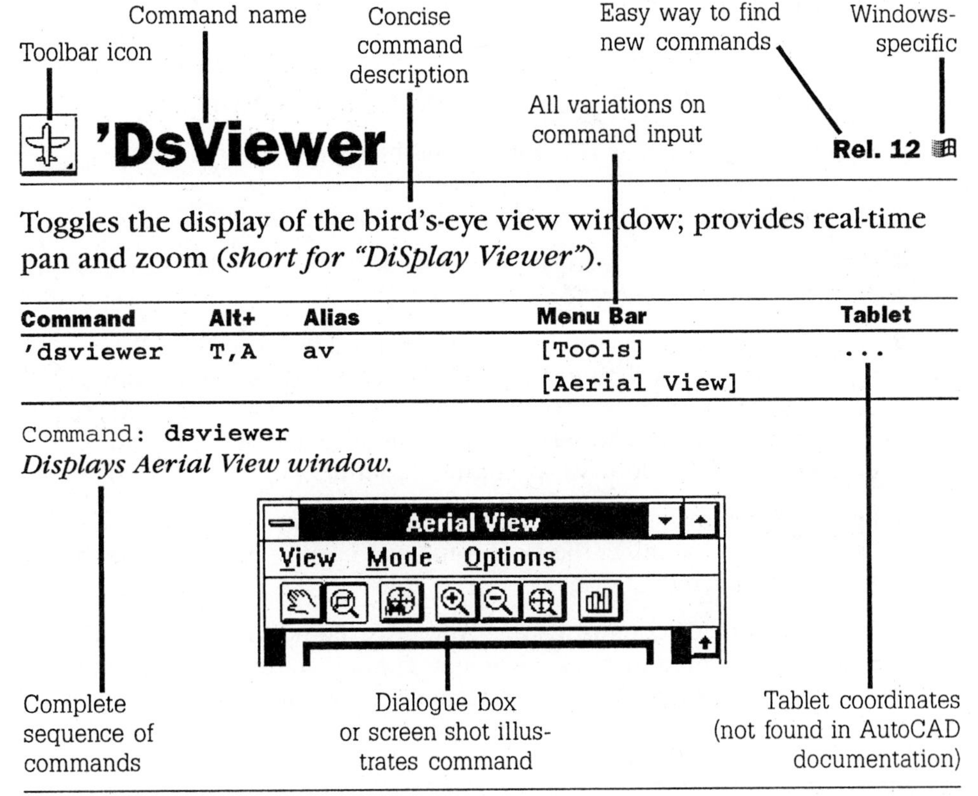

COMMAND NAME & INPUT OPTIONS

The name of the command is in mixed upper and lower case, such as **AseSqlEd**, to help understand the construction of the command name, which tend to be condensed.

All alternative methods of command input are listed for each command:

- Alternate command name spelling, such as **Donut** and **Doughnut**.
- ' (*the apostrophe prefix*) indicates transparent commands, such as **'BlipMode**. The list of transparent commands in this book is more accurate than found in the Release 13 documentation.
- Aliases, such as **L** for the **Line** command.
- Shortcut keystrokes, such as **[Alt]+T,A** for the **DsViewer** command.
- The ⊞ icon indicates AutoCAD commands specific to the Windows version.
- Side Menu picks, such as **[DRAW 1] [Line:]** for the **Line** command.
- Pull-down menu picks, such as **[Construct] [Region]** for the **Region** command.
- Table menu coordinates, such as **N 1** for the **Hide** command.
- Control-key combinations, such as **[Ctrl]+E** for the **Isoplane** toggle.
- Function keys, such as **[F7]** for the **Grid** toggle.

The brief command description includes the following notes:

- Explanation of condensed command names; for example, the **VpLayer** command name is short for ViewPort LAYER.
- "External" commands are defined by an AutoLISP, ADS, or ARx routine, rather than being part of the AutoCAD core. These commands do not operate when AutoCAD cannot access the external routines. For example, the **DdModify** command is an external command defined by the DdModify.Lsp AutoLISP routine.
- Renamed commands had a different name in an earlier release of AutoCAD; for example, the **Box** command used to be known as the **SolBox** command.
- "Undocumented" commands that Autodesk did not document in the Release 13 printed manuals or on-line documentation. For example, the **DlxHelp** command is not documented by Autodesk.

VERSION & RELEASE NUMBERS

The version or release number indicates when the command first appeared in AutoCAD, such as **Ver. 1.4** or **Rel. 13**. This is useful when working with older versions of AutoCAD. See Appendix B for the list of commands removed from AutoCAD up to and including Release 13.

RELATED COMMANDS & VARIABLES

Following the **Command Options** section, each command includes one or more of the following, where applicable:

- Related AutoCAD commands
- Related AutoLISP programs
- Related Autodesk programs
- Related system variables
- Related dimension variables
- Related Windows commands
- Related environment variables
- Related files
- Related blocks
- Input options

DEFINITIONS & TIPS

Many commands include one or more tips that help you use the command more efficiently or warn you of the command's limitations. Several commands include a list of definitions of acronyms and jargon words.

OPERATING SYSTEM

This edition of the *Illustrated AutoCAD Quick Reference* is specific to the Windows version of AutoCAD Release 13. An edition of this book (*ISBN 0-8273-6645-0*) is available for the DOS version, which has a different command structure and additional commands not found in the Windows version.

Ralph Grabowski
Abbotsford, British Columbia
March 7, 1995

Email: ralphg@haven.uniserve.com
or 72700,3205 via CompuServe

'About

Rel. 12

Displays the AutoCAD version, serial number, and the Acad.Msg file.

Command	Alt+	Side Menu	Menu Bar	Tablet
'about	H, A	[HELP]	[Help]	...
		[About]	[About AutoCAD]	

Command: **about**
Displays dialogue box:

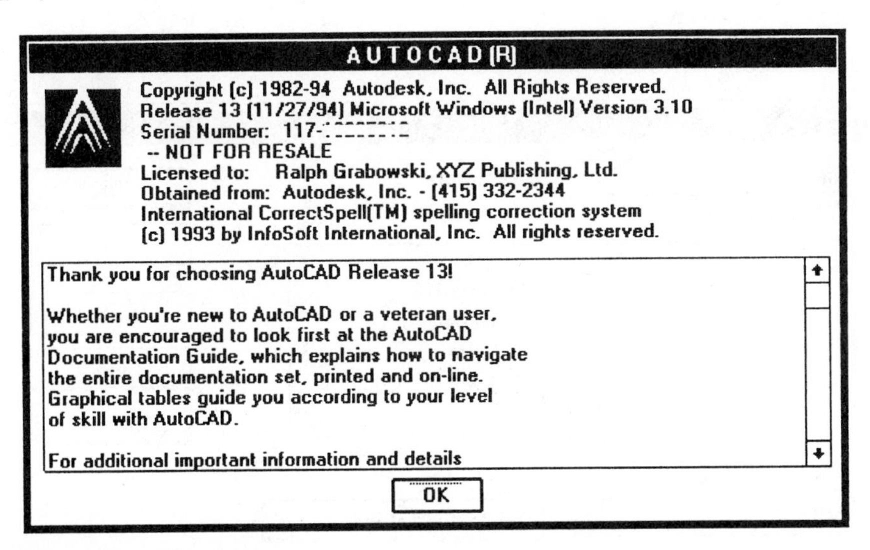

COMMAND OPTIONS

None

RELATED AUTOCAD COMMANDS

- **Status** Display information about the drawing and environment.
- **Stats** Display information about the rendering environment.

RELATED SYSTEM VARIABLES

- **_PkSer** The AutoCAD software serial number.
- **_Server** Network authorization code.

RELATED FILE

- **Acad.Msg** The ASCII text file displayed by the **About** command.

TIPS

- To change the message displayed by the **About** command, edit the \Acad13\Common\Support**Acad.Msg** file with a text editor.

- "**Serial**" is an alias for the **_PkSer** system variable.

AcisIn

Rel. 13

Imports an SAT file (short for Save As Text; an ASCII-format ACIS file) into the drawing, then creates 3D solids, 2D regions, and bodies.

Command	Alt+	Side Menu	Menu Bar	Tablet
acisin	F,I	[FILE]	[File]	...
		[IMPORT]	[Import]	
		[SATin:]		

Command: **acisin**

*Displays **Select ACIS File** dialogue box.*

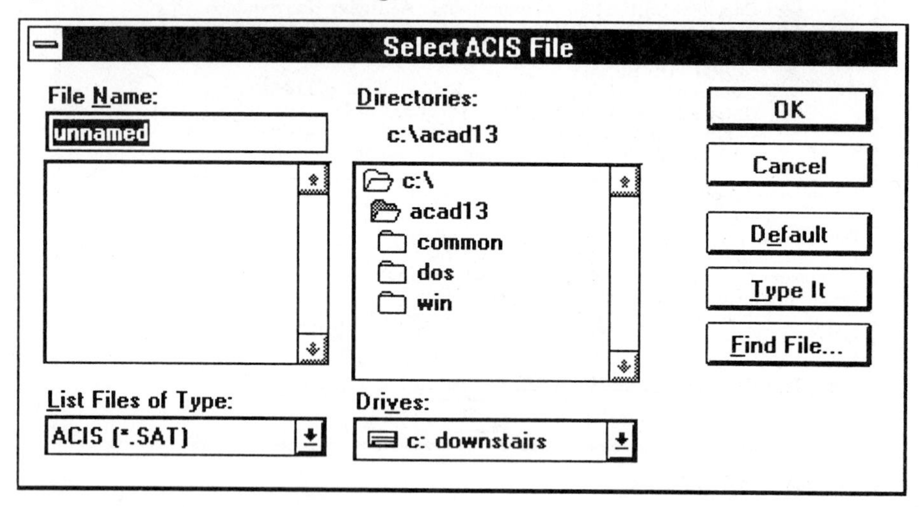

COMMAND OPTIONS

None

RELATED AUTOCAD COMMANDS

- **AcisOut** Exports ACIS objects (3D solids, 2D regions, and bodies) to a SAT file.
- **AMEconvert** Converts AME v2.0 and v2.1 solid models and regions into ACIS solids.

RELATED FILE

- ***.SAT** The ASCII format of ACIS model files.

TIP

- ACIS is short for the "Andrew, Charles, Ian's Solids," the solids modeling engine from Spatial Technologies used in Release 13 and Designer.

AcisOut

Rel. 13

Exports AutoCAD 3D solids, 2D regions, and bodies to a SAT file.

Command	Alt+	Side Menu	Menu Bar	Tablet
acisout	F, E	[FILE] [EXPORT] [ACISout:]	[File] [Export]	...

```
Command: acisout
Selection objects: [pick]
```

COMMAND OPTIONS

None

RELATED AUTOCAD COMMANDS

- **AcisIn** Imports a SAT file and creates 3D solids, 2D regions, and bodies.
- **StlOut** Exports ACIS solid model in STL format.
- **3dsOut** Exports ACIS solid models as 3D faces.

RELATED FILE

- ***.SAT** The ASCII format of ACIS model files. Sample output :

```
105 31 1 0
body $1 $2 $-1 $-1 #
f_body-lwd-attrib $-1 $3 $-1 $0 #
lump $4 $-1 $5 $0 #
ref_vt-lwd-attrib $-1 $-1 $1 $0 $6 $7 #
...
color-adesk-attrib $-1 $29 $-1 $23 256 #
vertex $-1 $23 $30 #
intcurve-curve $-1 0 { surfintcur nubs 3 periodic 41
...
2 18.996755648173917 2
19.187154530129618 3
8.2317032253755595 6.7400931724018447 0
8.2317032253755595 6.7400931724018447 -0.0634662939852331
8.2294598728158608 6.7508848507648285 -0.12180055063975684
...

} #
epar-lwd-attrib $-1 $-1 $26 $23 #
point $-1 8.2317032253755595 6.7400931724018447 0 #
```

TIP

- **AcisOut** does not export objects that are not 3D solids, 2D regions, or bodies.

Ai_Box, *etc.* Rel. 11

Draws nine basic 3D surface objects from polygon meshes: box, pyramid, wedge, dome, dish, mesh, sphere, cone, and torus (*an external command in 3d.Lsp*).

Command	Alias	Side Menu	Pulldown	Tablet
ai_box	...	[DRAW 2] [SURFACES] [Box:]	[Draw] [Surfaces] [3D Objects]	...
ai_cone	...	[DRAW 2] [SURFACES] [Cone:]	[Draw] [Surfaces] [3D Objects]	...
ai_dish	...	[DRAW 2] [SURFACES] [Dish:]	[Draw] [Surfaces] [3D Objects]	...
ai_dome	...	[DRAW 2] [SURFACES] [Dome:]	[Draw] [Surfaces] [3D Objects]	...
ai_mesh	...	[DRAW 2] [SURFACES] [Mesh:]	[Draw] [Surfaces] [3D Mesh]	...
ai_pyramid	...	[DRAW 2] [SURFACES] [Pyramid:]	[Draw] [Surfaces] [3D Objects]	...
ai_torus	...	[DRAW 2] [SURFACES] [Torus:]	[Draw] [Surfaces] [3D Objects]	...
ai_wedge	...	[DRAW 2] [SURFACES] [Wedge:]	[Draw] [Surfaces] [3D Objects]	...

```
Command: ai_box
Corner of box:
Length:
Cube/<Width:

Command: ai_cone
Base center point:
Diameter/<radius> of base:
Diameter/<radius> of top:
Height:
Number of segments <16>:
```

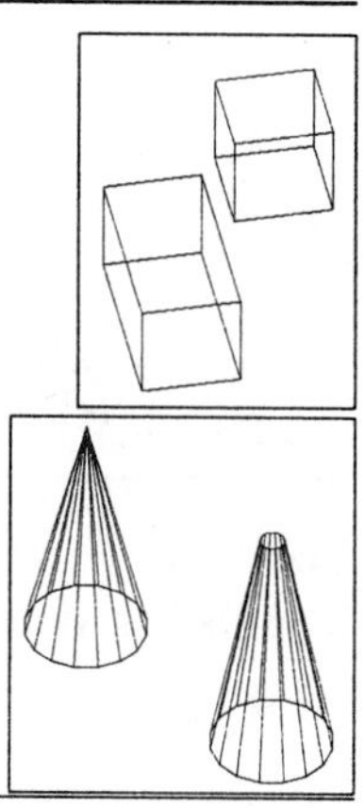

Command: **ai_dish**
Center of dish:
Diameter/<radius>:
Number of longitudinal segments <16>:
Number of latitudinal segments <16>:

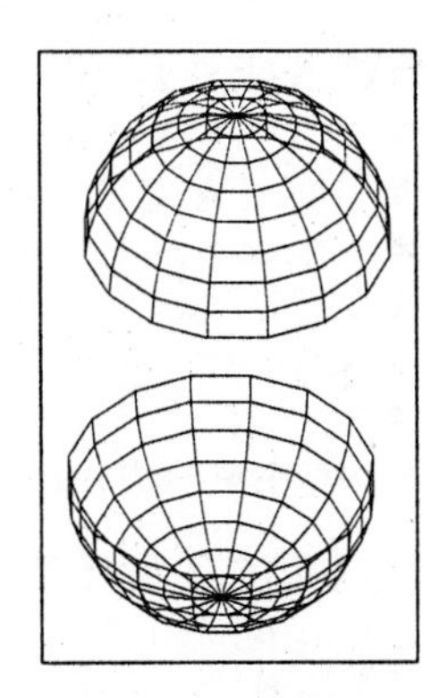

Command: **ai_dome**
Center of dome:
Diameter/<radius>:
Number of longitudinal segments <16>:
Number of latitudinal segments <16>:

Command: **ai_mesh**
First corner:
Second corner:
Third corner:
Fourth corner:
Mesh M size:
Mesh N size:

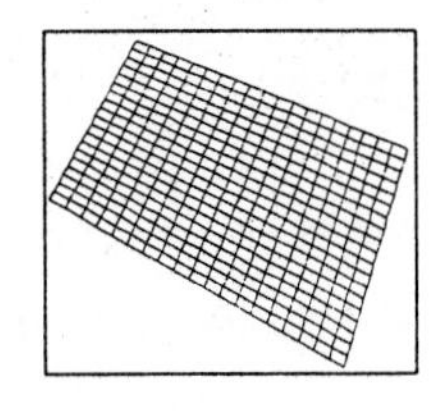

Command: **ai_pyramid**
First base point:
Second base point:
Third base point:
Tetrahedron/<Fourth base point>:
Ridge/Top/<Apex point>:

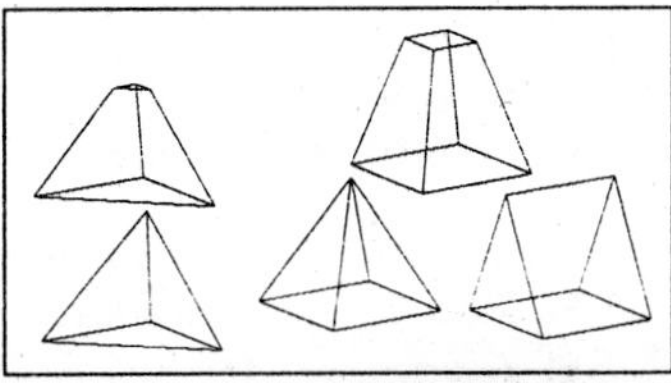

Command: **ai_sphere**
Diameter/<radius>:
Number of longitudinal segments <16>:
Number of latitudinal segments <16>:

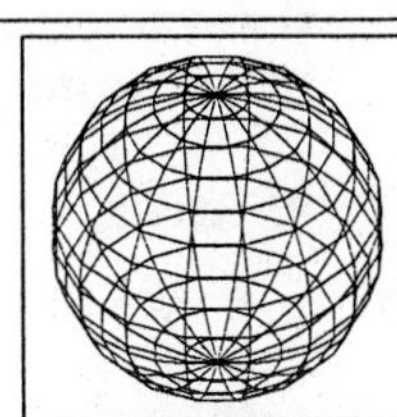

Command: **ai_torus**
Center of torus:
Diameter/<radius> of torus:
Diamter/<radius> of tube:
Segments around tube circumference <16>:
Segments around torus circumference <16>:

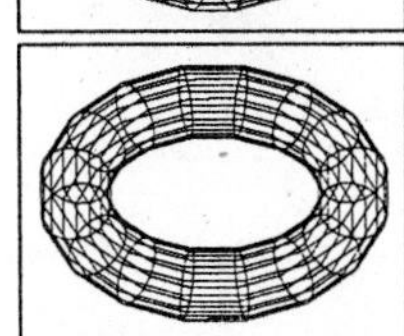

Command: **ai_wedge**
Corner of wedge:
Length:
Width:
Height:
Rotation angle about Z axis:

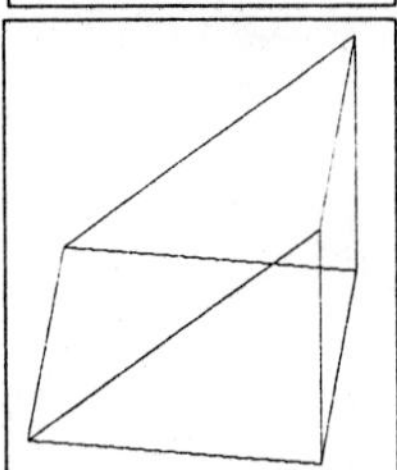

RELATED AUTOCAD COMMANDS

- **3D** Displays a dialogue box with all nine 3D surface objects.
- **Box** Draws a 3D solid box.
- **Cone** Draws a 3D solid cone.
- **Cylinder** Draws a 3D solid cylinder.
- **Sphere** Draws a 3D solid sphere.
- **Torus** Draws a 3D solid torus.
- **Wedge** Draws a 3D solid wedge.

RELATED SYSTEM VARIABLES

- **SurfU** Surface mesh density in the m-direction.
- **SurfV** Surface mesh density in the n-direction.

TIPS

- You canot perform Boolean operations (intersect, subtract, and union) on 3D surface objects.

- You cannot convert 3D surface objects into 3D solid objects.

- Convert 3D solid objects into 3D surface objects by exporting with the **3dsIn** command, then importing with the **3dsOut** command.

- By default, 3D solid models look sparse; increase the mesh density to 16 with the **IsoLines** system variable.

- Varients of objects drawn by the **Ai_** commands:
  - **Box** Draws rectangular box or cube.
  - **Cone** Draws pointy cone or truncated cone.
  - **Pyramid** Draws pyramid and truncated pyramid, tetrahedron and truncated tetrahedron, and roof shape.
  - **Torus** Draws donut or football.

- Mesh m- and n-sizes are limited to values between 2 and 256.

Align

Rel. 12

Moves, transforms, and rotates objects in three dimensions (*an external command in Geom3d.Exp*).

Command	Alias	Side Menu	Menu Bar	Tablet
align	...	[MODIFY] [Align:]	...	Y 19

```
Command: align
Select objects: [pick]
Select objects: [Enter]
1st source point: [pick]
1st destination point: [pick]
2nd source point: [pick]
2nd destination point: [pick]
3rd source point: [pick]
3rd destination point: [pick]
```

COMMAND OPTIONS

None

RELATED AUTOCAD COMMANDS

- **Move** Performs a move in two dimensions.
- **Mirror3d** Mirrors objects in three dimensions.
- **Rotate3d** Rotates objects in three dimensions.

TIPS

- Enter the first pair of points to define the move distance:

```
1st source point: [pick]
1st destination point: [pick]
2nd source point: [Enter]
```

- Enter two pairs of points to define a 2D (or 3D) transformation and rotation:

```
1st source point: [pick]
1st destination point: [pick]
2nd source point: [pick]
2nd destination point: [pick]
3rd source point: [Enter]
<2d> or 3d transformation:
```

- The third pair defines the 3D transformation.

AmeConvert

Rel. 13

Converts solid models and regions created by AME v2.0 and v2.1 (from Release 12) into ACIS solids models.

Command	Alias	Side Menu	Menu Bar	Tablet
ameconvert	...	[DRAW 2] [SOLIDS] [AMEconv:]	...	...

```
Command: ameconvert
Select objects: [pick]
Processing 1 of 17 Boolean operations. | / - \
```

COMMAND OPTIONS

None

RELATED AUTOCAD COMMAND

- **AcisIn** Imports ACIS models from a SAT file.

TIPS

- After conversion, the AME model remains in the drawing in the same location as the ACIS model. Erase, if necessary.
- AME holes may become blind holes in ACIS.
- AME fillets and chamfers may be placed higher or lower in ACIS.
- Once the **AMEconvert** command converts a Release 12 PADL drawing into Release 13 ACIS model, it cannot be converted back to PADL format.
- Old AME models are stored in Release 13 as an anonymous Block Reference.

DEFINITIONS

ACIS

- Andy, Charles, and Ian's Solids.
- The solids modeling system used in Release 13.

AME

- Advanced Modeling Extension.
- AutoCAD's solids modeling module.

PADL

- Parts and Description Language.
- The solids modeling system used in Releases 10 through 12.

'Aperture

Ver. 1.3

Sets the size, in pixels, of the object snap target height (or box cursor).

Command	Alt+	Side Menu	Menu Bar	Tablet
'aperture	O,O,R	...	...	...

```
Command: aperture
Object snap target height (1-50 pixels) <10>:
```

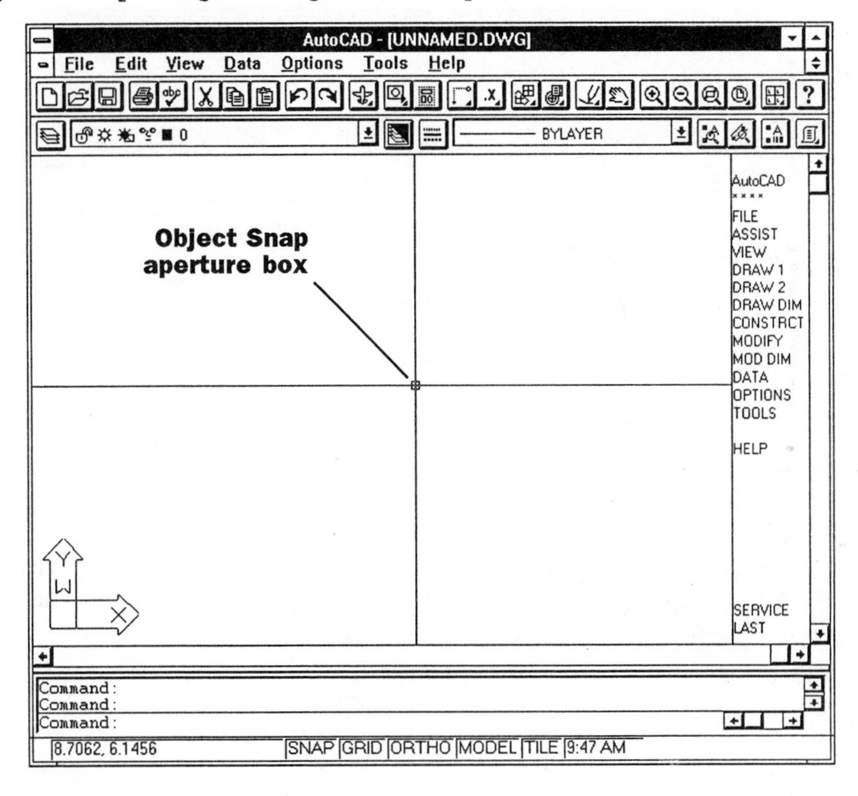

RELATED AUTOCAD COMMANDS

- **DdOSnap** Sets the aperture size interactively.
- **DdSelect** Sets the size of the object selection pickbox.
- **OSnap** Sets the object snap modes.

RELATED SYSTEM VARIABLES

- **Aperture** Contains the current target height:
  - **1** Minimum size.
  - **10** Default size.
  - **50** Maximum size.

'AppLoad

Rel. 12

Creates a list of AutoLISP, ADS, and ARx applications to load (*an external file in AcadApp.Exe; short for APPlication LOADer*).

Command	Alt+	Side Menu	Menu Bar	Tablet
'appload	T,P	[TOOLS]	[Tools]	V 25
		[Appload:]	[Applications]	

Command: **appload**

Display dialogue box.

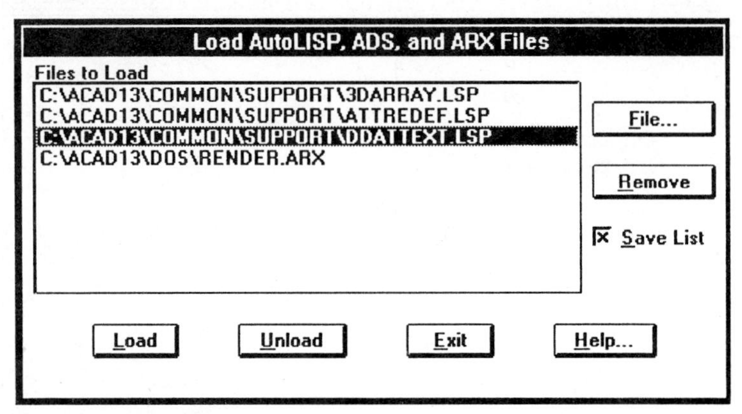

COMMAND OPTIONS

File Displays file dialogue box to select LSP (*AutoLISP*), EXE (*ADS*), and ARX and DLL (*ARx*) files.

Remove Removes selected filenames from the list.

Save List Saves the list to file AppLoad.Dfs.

Load Loads all or selected files into AutoCAD.

Unload Unloads all or selected files out of AutoCAD.

Exit Exits the dialogue box.

RELATED AUTOCAD COMMAND

- **Arx** Lists ARx programs currently loaded

RELATED AUTOLISP FUNCTIONS

- **(load)** Loads an AutoLISP program.
- **(xload)** Loads an ADS program.
- **(autoload)** Predefines commands to load AutoLISP program.
- **(autoxload)** Predefines commands to load ADS program.

RELATED FILES

- **AppLoad.Dfs** Contains list of programs to load.
- ***.Lsp,*.Exe,*.Arx** AutoLISP, ADS, and ARx programs loaded by **AppLoad**.

Arc

Ver. 1.0

Draws a 2D arc of less than 360 degrees, by eleven different methods.

Command	Alias	Side Menu	Menu Bar	Tablet
arc	a	[DRAW 1] [Arc:]	[Draw] [Arc >]	M 9

```
Command: arc
Center/<Start point>: [pick]
Center/End/<Second point>: [pick]
End point: [pick]
```

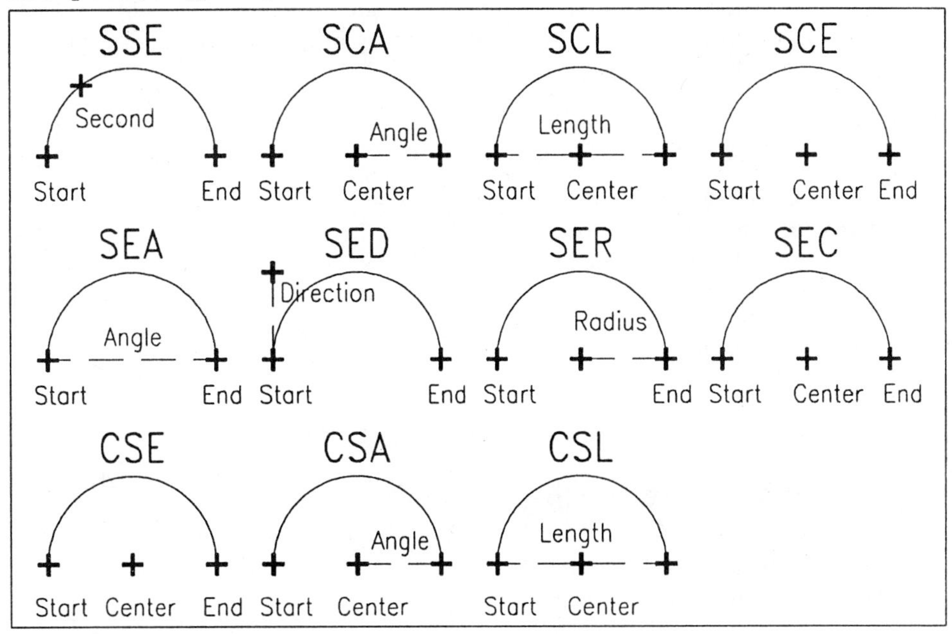

COMMAND OPTIONS

 *SSE arc:*

<Start point> Indicates the start point of a three-point arc (*see figure*):

<Second point> Indicates a second point anywhere along the arc.

End point Indicates the end point of the arc.

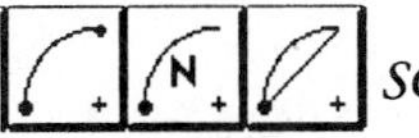 *SCE, SCA, and SCL arcs:*

<Start point> Indicates the start point of a two-point arc:

Center Indicates the center point of the arc:

Angle Indicates the arc's included angle.

Length of chord
Indicates the length of the arc's chord.
<End point> Indicates the end point of the arc.

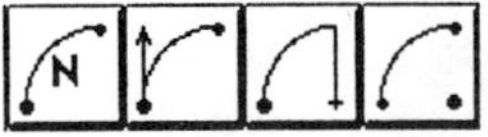 *SEA, SED, SER, and SEC arcs:*

<Start point> Indicates the start point of a two-point arc:
End Indicates the end point of the arc:
<Center point>
Indicates the arc's center point.
Angle Indicates the arc's included angle.
Direction Indicates the tangent direction from the arc's start point.
Radius Indicates the arc's radius.

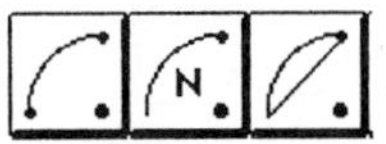 *CSE, CSA, and CSL arcs:*

Center Indicates the center point of a two-point arc:
Start point Indicates the arc's start point:
<End point> Indicates the arc's end point.
Angle Indicates the arc's included angle.
Length of chord
Indicates the length of the arc's chord.

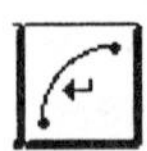 [Enter] Continues arc from endpoint of last-drawn line or arc.

RELATED AUTOCAD COMMANDS

- **Circle** Draws an arc of 360 degrees.
- **Ellipse** Draws elliptical arcs.
- **Polyline** Draws connected polyline arcs.
- **ViewRes** Controls the roundness of arcs.

RELATED SYSTEM VARIABLE

- **LastAngle** Saves the included angle of the last-drawn arc (read-only).

TIPS

- To precisely start an arc from the endpoint of the last line or arc, press [Enter] at the "Enter/<Start point>" prompt.

- You can only drag the arc during the last-entered option.

- Specifying an x,y,z-coordinate as the starting point of the arc draws the arc at the z-elevation.

- In some cases, it may be easier to draw a circle and use the **Break** and **Trim** commands to convert the circle into an arc.

Area

Ver. 1.0

Calculates the area and perimeter of areas, closed entities, or polylines.

Command	Alias	Side Menu	Menu Bar	Tablet
area	...	[ASSIST] [INQUIRY] [Area:]	...	R 1

```
Command: area
<First point>/Object/Add/Subtract: [pick]
Next point: [pick]
Next point: [Enter]
Area = 1.8398, Perimeter = 6.5245
```

COMMAND OPTIONS

<First point> Indicates the first point to begin measurement.
Entity Indicates the entity to be measured.
Add Switches to add-area mode.
Subtract Switchs to subtract-area mode.
[Enter] Indicates the end of the area outline.

RELATED AUTOCAD COMMANDS

- **DbList** Lists all information of the entire drawing.
- **List** Lists all information of the selected entity.

RELATED SYSTEM VARIABLES

- **Area** Contains the most recently calculated area.
- **Perimeter** Contains the most recently calculated perimeter.

TIPS

- AutoCAD automatically "closes the polygon" before measuring the area.

- You can specify 2D x,y-coordinates or 3D x,y,z-coordinates.

- The **Object** option returns the following information:
  - **Circle, ellipse** Area and circumference.
  - **Planar closed spline** Area and circumference.
  - **Closed polyline, polygon** Area and perimeter.
  - **Open objects** Area and length.
  - **Region** Area summed for all objects in region.
  - **Solid** Surface area.

- The area of a wide polyline is measured along its centerline; closed polylines must have only one closed area.

Array

Ver. 1.3

Creates a 2D rectangular or polar array of objects.

Command	Alias	Side Menu	Menu Bar	Tablet
array	...	[CONSTRCT] [Array:]	...	W 21

Rectangular array:

```
Command: array
Select objects: [pick]
Select objects: [Enter]
Rectangular or Polar array (R/P): r
Number of rows (—) <1>:
Number of columns (|||) <1>:
Unit cell or distance between rows (—):
Distance between columns (|||):
```

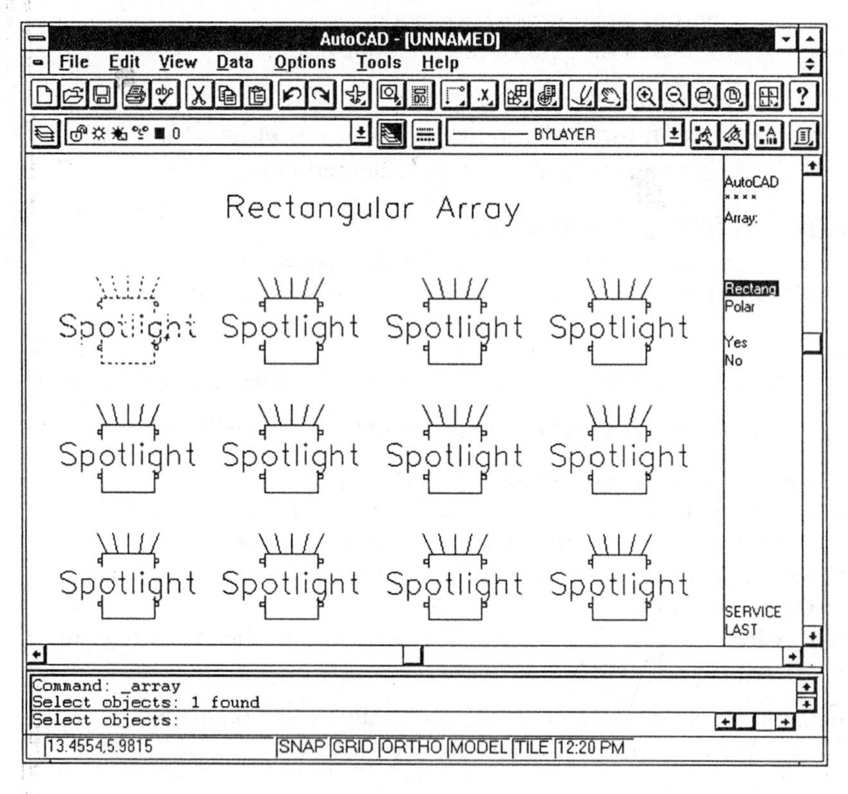

Polar array:

```
Command: array
Select objects: [pick]
Select objects: [Enter]
Rectangular or Polar array (R/P): p
Center point of array: [pick]
Number of items:
Angle to fill (+=ccw, -=cw) <360>:
Rotate objects as they are copied? <Y>
```

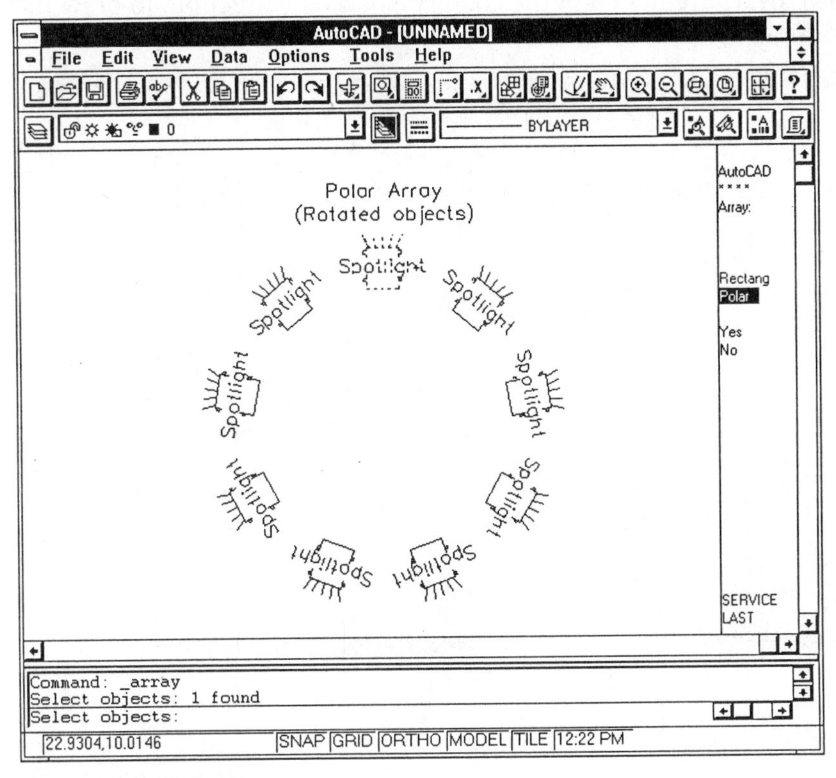

COMMAND OPTIONS

R Creates a rectangular array of the selected object.

P Creates a polar array of the selected object.

RELATED AUTOCAD COMMANDS

- **3dArray** Creates rectangular or polar array in 3D space.
- **Copy** Creates one or more copies of the selected object.
- **MInsert** Creates a rectangular block array of blocks.

RELATED SYSTEM VARIABLE

■ **SnapAng** Determines the angle of rectangular arrays.

TIPS

■ To create an array at an angle, use the **Rotation** option of the **Snap** command.

■ Rectangular array draws up in the positive x-direction, and right in the positive y-direction; to draw the array in the opposite direction, specify negative row and column distances.

■ Polar arrays are drawn in the counter-clockwise direction; to draw the array in the opposite direction, specify a negative angle.

■ Use the **Divide** or **Measure** commands to create an array along a path.

Arx

Rel. 13

Displays information regarding currently-loaded ARx programs.

Command	Alias	Side Menu	Menu Bar	Tablet
arx	...	...	...	...

```
Command: arx
Load/Popcmds/Unload/?/<eXit>: ?
What to List:  CLasses/<Commands>/Objects/Programs/Services:
```

COMMAND OPTIONS

Load Explicitly loads an ARx application.

Popcmds Specifies the name of the first subdirectory where AutoCAD looks for commands contained in ARx programs.

Unload Explicitly unloads an ARx application to recover memory.

? Lists further options:

- **CLasses** Lists the class hierarchy for ARx objects.
- **Commands** Lists the commands registered by ARx programs.
- **Objects** Lists the names of objects entered into the "system registry."
- **Programs** Lists the names of loaded ARx programs.
- **Services** Lists the names of services entered into the ARx "service dictionary."

RELATED AUTOCAD COMMAND

- **AppLoad** Loads AutoLISP, ADS, and ARx programs.

RELATED AUTOLISP FUNCTIONS

- **(arx)** Lists currently loaded ARx programs.
- **(arxload)** Loads an ARx application.
- **(autorxload)** Predefines commands that load the ARx program.
- **(arxunload)** Unloads an ARx application.
- **(load)** Loads an AutoLISP program.
- **(xload)** Loads an ADS program.
- **(ads)** Lists the ADS programs currently loaded.

RELATED FILE

- ***.Arx** ARx program files.

AseAdmin

Rel. 12

Administers links between the drawing and an external database (*short for Autocad Sql Extension ADMINistration; formerly the **AseSetDBMS, AseSetDB, AseSetTable, AseEraseTable, AseCloseTable, AseCloseDB, AseEraseDB, AseEraseDBMS, AseEraseAll, AsePost,** and **AseReloadDA** commands; an exernal command in Ase.Arx*).

Command	Alias	Side Menu	Menu Bar	Tablet
aseadmin	...	[TOOLS] [EXT DBMS] [Admin:]	...	...

```
Command: aseadmin
```

Displays dialogue box.

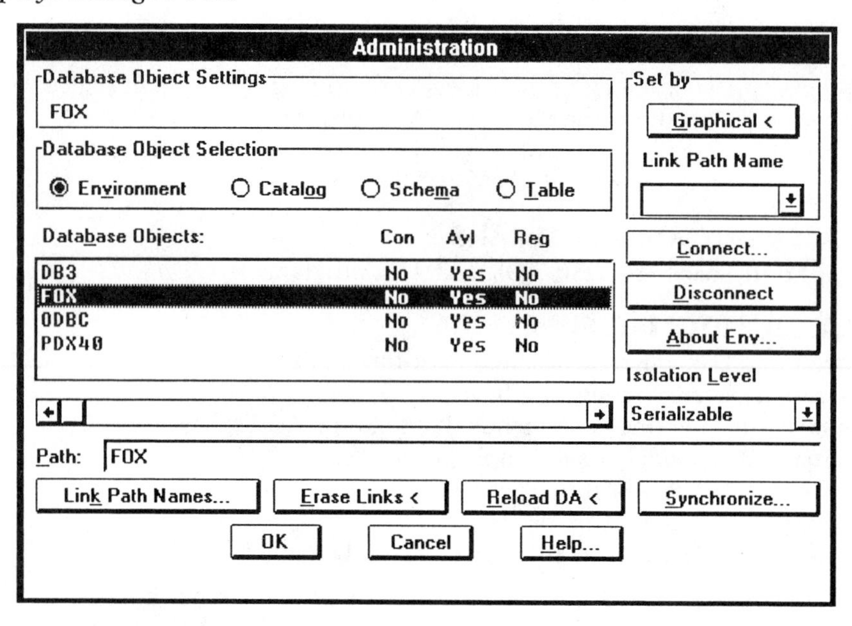

COMMAND OPTIONS

Database Object Selection:

Environment Displays list of environments.
Catalog Displays list of catalogues after connection with DBMS driver.
Schema Displays list of schemas; not supported by all databases.
Table Displays list of tables in current schema.

Database Objects:

Yes Available.

No Not available.

? Not detected by AutoCAD.

Path Name of database object or logical path name.

Set By Sets database object by:

Graphical Selecting object in drawing.

Link Path Name
Database object hierarchy with key column definitions.

Connect Loads database driver and connects it with AutoCAD; displays dialogue box:

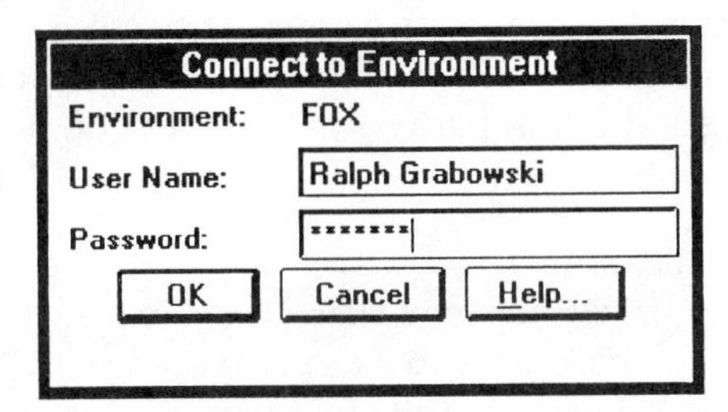

Disconnect Disconnect database driver from AutoCAD.

About Env Displays information about the ASE environment via dialogue box:

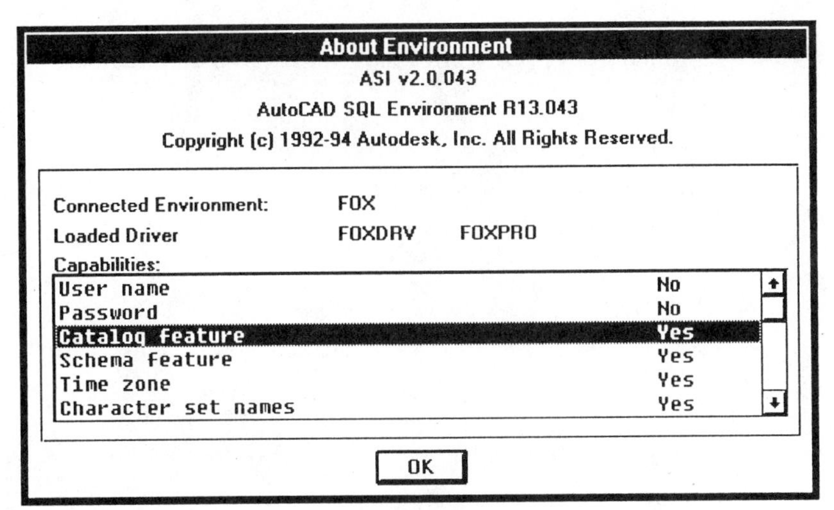

Isolation Level:

Serializable Concurrent SQL transactions have same result as sequential transactions.

Uncommitted Does not lock records in use.

Committed Locks out in-use records.

Repeatable Does not lock out selection sets being changed.

Erase Links Removes all links with selected database objects.
Reload DA Updates Displayable Attributes; deleted rows show ****.
Synchronize Resynchronizes link info between drawing and database.
Link Path Names
Displays **Link Path Names** dialogue box.

Key Selection:
- **On** Makes the selecte column a key column.
- **Off** Turns off key column status.

Link Path:
- **New** Specifies a new link path.
- **Existing** Lists the current link path.
- **Erase** Erases links to current database object.
- **Erase All** Erases all link paths and links to database objects.
- **Rename** Changes the link path; displays dialogue box:

RELATED FILES

- **Asi.Ini** ASI initialization file.
- **Asi*.Xmx** Database driver message files.
- ***.Dbf** Sample database files in \Acad13\Common\Sample\Dbf.
- **AseSmp.Dwg** Sample ASE drawing file in \Acad13\Common\Sample.
- ***.Xmx** International language message files in \Acad13\Dos\Ase\Lang.
- ***.H, *.Lib** ASI programming support files in \Acad13\Dos\Ase\Sample.

TIP

- The **Undo** command does not work reliably with ASE commands.

AseExport

Rel. 12

Exports link information to an external database file (*an external command in Ase.Arx*).

Command	Alias	Side Menu	Menu Bar	Tablet
aseexport	...	[TOOLS] [EXT DBMS] [Export:]	...	...

```
Command: aseexport
All/Environment/Catalog/Schema/<Table>/Lpn: [Enter]
Sdf/Cdf/<Native>/sKip:
Name of file:
```

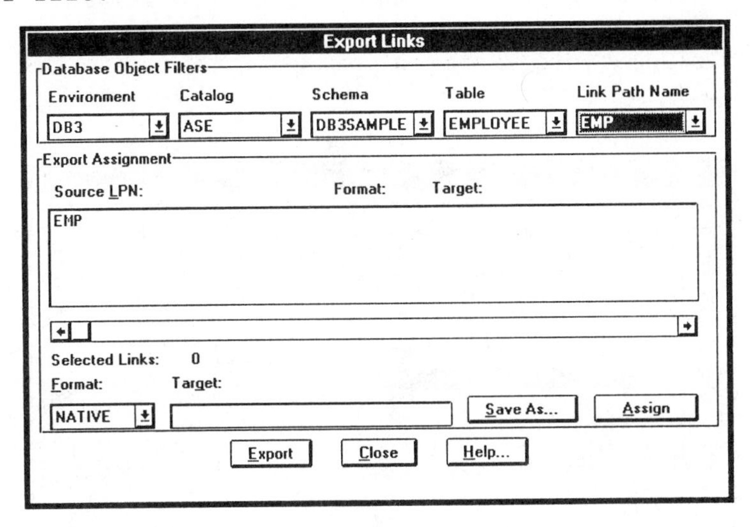

COMMAND OPTIONS

All Exports all link information.

Environment Exports environment-related link information.

Catalog Exports a catalog with links to selected objects.

Schema Exports a schema with links to selected objects.

<Table> Exports a table with links to selected objects:

Lpn Exports links for a single table registration (*LPN is short for Link Path Name*).

Sdf Space-deliminted file, 1line per link, space-padded to 16 characters.

Cdf Comma-delimited file, one line per link.

<Native> Native database file format, one file per source LPN and one record per link.

sKip Exits the command.

TIP

- Enter * to export all link information.

AseLinks

Rel. 12

Lists, edits, and displays link information between drawing and database (*formerly the **AseEditLink**, **AseViewLink**, and **AseDelLink** commands; an external command in Ase.Arx*).

Command	Alias	Side Menu	Menu Bar	Tablet
aselinks	...	[TOOLS] [EXT DBMS] [Links:]	...	...

Command: **aselinks**
All/Environment/Catalog/Schema/Table/Lpn/Object/<eXit>: **A**
Browse/Next/Prior/First/Del/delAll/Update/Rows/<eXit>: **R**
Cursor-state/Textual/Keys/<eXit>: **C**
Scrollable/Updatable/<Read-only>: **K**
Enter value for *key-column-name-n*: **[*enter value*]**
Browse/Next/Prior/First/Last/<eXit>:

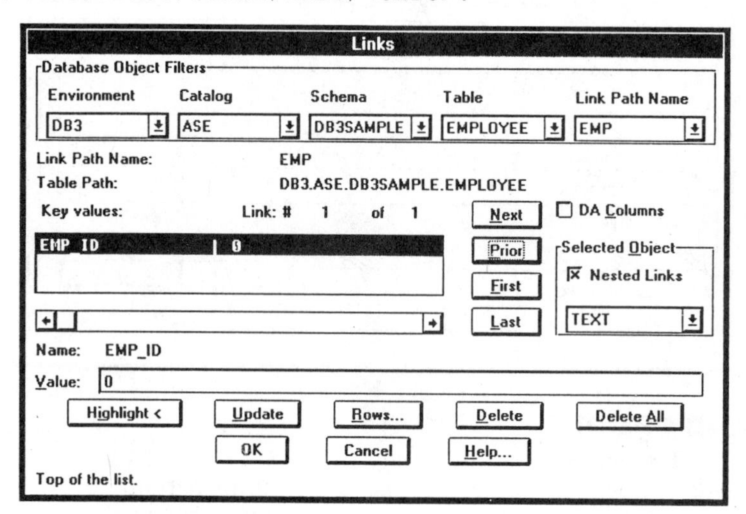

COMMAND OPTIONS

All Selects all links and displays the following options:

- **Browse** Displays all links.
- **Next** Displays next link.
- **Prior** Displays prior link.
- **First** Displays first link.
- **Last** Displays last link.
- **Del** Deletes current link.
- **delAll** Deletes all links to the selected object.
- **Update** Updates key values to the current link.

Rows Edits the current link; if a read-only cursor cannot be opened, displays the following options:

- **Textual** Selects a row or rows matching an SQL search condition.
- **Keys** Selects a row that matches the key values.
- **Cursor-state** Displays the following options:
  - **Scrollable** Makes a row the current row.
  - **Updateable** Row can be edited but not scrolled.
  - **<Read-only>** Rows can't be edited or deleted.

Environment Selects all links within the current environment.

Catalog Selects all links in the current catalogue.

Schema Selects all links in the current schema.

Table Selects all links in the current table.

Lpn Selects all links in the current link path name.

Object Filters our nested links in blocks and xrefs.

<eXit> Exits the command.

RELATED AUTOCAD COMMANDS

All ASE commands.

TIPS

■ Press **[Enter]** to the 'Enter text condition:' prompt to select all rows in the table.

■ A read-only cursor is nonscrollable.

AseRows

Rel. 12

Creates links and selection sets; displays and edits table data *(formerly the **AseAddRow, AseDelRow, AseEditRow, AseQEdit, AseQView, AseViewRow, AseMakeDA, AseMakeLink, AseQLink,** and **AseMakeDA** commands; an external command in Ase.Arx).*

Command	Alias	Side Menu	Menu Bar	Tablet
aserows	...	[TOOLS] [EXT DBMS] [Rows:]	...	...

```
Command: aserows
Settings/Insert/Cursor-state/Textual/Keys/<eXit>/Select
object: [pick]
Browse/Next/Prior/First/Last/Select/Unselect/Edit/Insert/Del/
Mlink/MDA/linKs/<eXit>: M
Enter a column name or ? for list:
Justify/Style/<Start point>:
```

*When system variable CmdDia is turned on (= 1), the **AseRows** command displays the following dialogue box:*

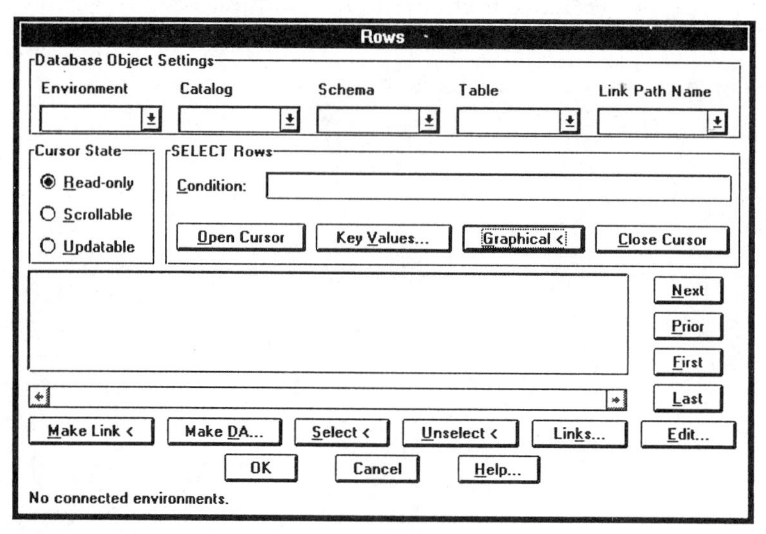

COMMAND OPTIONS

Settings Displays the following options:

- **Environment** Setss up an environment or selects a different one.
- **Catalog** Changes the catalogue.
- **Schema** Changes the current schema.
- **Table** Changes the current table.

Lpn Sets the current link path name.
Insert Inserts a new row in the current table.
Cursor-state Displays the following options:
Scrollable Makes a row the current row.
Updateable Row can be edited but not scrolled.
<Read-only> Rows cannot be edited or deleted.
Textual Selects a row or rows from a table.
Keys Selects a new row that matches key values.
Select object After you select an object, displays the following options:
Browse Displays all links.
Next Displays next link.
Prior Displays prior link.
First Displays first link.
Last Displays last link.
Select Highlights linked objects and adds them to selection set.
Unselect Removes objects from selection set.
Edit Changes the current row.
Insert Inserts row into table.
Del Deletes current link.
Mlink Makes link between row and selected object.
MDA Makes displayable attribute.
linKs Displays link information.
<eXit> Exits the command.

RELATED AUTOCAD COMMANDS

All ASE commands

TIPS

■ Press **[Enter]** to the 'Enter text condition:' prompt to select all rows in the table.

■ A read-only cursor is nonscrollable.

AseSelect

Rel. 12

Creates a selection set from rows linked to graphic and text selection sets (*an external command in Ase.Arx*).

Command	Alias	Side Menu	Menu Bar	Tablet
aseselect	...	[TOOLS] [EXT DBMS] [Select:]	...	...

```
Command: aseselect
Graphical/Textual/<eXit>: T
All/Environment/Catalog/Schema/<Table>/Lpn: L
Union/Intersect/subtractA/SubtractB/<eXit>:
```

When system variable CmdDia is turned on (= 1), the ***AseSelect*** *command displays the following dialogue box:*

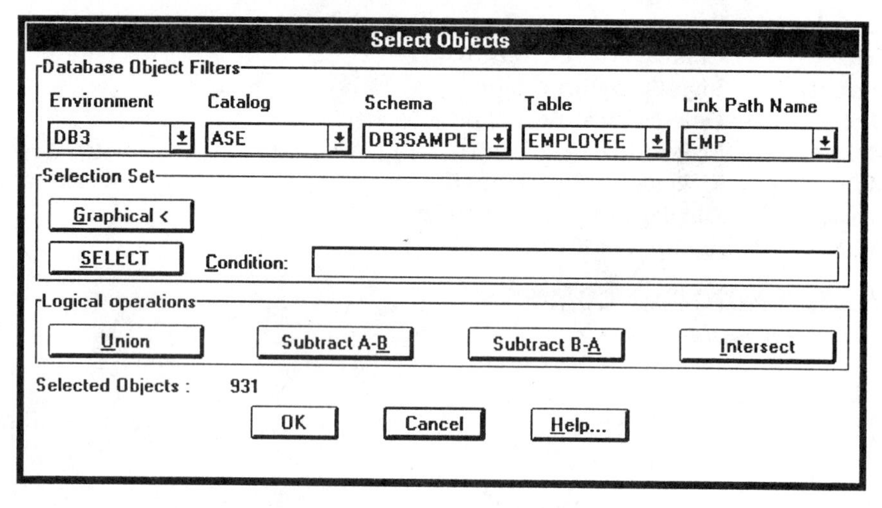

COMMAND OPTIONS

Graphical Selects objects from screen.

Textual Makes SQL selection; displays the following options:

- **All** Selects all objects linked to rows matching SQL condition.
- **Environment** Selects all objects within the current environment.
- **Catalog** Selects all objects in the current catalogue.
- **Schema** Selects all objects in the current schema.
- **<Table>** Selects all objects in the current table.
- **Lpn** Selects all objects in the current link path name.

Union Selection set contains all linked objects meeting criteria and selected graphic objects.

Intersect	Selection set contains objects in both sets.
subtractA	Subtracts first selection set from second set.
SubtractB	Subtracts second selection set from first set.
<eXit>	Exits the command.

RELATED AUTOCAD COMMANDS

All ASE commands

TIPS

- By default, all external database ojects are can be selected.
- Enter * (*asterisk*) to select all objects.
- Type an **SQL WHERE** sentance at the 'Enter text condition:' prompt.
- An **SQL WHERE** sentence operates on a single table at a time.

AseSqlEd

Rel. 12

Executes SQL statements *(short for SQL EDitor; an external command in Ase.Arx).*

Command	**Alias**	**Side Menu**	**Menu Bar**	**Tablet**
asesqled	...	[TOOLS] [EXT DBMS] [SQLedit:]	...	...

```
Command: asesqled
Settings/Options/Isolation/Autocommit/File/<SQL>/Native/Com-
mit/Rollback/eXit: I
Uncommitted/Committed/Repeatable/<Serializable>:
```

*When system variable CmdDia is turned on (= 1), the **AseSqlEd** command displays the following dialogue box:*

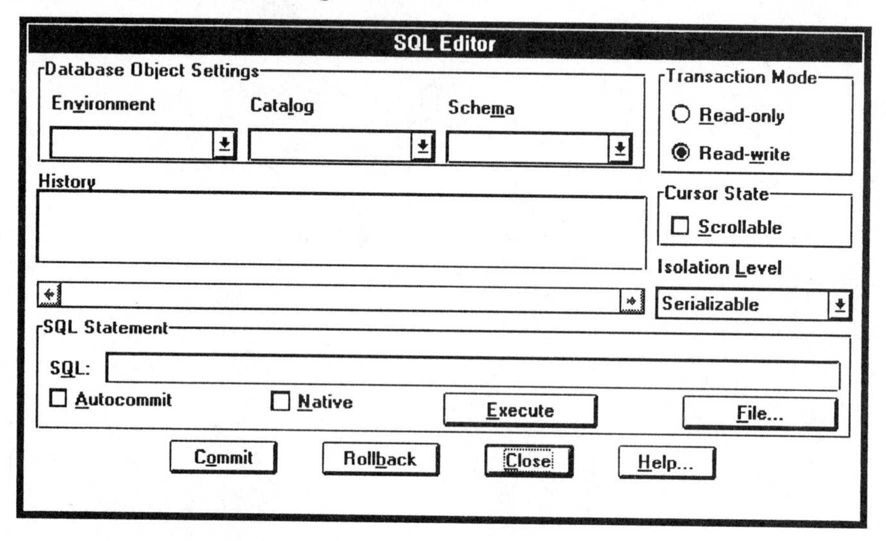

COMMAND OPTIONS:

Settings Changes database object settings:
- **Environment** Sets up an environment or selects a different one.
- **Catalog** Changes the catalogue.
- **Schema** Changes the current schema.

Options Transaction mode and cursor type:
- **Scrollable** Makes a row the current row.
- **read-Write** Row can be edited and deleted.
- **Read-only** Rows cannot be edited or deleted.

Isolation Selects level of SQL isolation:

- **Uncommitted** Other users can access any record.
- **Committed** Other users can access completed records.
- **Repeatable** Other users can access the current selection set.
- **<Serializable>** Concurrent transactions have same result as consecutive ones.

Autocommit Automatically makes multiple changes to the database.

File Runs SQL statements from a file in batch mode; displays the **Open SQL File** dialogue box..

<SQL> Executes an SQL statement at the **SQL>** prompt.

Native Allows use of non-SQL database commands.

Commit Saves changes made to database.

Rollback Cancels changes made to database.

eXit Exits the command.

RELATED AUTOCAD COMMANDS

All ASE commands

RELATED FILE

- ***.Txt** File containing SQL statements to be executed in batch mode.

TIPS

- Database column values (*except column names*) are case-sensitive.
- Search string character values are enclosed by ' ' (single quotes)
- Use a pair of single quotes ('') to create an apostrophe.
- Use the following metacharacters when entering column data at the Command: prompt:
  - A period (.) represents a null.
  - A pair of periods (..) represents a single period.
  - Pressing **[Enter]** leaves column data unchanged.
- The SQL batch file can use the following metacharacters:
  - ; Semi-colon prefix indicates a comment line.
  - $ Dollar prefix indicates a comment echoed to the command line.
  - & Ampersand suffix continues SQL command on the next line.

NON-EXISTANT COMMAND: AseUnload

The **AseUnload** command, documented by Autodesk, does not exist in Release 13. Instead, use the **(arxunload "ase")** AutoLISP function to unload ASE from memory.

AttDef

Ver. 2.0

Defines attribute modes and prompts (*short for ATTribute DEFinition*).

Command	**Alias**	**Side Menu**	**Menu Bar**	**Tablet**
`attdef`	...	...	...	...

```
Command: attdef
Attribute modes – Invisible:N  Constant:N  Verify:N  Preset:N
Enter (ICVP) to change, RETURN when done: [Enter]
Attribute tag:
Attribute prompt:
Default attribute value:
Justify/Style/<Start point>: [pick]
```

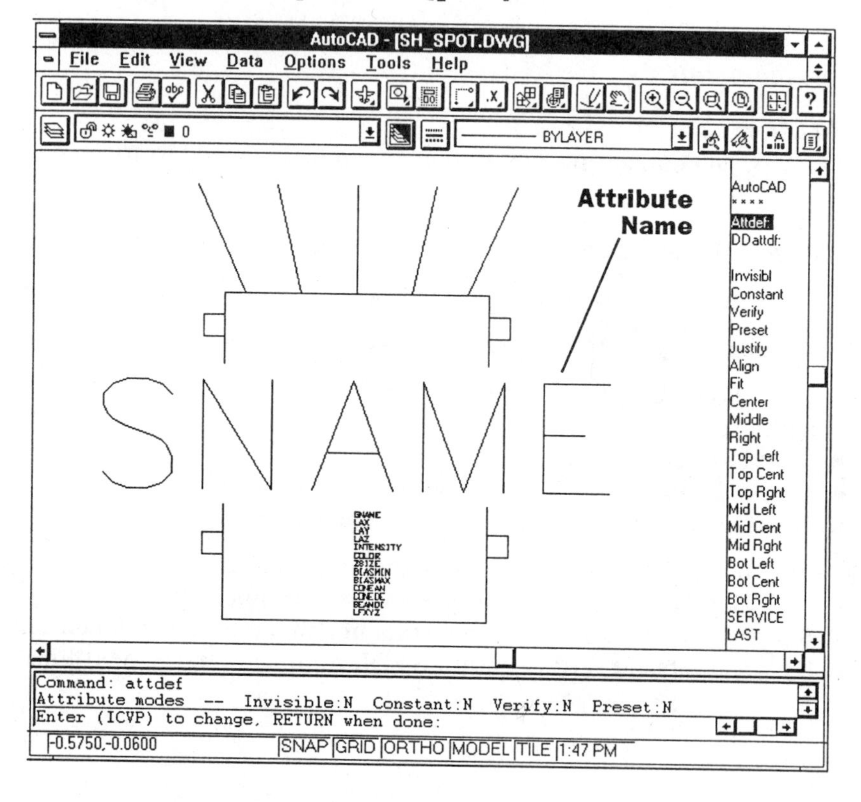

COMMAND OPTIONS

Attribute mode Selects the modes for the attribute:

- **I** Toggles visibility of attribute text in drawing (*short for Invisible*).
- **C** Toggles fixed or variable value of attribute (*short for Constant*).
- **V** Toggles confirmation prompt during input (*short for Verify*).
- **P** Toggles automatic insertion of default values (*short for Preset*).

Justify Selects the justification mode for the attribute text.

Style Selects the text style for the attribute text.

<Start point> Indicates the start point of the attribute text.

RELATED AUTOCAD COMMANDS

- **AttDisp** Controls the visibility of attributes.
- **AttEdit** Edits the values of attributes.
- **AttExt** Extracts attributes to disk.
- **AttRedef** Redefines an attribute or block.
- **Block** Binds attributes to objects.
- **DdAttDef** Defines attributes via a dialogue box.
- **DdAttE** Extracts the values of attributes via a dialogue box.
- **DdEdit** Edits the values of attributes via a dialogue box.
- **Insert** Inserts a block and prompts for attribute values.

RELATED SYSTEM VARIABLES

- **AFlags** Contains the value of modes in bit form:
  - **0** No attribute mode selected.
  - **1** Invisible.
  - **2** Constant.
  - **4** Verify.
  - **8** Preset.
- **AttDia** Toggles use of dialogue box during **Insert** command:
  - **0** Uses command-line prompts.
  - **1** Uses dialogue box.
- **AttReq** Toggles use of defaults or user prompts during **Insert** command:
  - **0** Assumes default values of all attributes.
  - **1** Prompts for attributes.

TIPS

- Constant attributes cannot be edited.
- Attribute tags cannot be null (have no value); attribute values may be null.
- When you press **[Enter]** at the 'Starting point:' prompt, **AttDef** automatically places the next attribute below the previous one.

'AttDisp

Ver. 2.0

Controls the display of all attributes in the drawing (*short for ATTribute DISPlay*).

Command	Alias	Side Menu	Menu Bar	Tablet
'attdisp	...	[OPTIONS]	[Options]	...
		[DISPLAY]	[Display]	
		[AttDisp:]	[Attribute Display]	

```
Command: attdisp
Normal/ON/OFF <Normal>:
```

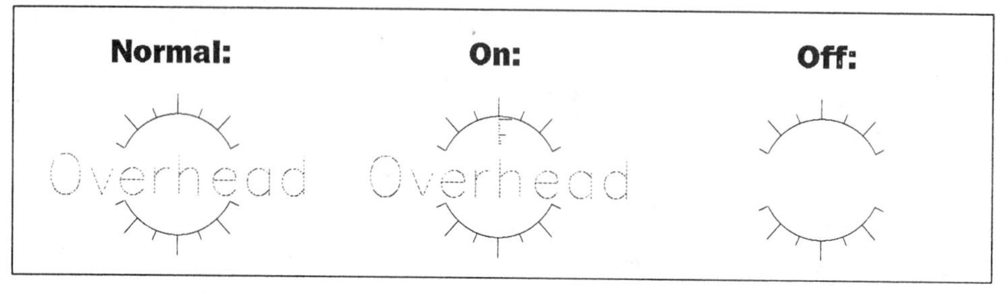

COMMAND OPTIONS

Normal Displays attributes according to Attdef setting.

ON Displays all attributes, regardless of **AttDef** setting.

OFF Displays no attribute, regardless of **AttDef** setting.

RELATED AUTOCAD COMMANDS

■ **AttDef** Defines new attributes, including their default visibility.

RELATED SYSTEM VARIABLES

■ **AttMode** Contains current setting of **AttDisp**:

0 Off.
1 Normal.
2 On.

TIPS

■ If **RegenAuto** is off, use the **Regen** command after **AttDisp** to see changes to attributes.

■ If you define invisible attributes, **AttDisp** lets you turn them on.

AttEdit

Ver. 2.0

Edits attributes in a drawing (*short for ATTribute EDIT*).

Command	Alias	Side Menu	Menu Bar	Tablet
attedit	...	...	...	...

```
Command: attedit
Edit attributes one at a time? <Y> [Enter]
Block name specification <*>:
Attribute tag specification <*>:
Attribute value specification <*>:
Select Attributes: [pick]
1 attributes selected.
Value/Position/Height/Angle/Style/Layer/Color/Next <N>:
```

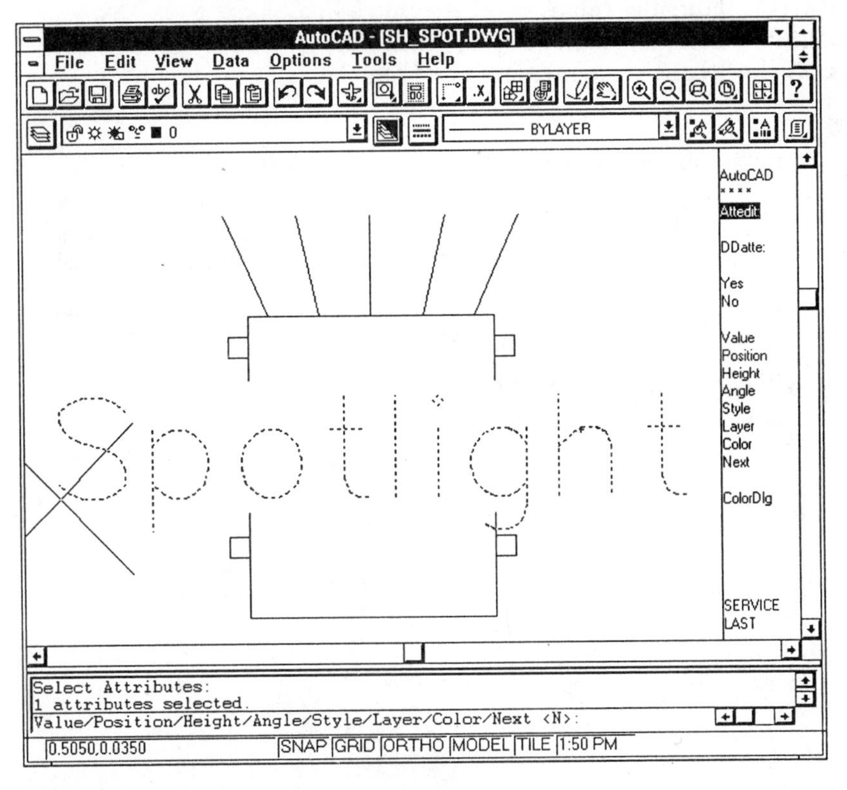

COMMAND OPTIONS

Value Changes or replaces the value of the attribute:

- **Change** Changes some of the attribute value.
- **Replace** Replaces attribute with a new value.

Position Moves the text insertion point of the attribute.

Height	Changse the attribute text height.
Angle	Changes the attribute text angle.
Style	Changes the text style of the attribute text.
Layer	Moves the attribute to a different layer.
Color	Changes the color of the attribute text.
Next	Edits the next attribute.

RELATED AUTOCAD COMMANDS

- **AttDef** Defines an attribute's original value and parameter.
- **AttDisp** Toggles an attributes visibility.
- **AttEdit** Edits the values of attributes.
- **AttRedef** Redefines attributes and blocks.
- **DdAttDef** Defines attributes via a dialogue box.
- **DdAttE** Edits the values of all attributes in a block.
- **DdEdit** Edits the values of one attribute.
- **Explode** Reduces an attribute to its tag.

TIPS

- Constant attributes cannot be edited with **AttEdit**.

- You can only edit attributes parallel to the current UCS.

- Unlike other text input to AutoCAD, attribute values are case-sensitive.

- To edit null attribute values, use **AttEdit**'s global edit option and enter \ (backslash) at the 'Attribute value specification:' prompt.

- The wildcard characters ? and * are interpreted literally at the 'String to change:' and 'New String:' prompts.

AttExt

Ver. 2.0

Extracts attribute data from the drawing to a file on disk (*short for ATTribute EXTract*).

Command	Alt+	Side Menu	Menu Bar	Tablet
attext	F, E	...	...	...

```
Command: attext
CDF, SDF or DXF Attribute extract (or Objects)? <C>:
```

*Displays the **Select Template File** and **Create Extract File** dialogue boxes.*

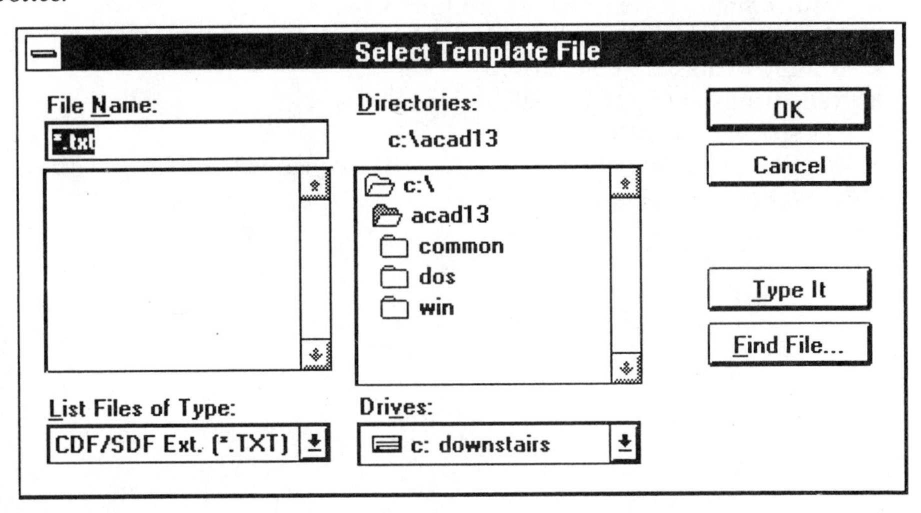

COMMAND OPTIONS

CDF Outputs attributes in comma-delimited format.
SDF Outputs attributes in space-delimited format.
DXF Outputs attributes in DXF format.
Objects Selects objects from which to extract attributes.

RELATED AUTOCAD COMMANDS

- **AttDef** Defines attributes.
- **DdAttExt** Defines attribute extraction via a dialogue box.

RELATED FILES

- ***.Txt** Required extension for the template file.
- ***.Txt** Extension for CDF and SDF extraction files.
- ***.Dxx** Extension for DXF extraction files.

TIPS

■ To output the attributes to the printer, specify:

CON	Appears on the text screen.
PRN *or* **LPT1**	Prints on parallel port 1.
LPT2 *or* **LPT3**	Prints to parallel port 2 or 3.

■ Before you can specify the SDF or CDF option, you must create the template file.

■ CDF files use the following conventions:

- ■ Specified field widths are the maximum width.
- ■ Positive number fields have a leading blank.
- ■ Character fields are enclosed in ' ' (single quote marks).
- ■ Trailing blanks are deleted.
- ■ Null strings are '' (two single quote marks).
- ■ Use spaces; do not uses tabs.
- ■ Use the C:DELIM and C:QUOTE records to change the field and string delimiters to another character.

QUICK START: Exporting Attributes

How to extract attribute data from a drawing:

1. CREATE TEMPLATE FILE

If you want the attribute extracted in CDF or SDF format, you must first create a template file; the DXF format does not use a template file. The *template file* is used by the **AttExt** (or **DdAttExt**) command to (1) determine which attributes to extract; and (2) the format of the extracted information. The template file uses the following format codes:

The *Type* describes the type of attribute data:

- C Alpha-numeric characters.
- N Numbers only.

The *Width* (www) describes the field width from 001 to 999 characters wide, padded with leading zeros.

The *Precision* (ddd) describes the number of decimal places from 001 to 999 (i.e., .1 to .00000...001), padded with leading zeros.

Field Name	Type,Width,Precision	Description
BL:NAME	Cwww000	Name of block
BL:NUMBER	Nwww000	Number of occurances
CHAR_ATTRIBUTE_TAG	Cwww000	Character attribute tag
NUMERIC_ATTR_TAG	Nwwwddd	Numeric attribute tag
BL:LAYER	Cwww000	Block's layer name
BL:ORIENT	Nwwwddd	Block rotation angle
BL:LEVEL	Nwww000	Block's nesting level
BL:X	Nwwwddd	Block insertion x-coordinate
BL:Y	Nwwwddd	Block insertion y-coordinate
BL:Z	Nwwwddd	Block insertion z-coordinate
BL:XSCALE	Nwwwddd	Block's x-scale factor
BL:YSCALE	Nwwwddd	Block's y-scale factor
BL:ZSCALE	Nwwwddd	Block's z-scale factor
BL:XEXTRUDE	Nwwwddd	Block's x-extrusion
BL:YEXTRUDE	Nwwwddd	Block's y-extrusion
BL:ZEXTRUDE	Nwwwddd	Block's z-extrusion
BL:HANDLE	Cwwwddd	Block's handle hex-number.

Shell out of AutoCAD and use a text editor to create a template file. Here is an example:

Field	Template	Description
BL:NAME	C008000	*8-character block name*
BL:NUMER	N004000	*Number of occurrences*
VENDOR	C016000	*Vendor attribute (16 chars)*
MODELNO	N012000	*Model # attr (12 digits)*

Save the file as ASCII text with the .TXT extension and return to AutoCAD.

2. SELECT OUTPUT FORMAT

Use either the **AttExt** or **DdAttExt** command to extract attribute data. Decide on the output format:

- **CDF** Comma-delimited format (the default), best for importing into a spreadsheet; sample output:
  `'Desk',55,'Steelcase',2248599597`
- **DXF** Drawing interchange format, similar to an entities-only DXF file.
- **SDF** Space-delimited format, best for importing into a database program; sample output:
  `Desk     55              Steelcase  2248599597`

3. SELECT OBJECTS

Either select the blocks you want to extract attributes from, or select all objects in the drawing. AutoCAD ignores all non-block objects and blocks with no attributes.

4. SPECIFY TEMPLATE FILE

Enter the name of the template file you created earlier.

5. SPECIFY OUTPUT FILENAME (Optional)

If you do not specify an output filename, AutoCAD uses the drawing's name, appending a .TXT to CDF and SDF files, or DXX to DXF files. Otherwise, specify any name except the template file's name.

6. CLICK [OK]

Click the **OK** button and AutoCAD places extracted attribute data into the output file. AutoCAD will stop if it finds any errors in the format of the template file, or if the selection set contains no attributes.

'AttRedef

Rel. 13

Redfines blocks and attributes (*short for ATTribute REDEFinition; an external command in AttRedef.Lsp*).

Command	Alias	Side Menu	Menu Bar	Tablet
'attredef	at	...	...	...

```
Command: redefine
Name of Block you wish to redefine:
Select objects for new Block...
Select objects: [pick]
Select objects: [Enter]
Insertion base point of new block: [pick]
```

RELATED AUTOCAD COMMANDS

- **AttDef** Defines an attribute's original value and parameter.
- **AttDisp** Toggles an attribute's visibility.
- **AttEdit** Edits the values of attributes.
- **DdAttDef** Defines attributes via a dialogue box.
- **DdAttE** Edits the values of several attributes.
- **DdEdit** Edits the value of one attribute.
- **Explode** Reduces an attribute to its tag.

TIPS

- Existing attributes retain their values.
- Existing attributes not included in the new block are erase.
- New attributes added to an existing block take on default values.

Audit

Rel. 11

Examines a drawing file for structural errors.

Command	Alt+	Side Menu	Menu Bar	Tablet
audit	F,M,A	[FILE]	[File]	...
		[MANAGE]	[Management]	
		[Audit:]	[Audit]	

```
Command: audit
Fix any errors detected? <N> y
```

Sample output:

```
0 Blocks audited
Pass 1 132 entities audited
Pass 2 132 entities audited
Total errors found 0 fixed 0
```

COMMAND OPTIONS

<N> Reports errors found; does not fix errors.

Y Reports and fixes errors found in drawing file.

RELATED AUTOCAD COMMANDS

- **Save** Saves a recovered drawing to disk.
- **Recover** Recovers a damaged drawing file.

RELATED SYSTEM VARIABLE

- **Auditctl** Controls the creation of the ADT audit log file:
  - **0** No log file written.
  - **1** ADT audit log file is written in drawing's directory.

RELATED FILES

- ***.ADT** Audit log file, reports the progress of the auditing process.

TIPS

- The **Audit** command is a diagnostic tool for validating and repairing the contents of a DWG file.

- Entities with error are placed in the previous selection set. Use an edit command, such as **Copy**, to view the entities.

- If **Audit** cannot fix a drawing file, use the **Recover** command.

'Base

Ver. 1.0

Changes the 2D or 3D insertion point of a drawing, located at (0,0,0) by default.

Command	Alias	Side Menu	Menu Bar	Tablet
'base	...	...	...	...

```
Command: base
Base point <0.0000,0.0000,0.0000>:
```

COMMAND OPTIONS

None

RELATED AUTOCAD COMMANDS

- **Block** Allows you to specify the insertion point of a new block.
- **Insert** Inserts another drawing into the current drawing.
- **Xref** References another drawing.

RELATED SYSTEM VARIABLE

- **InsBase** Contains the current setting of the drawing insertion point.

BHatch

Rel. 12

Automatically applies an associative hatch pattern within a boundary (*short for Boundary HATCH*).

Command	Alias	Side Menu	Menu Bar	Tablet
bhatch	...	[CONSTRCT]	[Draw]	W 16
		[Bhatch:]	[Hatch]	
			[Hatch]	

Command: **bhatch**

Displays dialogue box:

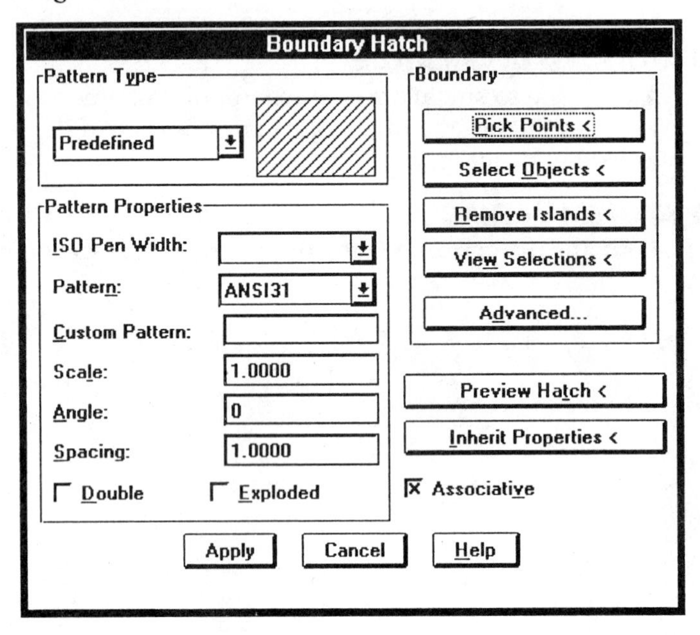

COMMAND OPTIONS

Pattern Type Select type of hatch pattern.

Pattern Properties:

- **ISO Pen Width** When an ISO hatch pattern is selected, scales pattern according to pen width.
- **Pattern** Selects name of hatch pattern defined in Acad.Pat.
- **Custom Pattern** Specifies name of a user-defined pattern.
- **Scale** Hatches pattern scale (*default: 1.0*).
- **Angle** Hatches pattern angle (*default: 0 degrees*).
- **Spacing** Spacing between lines of a user-defined hatch pattern.
- **Double** Toggles double hatching.
- **Exploded** Places hatch pattern as lines, rather than as a block.

Preview Hatch Previews the hatch pattern before it is applied.

Inherit Properties	Selects the hatch pattern parameters from an existing hatch pattern.
Default Properties	Changes pattern parameters back to default values.
Associative	Toggles between associative and non-associative pattern.

Boundary:

Pick Points	Picks points that define the hatch pattern boundary.
Select Objects	Selects objects to be hatched.
Remove Islands	Removes islands from the hatch pattern selection set.
View Selections	Views hatch pattern selection set.
Advanced...	Displays the **Advanced Options** dialogue box:

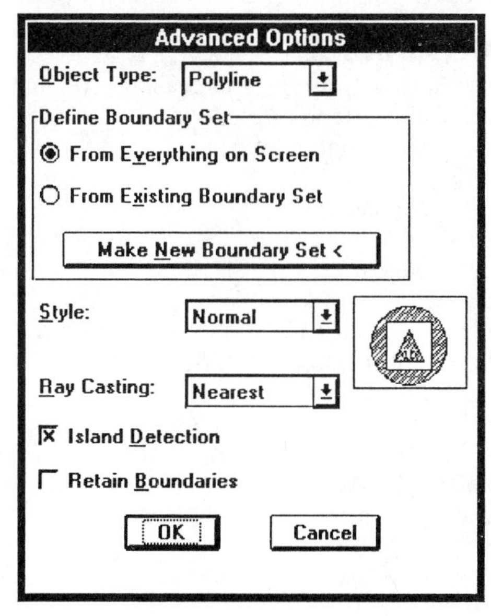

Object Type Boundary is made from a polyline or as a region.

Define Boundary Set:

From Everything on Screen	Selects all objects visible in current view.
From Existing Boundary Set	Selects current boundary selection set.
Make New Boundary Set	Dialogue box disappears to let you select the new boundary.

Style Hatching style:

Normal	Hatches alternate areas.
Outer	Hatches only the outermost area.
Ignore	Hatches everything within boundary.

Ray Casting Determines how AutoCAD searches for the hatch boundary:

Nearest	Searches from pick point to nearest object.
+X	Searches from pick point in +x direction.
-X	Searches from pick point in -x direction.
+Y	Searches from pick point in +y direction.

-Y Searches from pick point in -y direction.
Island Detection Toggles whether islands are detected.
Retain Boundaries Toggles whether the boundary is retained after hatch is placed.

RELATED AUTOCAD COMMANDS

- **Boundary** Automatically traces a polyline around a closed boundary.
- **Explode** Reduces a group of hatch patterns.
- **Hatch** Creates a non-associative hatch within a manually-selected perimeter.
- **PsFill** Fills a closed polyline with a PostScript pattern.

RELATED SYSTEM VARIABLES

- **DelObj** Toggles whether boundary is erased after hatch is place.
- **HpAng** Current hatch pattern angle.
- **HpBound** Hatch boundary made from:
  - **0** Polyline.
  - **1** Region (*default*)
- **HpDouble** Single or double hatching:.
  - **0** Single (*default*).
  - **1** Double.
- **HpName** Current hatch pattern name (*up to 31 characters long*):
  - "" No current hatch pattern name.
  - "**.**" Eliminate current name.
- **HpScale** Current hatch pattern scale factor.
- **HpSpace** Current hatch pattern spacing factor.
- **PickStyle** Controls hatch selection:
  - **0** Groups and associative hatches not selected.
  - **1** Groups selected.
  - **2** Associative hatches selected.
  - **3** Both selected
- **SnapBase** Starting coordinates of hatch pattern.

RELATED FILES

- **Acad.Pat** Hatch pattern definition file.

TIPS

- The **BHatch** command first generates a boundary polyline, then hatches the inside area. Use the **Boundary** command to obtain just the bounding polyline.

- **BHatch** stores hatching parameters in the pattern's extended entity data; use the **List** command to display parameters stored in extended entity data.

- See the **Hatch** command for a list of hatch patterns supplied with AutoCAD.

'Blipmode

Ver. 2.1

Turns the display of pick-point markers, known as "blips," off and on.

Command	Alias	Side Menu	Menu Bar	Tablet
blipmode	...	...	...	...

```
Command: blipmode
ON/OFF <On>: off
```

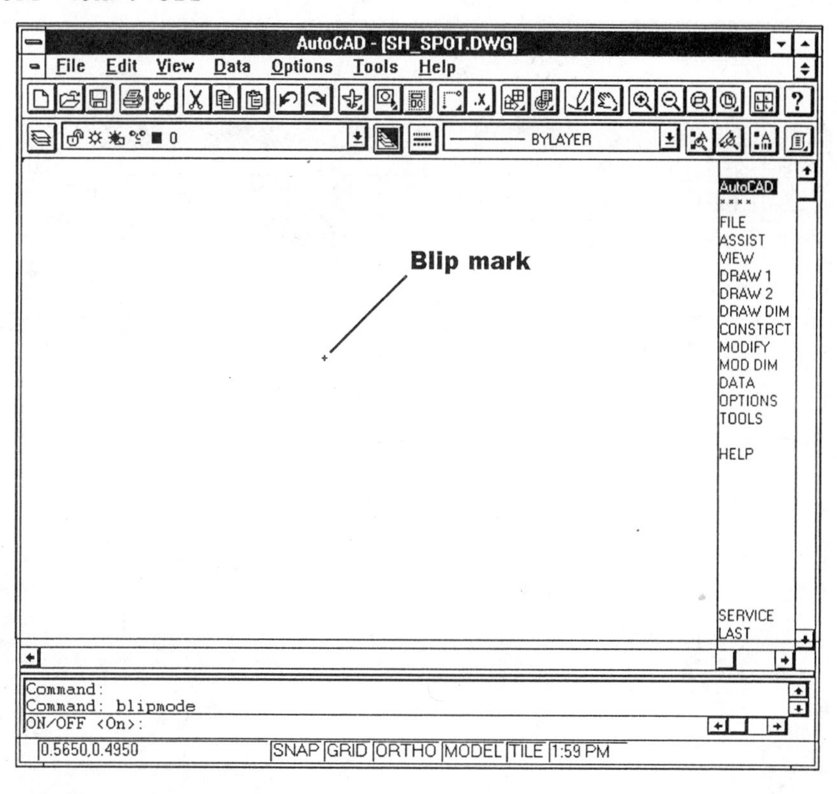

COMMAND OPTIONS

ON Turns on display of pick point markers.

OFF Turns off display of pick point markers.

RELATED AUTOCAD COMMANDS

- **DdRModes** Allows blipmode toggling via a dialogue box.
- **Redraw** Cleans blips off the screen.

RELATED SYSTEM VARIABLE

- **Blipmode** Contains the current setting of blipmode.

TIP

- You cannot change the size of the blipmark.

Block

Ver. 1.0

Defines a group of objects as a single named object; creates symbols.

Command	Alias	Side Menu	Menu Bar	Tablet
block	...	...	[Construct] [Block]	W 9

```
Command: block
Block name (or ?):
Insertion base point: [pick]
Select objects:
```

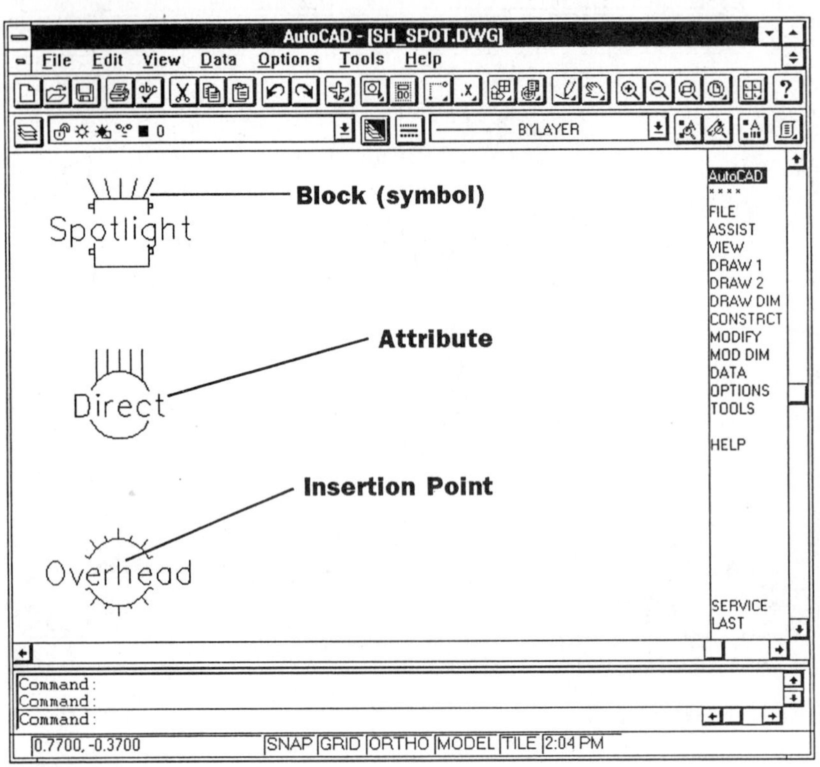

COMMAND OPTIONS

? Lists the blocks stored in the drawing.

RELATED AUTOCAD COMMANDS

- **Explode** Reduces a block into its original objects.
- **Insert** Adds a block or another drawing to the current drawing.
- **Oops** Returns objects to the screen after creating the block.
- **WBlock** Writes a block to a file on disk as a drawing.
- **XRef** Displays another drawing in the current drawing.

RELATED FILES

■ *.DWG All drawing files are insertable as blocks.

TIPS

■ A block name has up to 31 alpha-numeric characters, including $, -, and _. However, blocks stored on disk are limited to eight characters.

■ Release 13 has five types of blocks:

User Blocks	Blocks created by you.
Nested Blocks	A block within a block.
Unnamed Blocks	Blocks created by AutoCAD commands, such as hatch patterns created by the **Hatch** command.
External References	Externally referenced drawings.
Dependent Blocks	Blocks in an externally referenced drawing.

■ Use the **INSertion** object snap to select the block's insertion point.

■ A block created on a layer other than layer 0 is always inserted on that other layer.

■ A block created on layer 0 is inserted on the current layer.

BmpOut

Rel. 13

Exports selected objects in the current viewport to a BMP file.

Command	Alias	Side Menu	Menu Bar	Tablet
bmpout	...	...	???	...

```
Command: bmpout
Select objects:
```

*Displays **Create BMP File** dialogue box.*

COMMAND OPTIONS

None

RELATED AUTOCAD COMMANDS

- **Export** Exports the drawing in several formats.
- **SaveImg** Saves rendered view in three formats.
- **WmfOut** Exports selected objects to a WMF file.

RELATED WINDOWS COMMANDS

- **[Prt Scr]** Saves screen to Windows Clipboard.
- **[Alt]+[PrtScr]** Saves the optmost window to the Clipboard.

TIPS

- BMP is short for bitmap, the raster file standard for Windows.

- When responding All to the 'Select objects:' prompt, the **BmpOut** command does not select all objects in the drawing; rather, it selects all objects visible in the current viewport:

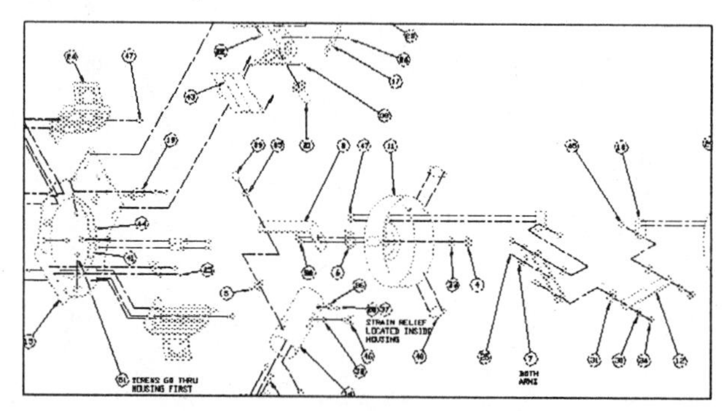

- The BMP format used by Release 13 is an unusual format incompatible with most graphics programs, including Windows Paintbrush, Word v2, WordPerfect, PageMaker, Publisher, Image-In, and Fractal Painter. I found that PaintShop Pro and Word 6 read AutoCAD's BMP files.

Boundary

Rel. 12

Creates a boundary as a polyline or 2D region (*formerly the BPoly command*).

Command	Alias	Side Menu	Menu Bar	Tablet
boundary	bpoly	[CONSTRCT] [Boundar:]	[Construct]]Boundary]	...

```
Command: -boundary
Advanced options/<Internal point>: A
Boundary set/Island detection/Object type/<eXit>: O
Region/Polyline: R
Boolean subtract inner islands?
```

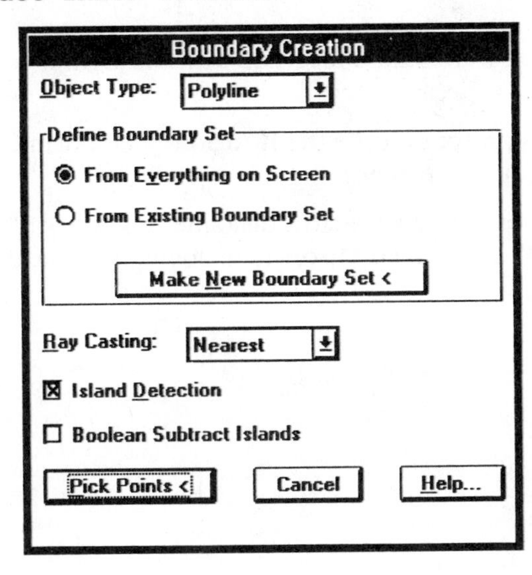

COMMAND OPTIONS

Object Type Creates the boundary out of a polyline or as a region.

Define Boundary Set:

- **From Everything on Screen** Selects all objects visible in current view.
- **From Existing Boundary Set** Selects the current boundary selection set.
- **Make New Boundary Set** Dialogue box disappears to allow you to select the boundary.

Ray Casting Determines how AutoCAD searches for the hatch boundary:

- **Nearest** Searches from pick point to nearest object.
- **+X** Searches from pick point in +x direction.
- **-X** Searches from pick point in -x direction.
- **+Y** Searches from pick point in +y direction.
- **-Y** Searches from pick point in -y direction.

Island Detection	Toggles whether islands are detected.
Boolean Subtract Island	Deletes islands from the boundary.

RELATED AUTOCAD COMMANDS

- **Polyline** Draws a polyline.
- **PEdit** Edits a polyline.
- **Region** Creates a 2D region from a collection of objects.

RELATED SYSTEM VARIABLES

- **HpBound** Object used to create boundary:
  0 Draw as polyline.
  1 Draw as region (*default*).

TIPS

- Use the **Boundary** command together with the **Offset** command to help create poching.

- Use the - (*dash*) prefix to force the **Boundary** command to display its prompt at the Command: prompt.

- Although the **Boundary Creation** dialogue box looks identical to the **BHatch** command's **Advanced Options** dialogue box, be aware there are differences.

Rel. 13

Draws a 3D box as an ACIS solid model (*an external command in Acis.Dll; formerly the SolBox command*).

Command	Alias	Side Menu	Menu Bar	Tablet
box	...	[DRAW 2]	[Draw]	M 7
		[SOLIDS]	[Solids]	
		[Box:]	[Box]	

```
Command: box
Center/<Corner of box> <0,0,0>: [pick]
Cube/Length/<other corner>: [pick]
Height:
```

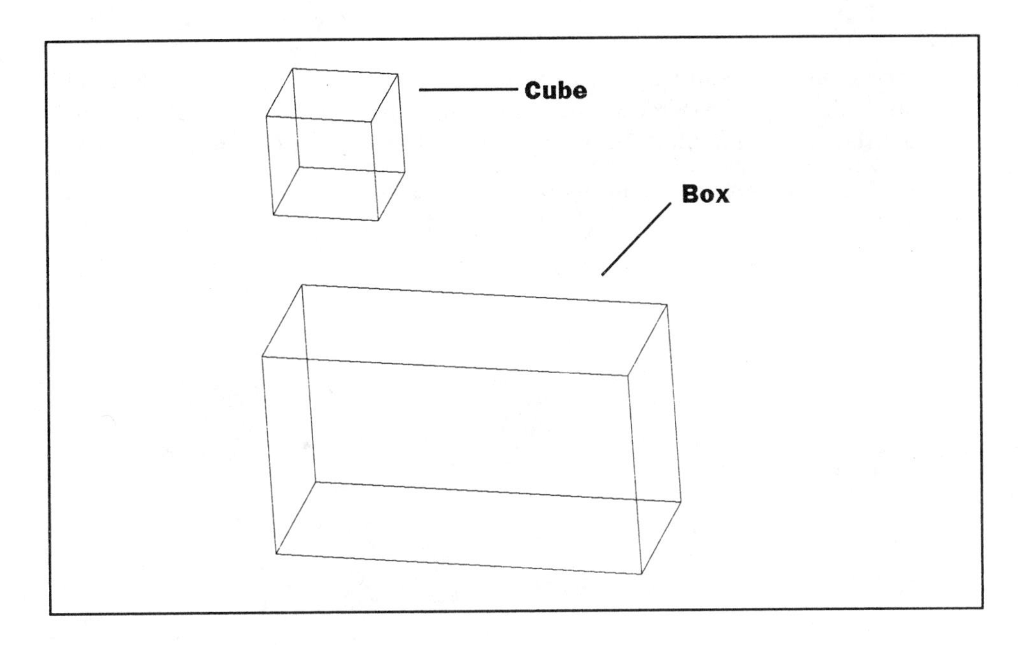

COMMAND OPTIONS

Center	Draws the box about a center point.
<Corner of box>:	Specifies one corner for the base of box.
Cube	Draws a cube box (all sides are the same length).
Length	Specifies the x, y, z-lengths.
Height	Specifies the height of the box.

RELATED AUTOCAD COMMANDS

- **Ai_Box** Draws a 3D wireframe box.
- **Cone** Draws a 3D solid cone.

- **Cylinder** Draws a 3D solid tube.
- **Sphere** Draws a 3D solid ball.
- **Torus** Draws a 3D solid donut.
- **Wedge** Draws a 3D solid wedge.

RELATED SYSTEM VARIABLES

- **DispSilh** Toggles display of 3D objects as silhouette after hidden-line removal and shading.
- **IsoLines** Number of isolines on solid surfaces:
  - **0** Minimum (*no isolines*).
  - **4** Default.
  - **16** A reasonable value.
  - **2,047** Maximum value.

TIPS

■ To bring an ACIS solid model into Release 12, do not use the **SaveAsR12** command, since solid models are decomposed to polylines and circles. Instead, use the **3dsOut** command, which converts ACIS solid models into 3D faces.

■ Once the **AMEconvert** command converts a Release 12 AME drawing into Release 13 ACIS model, it cannot be converted back.

Break

Ver. 1.4

Removes a portion of a line, trace, 2D polyline, arc, or circle.

Command	Alias	Side Menu	Menu Bar	Tablet
break	...	...	[Modify] [Break]	X 13
				X 14

```
Command: break
Select object: [pick]
Enter second point (or F for first point): f
Enter first point: [pick]
Enter second point: [pick]
```

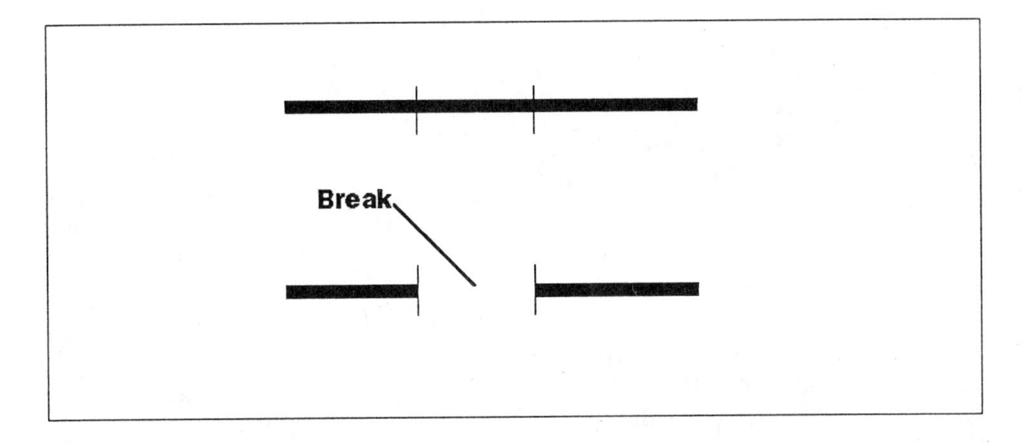

COMMAND OPTIONS

F Specifies the first break point.

@ Uses the first break point coordinates for the second break point.

RELATED AUTOCAD COMMANDS

- **Change** Changes the length lines.
- **PEdit** Removes and relocates vertices of polylines.
- **Trim** Shortens the lengths of lines.

'Cal

Rel. 12

The command-line algebraic and vector geometry calculator (*an external command in GeomCal.Exp*).

Command	Alias	Side Menu	Menu Bar	Tablet
'cal	...	[TOOLS]	[Assist]	P 3
		[GeomCal:]	[Calculator]	

```
Command: cal
>>Expression:
```

COMMAND OPTIONS

( )	Grouping of expressions.
[ ]	Vector expressions.
+	Addition.
-	Subtraction.
*	Multiplication.
/	Division.
^	Exponentiation.
&	Vector product of vectors.
sin	Sine.
cos	Cosine.
tang	Tangent.
asin	Arc sine.
acos	Arc cosine.
atan	Arc tangent.
ln	Natural logarithm.
log	Logarithm.
exp	Natural exponent.
exp10	Exponent.
sqr	Square.
sqrt	Square root.
abs	Absolute value.
round	Round off.
trunc	Truncate.
cvunit	Converts units using Acad.Unt.
w2u	WCS to UCS conversion.
u2w	UCS to WCS conversion.
r2d	Radians-to-degrees conversion.
d2r	Degrees-to-radians conversion.
pi	The value PI.
xyof	x- and y-coordinates of a point.

xzof	x- and z-coordinates of a point.
yzof	y- and z-coordinates of a point.
xof	x-coordinate of a point.
yof	y-coordinate of a point.
zof	z-coordinate of a point.
rxof	Real x-coordinate of a point.
ryof	Real y-coordinate of a point.
rzof	Real z-coordinate of a point.
cur	x,y,z-coordinates of picked point.
rad	Radius of entity
pld	Point on line, distance from.
plt	Point on line, using parameter *t*.
rot	Rotate point through angle about origin.
ill	Intersection of two lines.
ilp	Intersection of line and plane.
dist	Distance between two points.
dpl	Distance between point and line.
dpp	Distance between point and plane.
ang	Angle between lines.
nor	Unit vector normal.

RELATED AUTOCAD COMMANDS

- *All*

RELATED SYSTEM VARIABLES

- **UserI1 — UserI5** User-definable integer variables.
- **UserR1 — UserR5** User-definable real variables.

TIPS

- Since **'Cal** is a transparent command, it can be used to perform a calculation in the middle of another command.

- **Cal** understands the following prefixes:
  - * Scalar product of vectors.
  - & Vector product of vectors.

- And the following suffixes:
  - **r** Radian (*degrees is the default*).
  - **g** Grad.
  - ' Feet (*unitless distance is the default*).
  - " Inches.

Chamfer

Ver. 2.1

Bevels the intersection of two lines, all vertices of a 2D polyline, or the faces of 3D solid models.

Command	Alias	Side Menu	Menu Bar	Tablet
chamfer	...	[CONSTRCT] [Chamfer:]	[Construct] [Chamfer]	X 21

```
Command: chamfer
Polyline/Distances/Angle/Trim/Method<Select first line>: D
Enter first chamfer distance:
Enter second chamfer distance:
Polyline/Distances/Angle/Trim/Method<Select first line>: [pick]
Select second line: [pick]
```

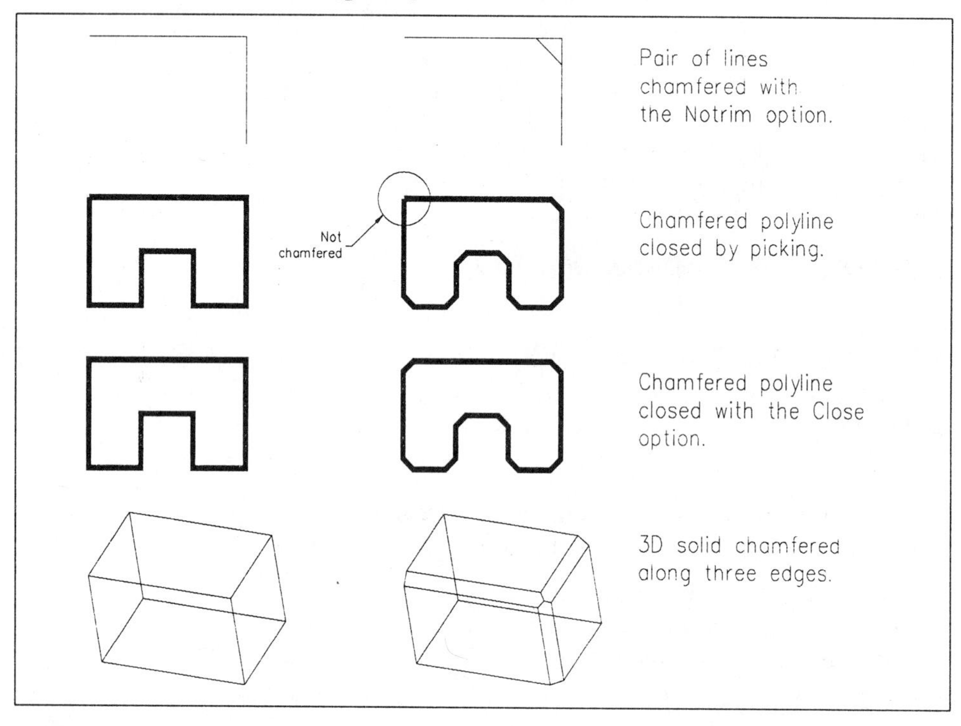

COMMAND OPTIONS

Polyline Chamfers all vertices of a polyline.
Distances Specifies the chamfer distances.
Angle Specifies the chamfer by a distance and an angle.
Trim Toggles whether lines/edges are trimmed after chamfer.
Method Toggles whether distance or angle are used.

RELATED AUTOCAD COMMANDS

- **Fillet** Rounds the intersection with a radius.

RELATED SYSTEM VARIABLES

- **ChamferA** First chamfer distance.
- **ChamferB** Second chamfer distance.
- **ChamferC** Length of chamfer.
- **ChamferD** Chamfer angle.
- **ChamMode** Toggles chamfer measurement:
  - **0** Chamfer by two distances.
  - **1** Chamfer by distance and angle.
- **TrimMode** Toggles whether lines/edges are trimmed after chamfer.

Change

Ver. 1.0

Modifies the color, elevation, layer, linetype, linetype scale, and thickness of any object, and certain properties of lines, circles, blocks, text, and attributes.

Command	Alias	Side Menu	Menu Bar	Tablet
change	...	[MODIFY] [Change:]	...	...

```
Command: change
Select objects: [pick]
Select objects: [Enter]
Properties/<Change point>: p
Change what property (Color/Elev/LAyer/LType/ltScale/
Thickness)?
```

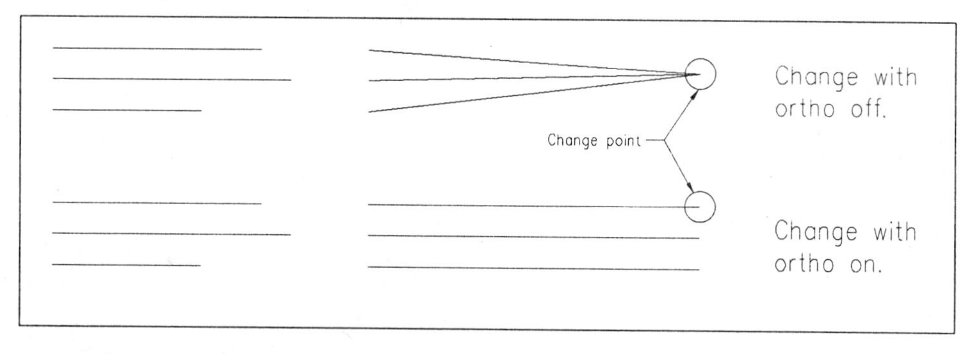

COMMAND OPTIONS

<Change point> Picks an entity to change:

- **<pick line>** Indicates the new length of line.
- **<pick circle>** Indicates the new radius of circle.
- **<pick block>** Indicates the new insertion point or rotation angle of a block.
- **<pick text>** Indicates the new location of text.
- **<pick block>** Indicates an attribute's new text insertion point, text style, height, rotation angle, text, tag, prompt, or default value.

[Enter] Changes the insertion point, style, height, rotation angle, and string of text.

Properties Changes properties of the object, as follows:

- **Color** Changes the color of the entity.
- **Elev** Changes the elevation of the entity.
- **LAyer** Moves the entity to a different layer.
- **LType** Changes the linetype of the entity.
- **ltScale** Changes the scale of the linetype.
- **Thickness** Changes the thickness of any entity, except blocks.

RELATED AUTOCAD COMMANDS

- **AttReDef** Changes a block or attributes.
- **ChProp** Contains the properties portion of the **Change** command.
- **Color** Changes the current color setting.
- **Colour** Changes the current colour setting.
- **DdChProp** Displays dialogue box for changing entity properties.
- **Elev** Changes the working elevation and thickness.
- **Modify** Changes most aspects of all objects.

RELATED SYSTEM VARIABLES

- **CeColor** The current color setting.
- **CeLType** The current linetype setting.
- **CircleRad** The current circle radius.
- **CLayer** The name of the current layer.
- **Elevation** The current elevation setting.
- **LtScale** The current linetype scale.
- **TextSize** The current height of text.
- **TextStyle** The current text style.
- **Thickness** The current thickness setting.

TIPS

■ The **Change** command cannot change the size of donuts, the radius or length of arcs, the length of polylines, or the justification of text.

■ Use the **Change** command to change the endpoints of a group of lines to a common vertex.

■ Turn ortho mode on to extend or trim a group of lines, without needing a cutting edge (as the **Extend** and **Trim** commands).

ChProp

Rel. 10

Modifies the color, layer, linetype, linetype scale, and thickness of most entities.

Command	Alias	Side Menu	Menu Bar	Tablet
chprop	...	...	...	...

```
Command: chprop
Select objects: [pick]
Select objects: [Enter]
Change what property (Color/LAyer/LType/ltScale/Thickness) ?
```

COMMAND OPTIONS

Color Changes the color of the object.
LAyer Moves the object to a different layer.
LType Changes the linetype of the object.
ltScale Changes the linetype scale.
Thickness Changes the thickness of any object except blocks.

RELATED AUTOCAD COMMANDS

- **Change** Allows changes to lines, circles, blocks, text, and attributes.
- **Color** Changes the current color setting.
- **Colour** Changes the current colour setting.
- **DdChProp** Dialogue box version of the **ChProp** command.
- **Elev** Changes the working elevation.
- **LtScale** Sets the linetype scale.
- **Modify** Changes most aspects of all objects.

RELATED SYSTEM VARIABLES

- **CeColor** The current color setting.
- **CeLtype** The current linetype name.
- **CLayer** The name of the current layer.
- **LtScale** The current linetype scale.
- **Thickness** The current thickness setting.

TIP

- Use the **Change** command to change the elevation of an object.

Circle

Ver. 1.0

Draws 2D circles by five different methods.

Command	Alias	Side Menu	Menu Bar	Tablet
circle	...	[DRAW 1] [Circle:]	[Draw] [Circle]	M 10

```
Command: circle
3P/2P/TTR/<Center point>: [pick]
Diameter/<Radius>: [pick]
```

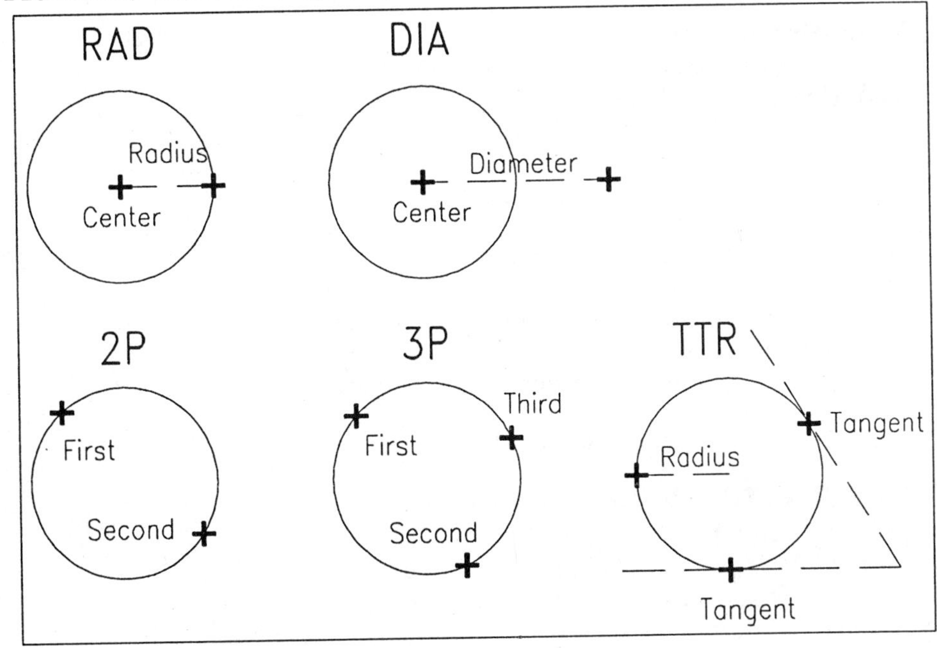

COMMAND OPTIONS

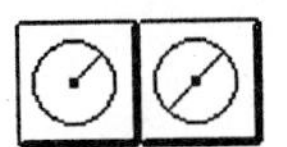

Center; Radius or Diameter:

<Center point> Indicates the circle's center point:

<Radius> Indicates the circle's radius.

Diameter Indicates the circle's diameter.

Three-point circle:

3P Draws a three-point circle:

First point: Indicates first point on circle; and
Second point: Indicates second point on circle; and
Third point: Indicates third point on circle.

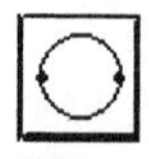

Two-point circle:

2P Draws a two-point circle:

First point on diameter:
Indicates first point on circle; and
Second point on diameter:
Indicates second point on circle.

Tangent-tangent-radius circle:

TTR Draws a circle tangent to two lines:

Enter Tangent spec:
Indicates first point of tangency; and
Enter second Tangent spec:
Indicates second point of tangency; and
Radius: Indicates first point of radius; and
Second point: Indicates second point of radius.

RELATED AUTOCAD COMMANDS

- **Arc** Draws an arc.
- **Donut** Draws a solid-filled circle or donut.
- **Ellipse** Draws an elliptical circle or arc.
- **Sphere** Draws a 3D solid ball.
- **ViewRes** Controls the visual roundness of circles.

RELATED SYSTEM VARIABLE

- **CircleRad** The current circle radius.

TIPS

- Sometimes it is easier to create an arc by drawing a circle, then using the **Break** or **Trim** commands to convert the circle into an arc

- Giving a circle thickness turns it into a cylinder.

'Color *or* 'Colour

Ver. 2.5

Sets the new working color.

Command	Alias	Side Menu	Menu Bar	Tablet
'color	...	...	...	...
'colour				

```
Command: color
New object color <BYLAYER>:
```

COMMAND OPTIONS

BYLAYER Sets working color to color of current layer.
BYBLOCK Sets working color of inserted blocks.
Color Number Sets working color using number (*1 to 255*), or name or abbreviation:

Color name	Number	Abbreviation
Red	1	R
Yellow	2	Y
Green	3	G
Cyan	4	C
Blue	5	B
Magenta	6	M
White	7	W
Greys	250 - 255	...

RELATED AUTOCAD COMMANDS

- **Change** Changes the color of entities.
- **ChProp** Changes the color of entities.
- **DdEModes** Sets new working color using a dialogue box.
- **DdChProp** Changes the color of entities via a dialogue box.

RELATED SYSTEM VARIABLES

- **CeColor** The current entity color setting:

1	(*red*) minimum value.
7	(*white*) default value.
255	maximum value.

TIPS

- 'BYLAYER' means that objects take on the color assigned to that layer.
- 'BYBLOCK' means that objects take on the color in effect at the time the block is inserted.
- White entities display as black when the background color is white.

Compile

Rel. 12

Compiles SHP shape, SHP font, and PFB PostScript font definition files into SHX format; TTF TrueType fonts cannot be compiled.

Command	Alias	Side Menu	Menu Bar	Tablet
`compile`	...	[TOOLS] [Compile:]	[Tools] [Compile]	...

Command: **compile**

COMMAND OPTIONS

None

RELATED AUTOCAD COMMANDS

- **Load** Loads a compiled SHX shape file into the current drawing.
- **Style** Loads SHP, SHX, TTF, PFA, and PFB font files into the current drawing.

RELATED SYSTEM VARIABLE

- **ShpName** The current SHP filename.

TIPS

- As of Release 12, the **Style** command converts SHP and PFB font files on-the-fly; it is only necessary to use the **Compile** command to obtain an SHX font file.

- The **Compile** command lets you convert any of the 10,000 PostScript fonts into AutoCAD's SHX format.

- TrueType fonts cannot be compiled.

Cone

Rel. 11

Draws a 3D ACIS cone with a circular or elliptical base (*formerly the SolCone command; an external command in Acis.Dll*).

Command	Alias	Side Menu	Menu Bar	Tablet
cone	...	[DRAW 2]	[Draw]	J 7
		[SOLIDS]	[Solids]	
		[Cone:]	[Cone]	

```
Command: cone
Elliptical/<center point> <0,0,0>: E
<Axis endpoint>/Center: C
Center of ellipse <0,0,0>: [pick]
Axis endpoint: [pick]
Other axis endpoint: [pick]
Apex/<Height>: A
Apex: [pick]
```

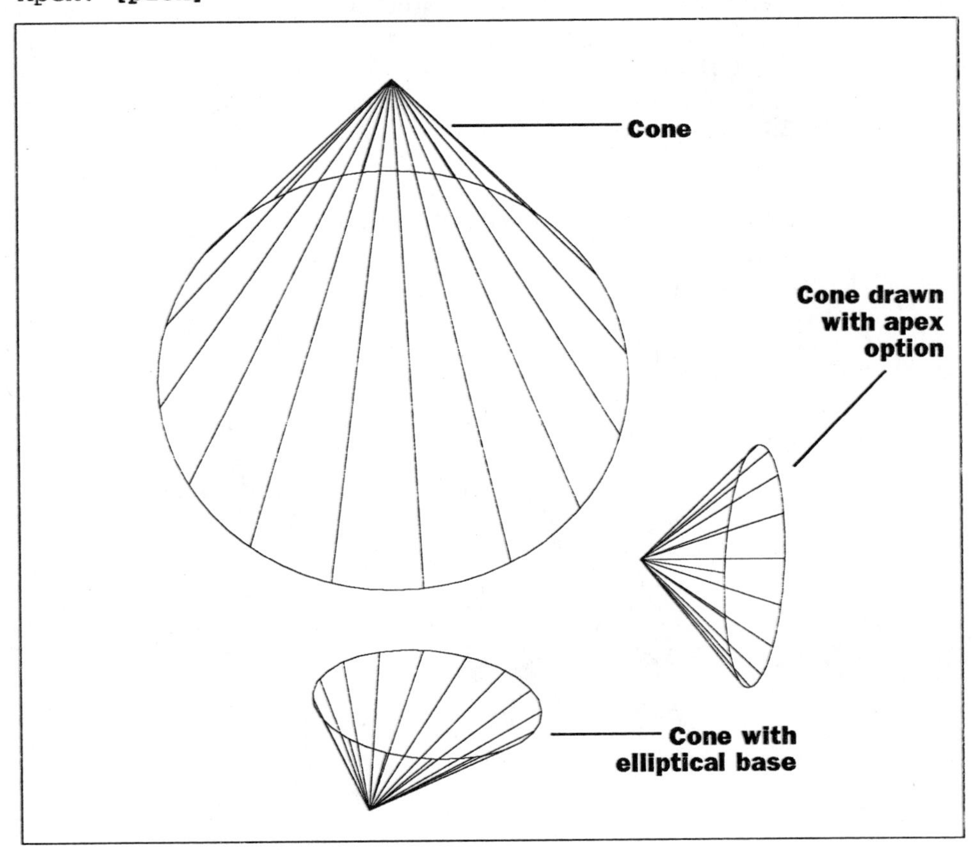

COMMAND OPTIONS

<Center point>	Indicates center of the cone's base.
Elliptical	Draws cone with an elliptical base.
Apex	Determines height and orientation.

RELATED AUTOCAD COMMANDS

- **Ai_Cone** Draws a 3D wireframe cone.
- **Box** Draws a 3D solid box.
- **Cylinder** Draws a 3D solid tube.
- **Sphere** Draws a 3D solid ball.
- **Torus** Draws a 3D solid donut.
- **Wedge** Draws a 3D solid wedge.

RELATED SYSTEM VARIABLES

- **DispSilh** Toggles display of 3D objects as silhouette after hidden-line removal and shading.
- **IsoLines** Number of isolines on solid surfaces:
  - **0** Minimum (*no isolines*).
  - **4** Default.
  - **16** A reasonable value.
  - **2,047** Maximum value.

TIP

- To draw a cone at an angle, use the **Apex** option.

Config

Rel. 12

Reconfigures AutoCAD for the graphics board, digitizer, plotter, and operating parameters (*short for CONFIGuration*).

Command	Alias	Side Menu	Menu Bar	Tablet
config	...	[OPTIONS]	[Options]	...
		[Config:]	[Configure]	

Command: **config**

COMMAND OPTIONS

<0> Exits **Config** back to the drawing screen (*default*).
1 Shows the current configuration of AutoCAD.
2 Allows detailed configuration.
3 Configures video display.
4 Configures digitizer.
5 Configures plotter.
6 Configures system console.
7 Configures operating parameters.

RELATED DOS COMMANDS

- **acad -r** Start AutoCAD with the **-r** parameter to reconfigure.
- **Set** Sets environment variables prior to starting AutoCAD for DOS.

RELATED AUTOCAD COMMANDS

- **DxlConfig** Configures the accelerated display driver.
- **RConfig** Configures **Render** with rendering graphics board and printer.
- **ReInit** Reinitializes peripherals.

RELATED ENVIRONMENT VARIABLES

- **AcadCfg** Points to the subdirectory containing the Acad.Cfg configuration file.
- **Acad** Points to the subdirectories AutoCAD should search for support files.
- **AcadAltMenu** Points to name of alternate tablet menus.
- **AcadDrv** Points to subdirectories containing device drivers.
- **AcadMaxMem** Maximum bytes of memory pager requests from DOS.
- **AcadPageDir** Points to subdirectory for paging file.
- **AcadMaxPage** Maximum bytes written to first pager file.
- **AcadPlCmd** Command to launch plot queue software.
- **AcadServer** Network license server (*default: Netinel*).
- **RenderCfg** Points to subdirectory containing the Render configuration file.
- **RdpAdi** Points to protected-mode ADI rendering display driver.
- **RhpAdi** Points to protected-mode ADI hardcopy rendering driver.

- **AveFaceDir** Points to subdirectory containing the temporary face file.
- **Ignore_Big_Screen**
  Allows pre-ADI v.42 display drivers to use Release 12 and 13's virtual screen.
- **Ignore_Dragg**
  Digitizer driver does not use mouse-button-down drag mode.

RELATED FILES

- **Acad.Cfg** Configuration file for AutoCAD.
- **Acad.Ini** Initialization file for AutoCAD.
- **Asi.Ini** Initialization file for ASE module.
- **Render.Cfg** Configuration file for Render module.

RELATED SYSTEM VARIABLES

- **DctCust** Name of user spelling dictionary.
- **DctMain** Name of current spelling dictionary.
- **Platform** Reports the AutoCAD hardware platform version: "Microsoft Windows (Intel) Version 3.10" for the Windows version.

TIPS

- Option 6, Configure system console, displays the following dialogue box:

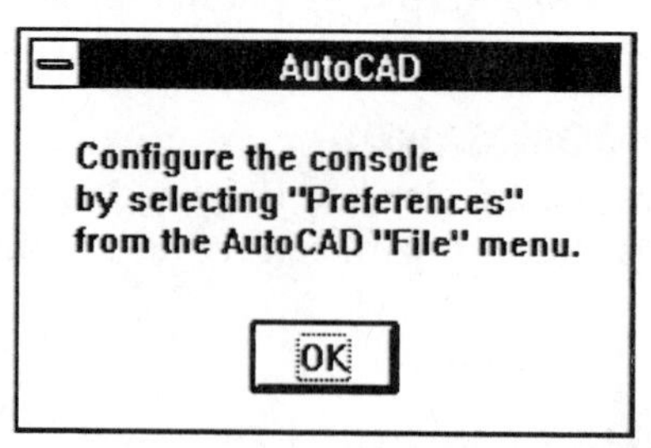

- Option 2, Allow detailed configuration, is now incorporated in the other options.

Copy

Ver. 1.0

Creates one or more copies of an object.

Command	Alias	Side Menu	Menu Bar	Tablet
copy	cp	[EDIT] [COPY:]	[Modify] [Copy]	X 15

Command: **copy**
Select objects: **[pick]**
Select objects: **[Enter]**
<Base point or displacement>/Multiple: **[pick]**
Second point of displacement: **[pick]**

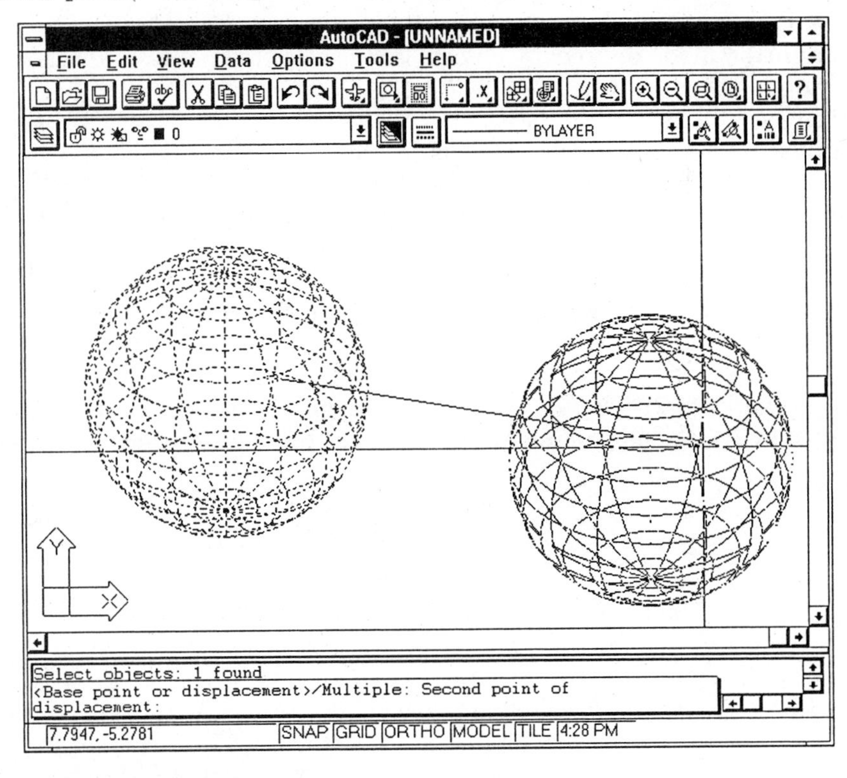

COMMAND OPTIONS

<Base point or displacement>
Indicates the starting point, or the distance to move.

Second point of displacement
Indicates the point to move.

Multiple Allows an object to be copied more than once.

[Esc] Cancels multiple object copying.

RELATED AUTOCAD COMMANDS

- **Array** Draws a rectangular or polar array of objects.
- **MInsert** Places an array of blocks.
- **Move** Moves an object to a new location.
- **Offset** Draws parallel lines, polylines, circles and arcs.

TIPS

■ Use the **M** (*multiple*) option to quickly place several copies of the original object.

■ Inserting a block multiple times is more efficient than placing multiple copies.

■ Turn ortho mode on to copy objects in a precise horizontal and vertical direction.

■ Turn snap mode on to copy objects in precise increments.

■ Use object snap modes to precisely copy objects from one geometric feature to another.

■ To copy an object by a known distance, enter 0,0 as the 'Base point.' Then, enter the known distance as the 'Second point.'

■ **Copy** works in 2D (*supply coordinate pairs*) and 3D (*supply coordinate triplets*). In 2D, the current elevation is used as the z-coordinate.

CopyClip

Rel. 12

Copies selected objects to the Windows Clipboard (*short for COPY to CLIPboard*).

Command	Ctrl+	Side Menu	Menu Bar	Tablet
copyclip	C	...	[Edit] [Copy]	...

```
Command: copyclip
Select objects: [pick]
Select objects: [Enter]
```

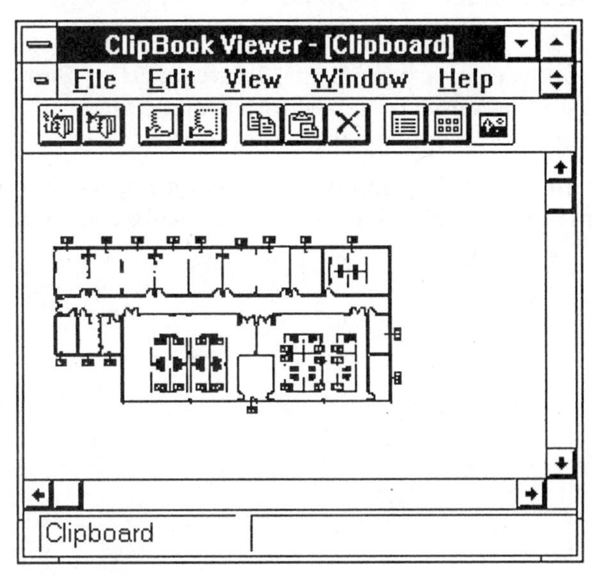

COMMAND OPTIONS

None

RELATED AUTOCAD COMMANDS

- **BmpOut** Exports selected objects in the current view to a BMP file.
- **CopyHist** Copies Text window text to the Windows Clipboard.
- **CopyLink** Copies current viewport to the Clipboard.
- **CutClip** Cuts selected objects to the Clipboard.
- **Export** Exports the drawing in several formats.
- **SaveImg** Saves rendered images in three formats.
- **WmfOut** Exports selected objects to a WMF file.

RELATED WINDOWS COMMANDS

- **[Prt Scr]** Copies the entire screen to the Windows Clipboard
- **[Alt]+[PrtScr]** Copies the topmost window to the Clipboard.

TIPS

■ AutoCAD's **CopyClip** command sends the selected objects to the Windows Clipboard in the following formats:

- **Bitmap** BMP (*bitmap*) raster format.
- **Native** AutoCAD R13 drawing format.
- **ObjectLink** Embedding information for OLE objects.
- **OwnerLink** Name of object's source application and filename.
- **Palette** Color palette used by object.
- **Picture** WMF (*Windows metafile*) vector format.

■ Contrary to the AutoCAD *Command Reference*, text objects are *not* copied to the Clipboard in text format.

■ In the other application, use the **Edit | Paste** or **Edit | Paste Special** commands to paste the AutoCAD image into the document; the **Paste Special** command lets you specify the pasted format.

■ When specifying the **All** option to the 'Select objects:' prompt, the **CopyClip** command only selects all objects visible in the current viewport.

CopyHist

Copies Text window text to the Windows Clipboard (*short for COPY HISTory*).

Command	Alias	Side Menu	Menu Bar	Tablet
copyhist	...	...	[Edit] [Copy]	...

Command: **copyhist**

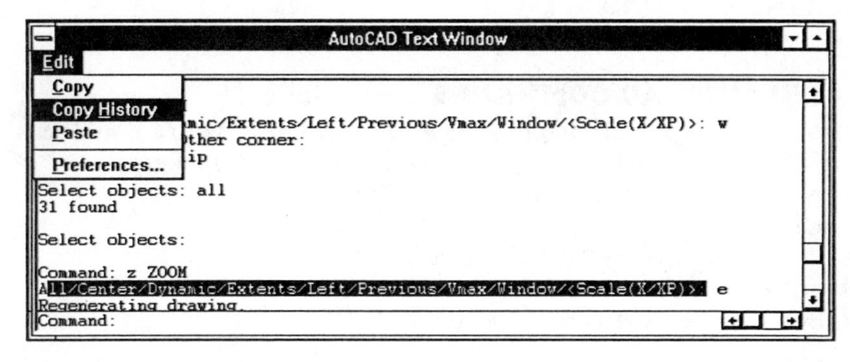

COMMAND OPTIONS

None

RELATED AUTOCAD COMMANDS

- **CopyClip** Copies selected text from the drawing to the Clipboard in text format.
- **CopyLink** Copies current viewport to Clipboard.
- **CutClip** Cuts selected objects to the Clipboard.

RELATED WINDOWS COMMAND

- **[Alt]+[PrtScr]** Copies the Text window to the Clipboard in graphics format.

TIPS

- AutoCAD's **CopyHist** command sends the text to the Windows Clipboard in the following formats:
  - **OemText** Identical to text format except that the Clipboard displays text with a proportional font; not treated differently when pasted into another applications.
  - **Text** Each line of text ends with a linefeed; does not support fonts or formatting.

- To copy a selected portion of Text window text to the Clipboard, highlight the text first, then use the **Edit | Copy** command.

CopyLink

Rel. 13

Copiess the current viewport to the Windows Clipboard.

Command	Alias	Side Menu	Menu Bar	Tablet
copylink	...	...	[Edit] [Copy Link]	...

Command: **copyclip**

COMMAND OPTIONS

None

RELATED AUTOCAD COMMANDS

- **BmpOut** Exports selected objects in the current view to a BMP file.
- **CopyClip** Copies selected objects to the Windows Clipboard.
- **CopyHist** Copies Text window text to the Clipboard.
- **CutClip** Cuts selected objects to the Clipboard.
- **Export** Exports the drawing in several formats.
- **SaveImg** Saves rendered images in three formats.
- **WmfOut** Exports selected objects to a WMF file.

RELATED WINDOWS COMMANDS

- **[Prt Scr]** Copies the entire screen to the Windows Clipboard
- **[Alt]+[PrtScr]** Copies the topmost window to the Clipboard.

TIPS

- AutoCAD's **CopyClip** command sends the selected objects to the Windows Clipboard in the following formats:
  - **Bitmap** BMP (*bitmap*) raster format.
  - **ObjectLink** Embedding information for OLE objects.
  - **OwnerLink** Name of object's source application and filename.
  - **Palette** Color palette used by object.
  - **Picture** WMF (*Windows metafile*) vector format.

- In the other application, use the **Edit | Paste** or **Edit | Paste Special** commands to paste the AutoCAD image into the document; the **Paste Special** command lets you specify the pasted format.

CutClip

Rel. 12

Copies selected objects to the Windows Clipboard (*short for COPY to CLIPboard*).

Command	Ctrl+	Side Menu	Menu Bar	Tablet
cutclip	X	...	[Edit] [Cut]	...

```
Command: cutclip
Select objects: [pick]
Select objects: [Enter]
```

COMMAND OPTIONS

None

RELATED AUTOCAD COMMANDS

- **BmpOut** Exports selected objects in the current view to a BMP file.
- **CopyClip** Copies selected objects to the Clipboard.
- **CopyHist** Copies Text window text to the Windows Clipboard.
- **CopyLink** Copies current viewport to the Clipboard.
- **Export** Exports the drawing in several formats.
- **SaveImg** Saves rendered images in three formats.
- **WmfOut** Exports selected obejcts to a WMF file.

RELATED WINDOWS COMMANDS

- **[Prt Scr]** Copies the entire screen to the Windows Clipboard
- **[Alt]+[PrtScr]** Copies the topmost window to the Clipboard.

TIPS

- AutoCAD's **CopyClip** command sends the selected objects to the Windows Clipboard in the following formats:
  - **Bitmap** BMP (*bitmap*) raster format.
  - **Native** AutoCAD R13 drawing format.
  - **OwnerLink** Name of object's source application and filename.
  - **Palette** Color palette used by object.
  - **Picture** WMF (*Windows metafile*) vector format.

- Contrary to the AutoCAD *Command Reference*, text objects are *not* copied to the Clipboard in text format.

- In the other application, use the **Edit | Paste** or **Edit | Paste Special** commands to paste the AutoCAD image into the document; the **Paste Special** command lets you specify the pasted format.

- When specifying the **All** option to the 'Select objects:' prompt, the **CutClip** command only selects all objects visible in the current viewport.

Cylinder

Rel. 12

Draws a 3D ACIS cylinder with a circular or elliptical cross section (*formerly the **SolCyl** command; an external command in Acis.Dll*).

Command	Alias	Side Menu	Menu Bar	Tablet
cylinder	...	[DRAW 2]	[Draw]	L 7
		[SOLIDS]	[Solids]	
		[Cylindr:]	[Cylinder]	

```
Command: cylinder
Elliptical/<center point> <0,0,0>: [pick]
Diameter/<Radius>: [pick]
Center of other end/<Height>: [pick]
```

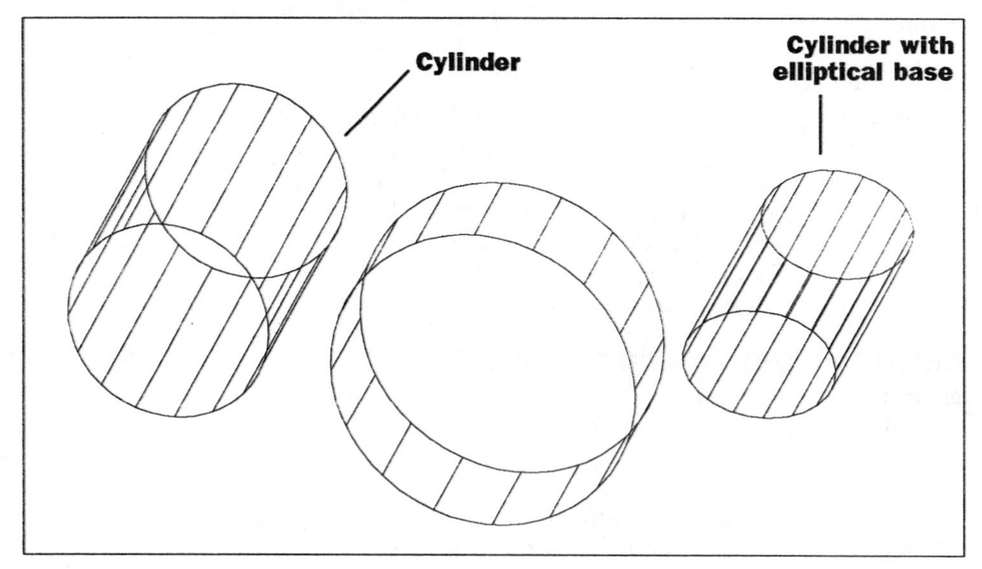

RELATED AUTOCAD COMMANDS

- **Ai_Cyl** Draws a 3D wireframe cylinder.
- **Box** Draws a 3D solid box.
- **Cone** Draws a 3D solid cone.
- **Extrude** Creates a cylinder with an arbitrary crosssection and sloped walls.
- **Sphere** Draws a 3D solid ball.
- **Torus** Draws a 3D solid donut.
- **Wedge** Draws a 3D solid wedge.

RELATED SYSTEM VARIABLES

- **DispSilh** Toggles display of 3D objects as silhouette after hidden-line removal and shading.

■ **IsoLines** Number of isolines on solid surfaces:

0 Minimum (*no isolines*).

4 Default.

16 A reasonable value.

2,047 Maximum value.

TIP

■ The **Ellipse** option draws a cylinder with an elliptical cross-section.

DbList

Ver. 1.0

Lists information on all entities in the drawing (*short for Data Base LISTing*).

Command	Alias	Side Menu	Menu Bar	Tablet
dblist	...	...	...	...

Command: **dblist**

Example listing:

```
          LINE        Layer: 1
                      Space: Model space
     from point, X=    3.7840   Y=    4.7169   Z=    0.0000
       to point, X=    4.1440   Y=    4.7169   Z=    0.0000
  Length =    0.3600,   Angle in X-Y Plane =       0
 Delta X =    0.3600, Delta Y = 0.0000, Delta Z = 0.0000

          ARC         Layer: 2
                      Space: Model space
   center point, X=    2.1000   Y=    7.0000   Z=    0.0000
   radius     1.2000
    start angle      0
      end angle    180
```

COMMAND OPTIONS

[Enter] Continues display after pause.
[[Esc] Cancels database listing.

RELATED AUTOCAD COMMANDS

- **Area** Lists the area and perimeter of objects.
- **Dist** Lists the 3D distance and angle between two points.
- **Id** Lists the 3D coordinates of a point.
- **List** Lists information about selected entities in the drawing.

RELATED SYSTEM VARIABLES

- *None*

DdAttDef

Rel. 12

Define an attribute definition via a dialogue box (*short for Dynamic Dialogue ATTribute DEFinition; an external file in DdAttDef.Lsp*).

Command	Alias	Side Menu	Toolbar	Tablet
`ddattdef`	...	`[CONSTRCT]`	`[Attrbibute]`	`W 7`
		`[DDatDef:]`	`[DdAttDef]`	

`Command:` **`ddattdef`**

Displays dialogue box.

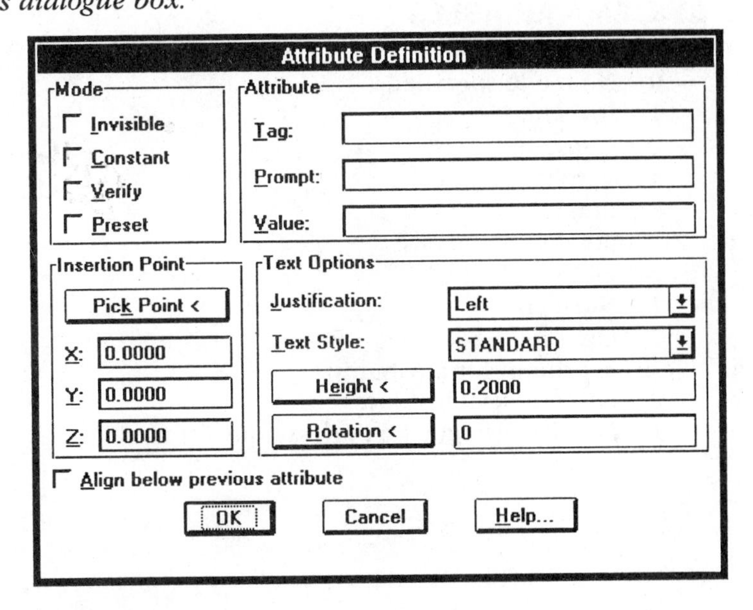

COMMAND OPTIONS

Mode Sets the attribute text modes:

- **Invisible** Makes the attribute text invisible.
- **Constant** Uses constant values for the attributes.
- **Verify** Verifies the text after input.
- **Preset** Presets the variable attribute text.

Attribute Sets the attribute text:

- **Tag** Identifies the attribute.
- **Prompt** Prompts the user for input.
- **Value** Specifies the default value for the attribute.

Insertion point Specifies the attributes insertion point:

- **Pick point** Picks insertion point with cursor.
- **X** X-coordinate insertion point.
- **Y** Y-coordinate insertion point.
- **Z** Z-coordinate insertion point.

Text options Specifies the attribute text options:

- **Justification** Sets the justification.
- **Text style** Selects a style.
- **Height** Specifies the height.
- **Rotation** Sets the rotation angle.

Align Automatically places the text below the previous attribute.

RELATED AUTOCAD COMMAND

■ **AttDef** Defines attribute definitions from the command line.

RELATED SYSTEM VARIABLES

■ **AFlags** Attribute mode:

- 0 No mode specified.
- 1 Invisible.
- 2 Constant.
- 4 Verify.
- 8 Preset.

■ **AttMode** Attribute display modes:

- 0 Off.
- 1 Normal.
- 2 On.

■ **AttReq** Toggles prompt for attributes:

- 0 Assume default values.
- 1 Enables dialogue box or prompts for attributes.

DdAttE

Rel. 9

Edits attribute data via a dialogue box (*short for Dynamic Dialogue ATTribute Editor*).

Command	Alias	Side Menu	Toolbar	Tablet
`ddatte`	...	[MODIFY] [AttEd:]	[Attribute] [DdAttE]	X 7

```
Command: ddatte
Select block: [pick]
```

Displays dialogue box.

Edit Attributes
Block Name: OVERHEAD
Light Name — Overhead
GNAME — Heather
Look-at X — 1
Look-at Y — 2
Look-at Z — 3
Light Intensity — 1
Light Color (RGB) — 1,1,1
Depth Map Size — 0
OK | Cancel | Previous | Next | Help...

COMMAND OPTIONS

None

RELATED AUTOCAD COMMANDS

- **AttEdit** Global attribute editor.
- **AttReDef** Changes the definition of attributes.
- **DdEdit** Edits attribute definitions.

RELATED SYSTEM VARIABLE

- **AttDia** Toggles use of **DdAttE** during **Insert** command.

TIP

- The **DdEdit** command edits attribute definitions.

DdAttExt

Rel. 12

Extracts attribute information to a file (*short for Dynamic Dialogue ATTribute EXTraction; an external command in Ddattext.Lsp*).

Command	Alt+	Side Menu	Menu Bar	Tablet
ddattext	F,E	[FILE]	[File]	...
		[EXPORT]	[Export]	
		[DDattEx:]	[DXX]	

```
Command: ddattext
```

Displays dialogue box.

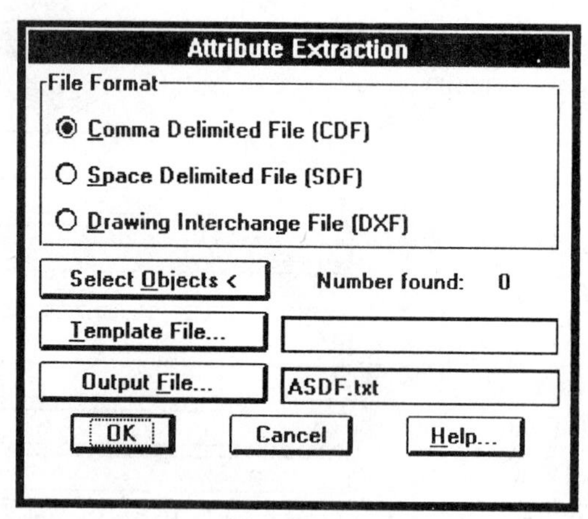

COMMAND OPTIONS

Comma delimited
Creates a CDF text file, where commas separate fields.

Space delimited
Creates a SDF text file, where spaces separate fields.

Drawing interchange
Creates an ASCII DXF file.

Select Objects
Returns to the graphics screen to select attributes for export.

Template file Specifies the name of the TXT template file for CDF and SDF files.

Output File Specifies the name of the attribute output file:

- **TXT** CDF and SDF formats.
- **DXX** DXF format.

RELATED AUTOCAD COMMAND

■ **AttExt** Attribute extraction via command-line interface.

DdChProp

Rel. 12

Modifies the color, layer, linetype, linetype scale, and thickness of most objects via a dialogue box (*short for Dynamic Dialogue CHange PROPerties; an external command in DdChProp.Lsp*).

Command	Alias	Side Menu	Menu Bar	Tablet
ddchprop	...	[MODIFY] [Ddchpro:]	...	...

```
Command: ddchprop
Select objects: [pick]
```

Displays dialogue box:

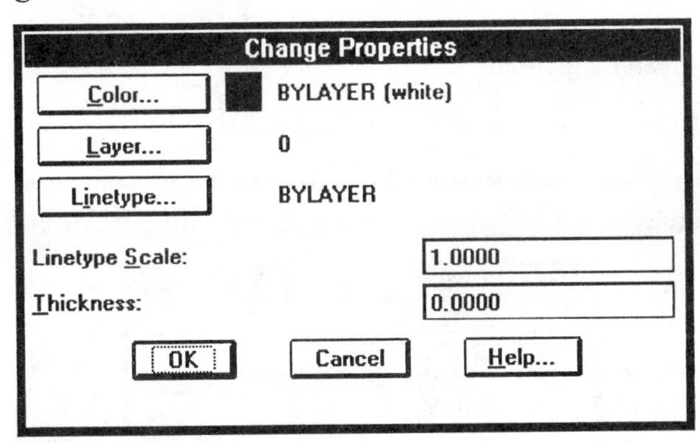

COMMAND OPTIONS

Color Changes the color of the selected objects by dialogue box:

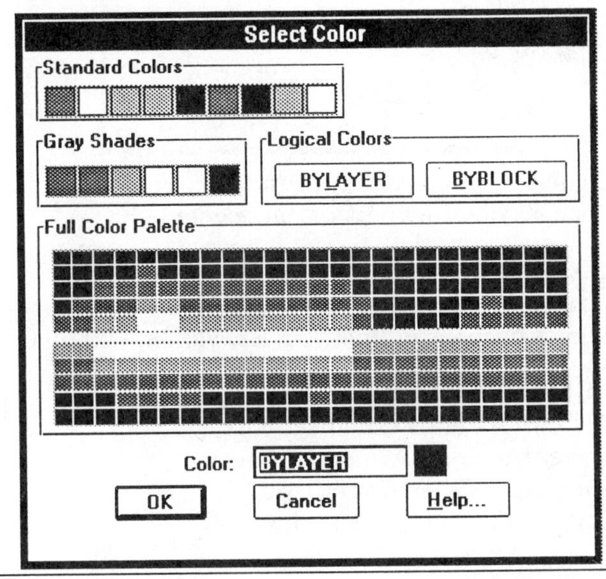

Layer name Moves the selected objects to a different layer by dialogue box:

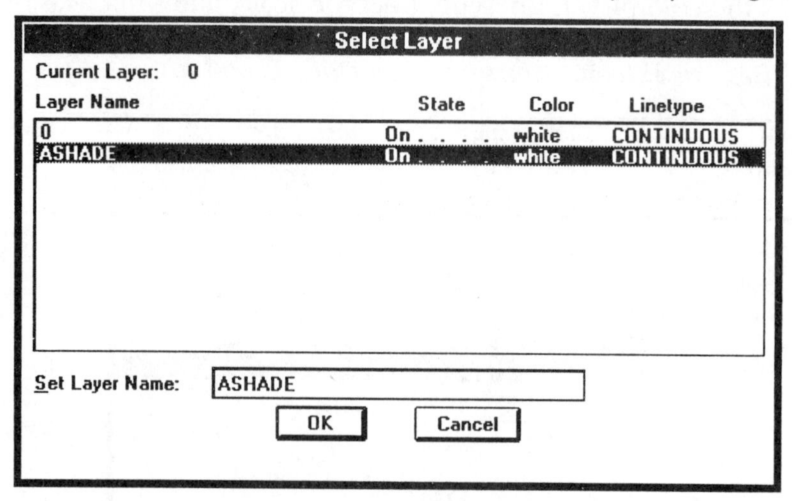

LInetype Changes the linetype of the selected objects by dialogue box:

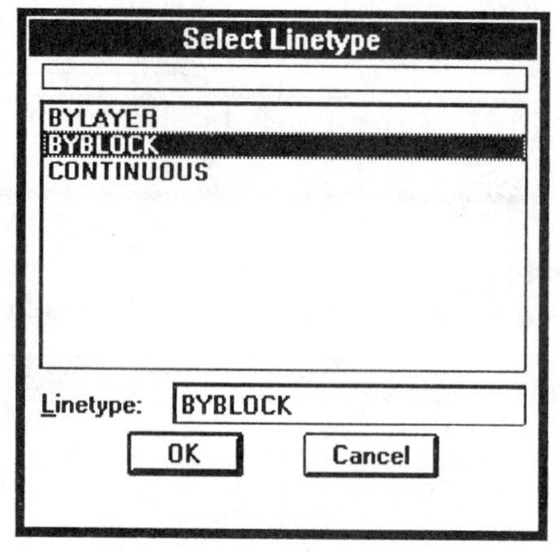

Linetype Scale Changes the linetype scale of the selected objects.
Thickness Changes the thickness of the selected objects.

RELATED AUTOCAD COMMANDS

- **ChProp** Change properties via command line.
- **Change** Allows changes to lines, circles, blocks, text, and attributes.

RELATED SYSTEM VARIABLES

- **CeColor** The current entity color setting.
- **CeLtype** The current entity linetype setting.
- **CLayer** The name of the current layer.
- **Thickness** The current thickness setting.

'DdColor

Rel. 13

Set the current working color by dialogue box (*an external command in DdColor.Lsp*).

Command	Alt+	Side Menu	Menu Bar	Tablet
'ddcolor	D,C	[DATA]	[Data]	...
		[Color:]	[Color]	
		[ColorDlg]		

Command: **ddcolor**

Displays dialogue box:

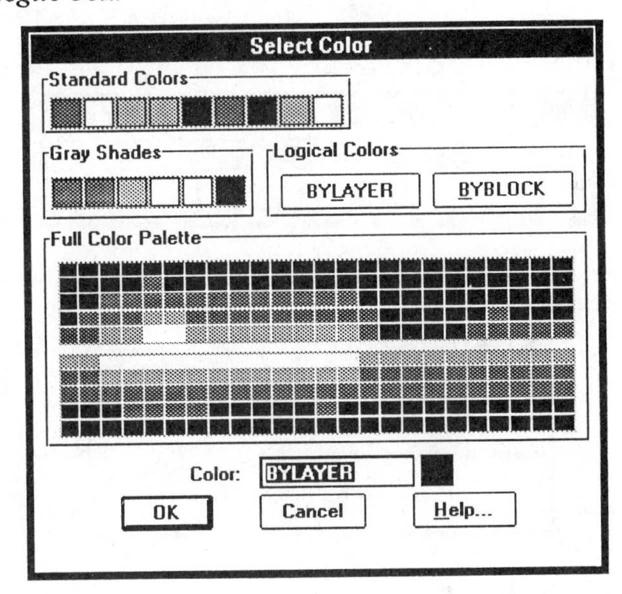

COMMAND OPTION

Color Enters color number, BYBLOCK, or BYLAYER.

RELATED AUTOCAD COMMANDS

- **Color** Changes the color from the 'Command' prompt.
- **DdChProp** Changes the color of selected objects.

RELATED SYSTEM VARIABLE

- **CeColor** Contains the number for the current working color.

DdEdit

Rel. 11

Edits a single line of text or a single attribute using a dialogue box; launches the text editor for editing multiline text (*short for Dynamic Dialogue EDITor*).

Command	Alias	Side Menu	Toolbar	Tablet
ddedit	...	[MODIFY] [Ddedit:]	[Modify] [Special Edit] [DdEdit]	U 5

Command: **ddedit**
<Select a TEXT or ATTDEF object>/Undo: **[pick text]**

Displays dialogue box.

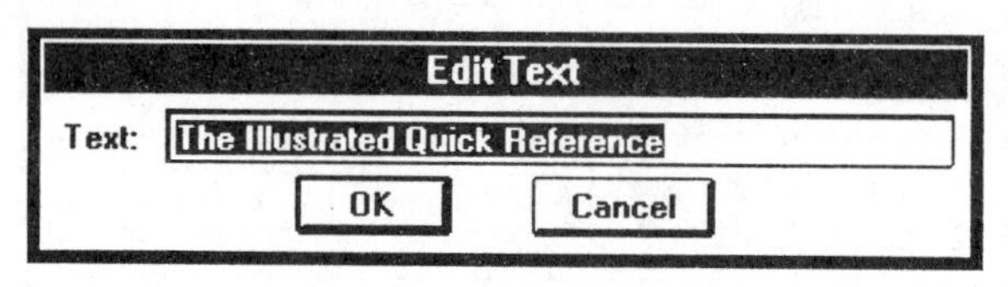

<Select a TEXT or ATTDEF object>/Undo: **[pick attribute definition]**
Displays dialogue box.

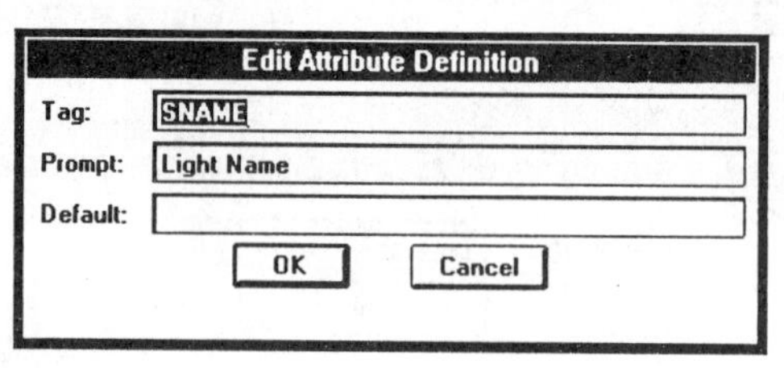

COMMAND OPTIONS

U Undoes editing operation.
[Esc] Cancels to end the command.

RELATED AUTOCAD COMMANDS

- **DdAttE** Edits all text attributes connected with a block.
- **Change** Edits some text attributes.

RELATED SYSTEM VARIABLE

- **MTextEd** Name of the text editor used for editing multiline text.

TIPS

- The **DdEdit** command automatically repeats; press **[Esc]** to cancel the command.

- Use the **DdAttE** command to edit attribute values.

'DdEModes

Rel. 9

Sets the working parameters (*short for Dynamic Dialogue Entity MODES*).

Command	Alt+	Side Menu	Menu Bar	Tablet
'ddemodes	D,O	[DATA]	[Data]	Y 9
		[DDemode:]	[Object Creation]	

```
Command: ddemodes
```

Displays dialogue box.

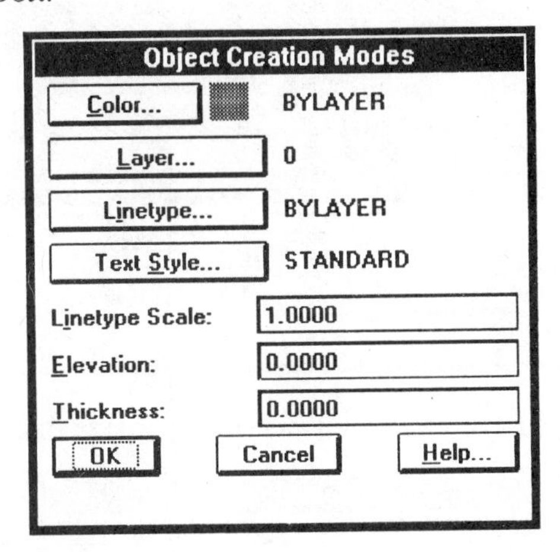

COMMAND OPTIONS

Elevation Sets the new working elevation.

Thickness Sets the new working thickness.

Layer Sets the new working layer; displays dialogue box.

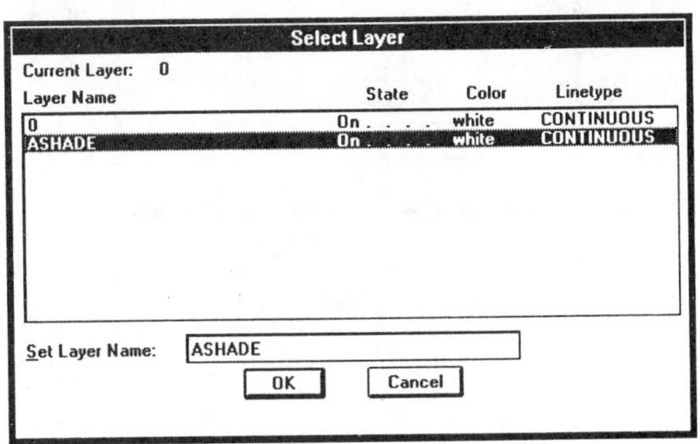

Color Sets the new working color; displays dialogue box.

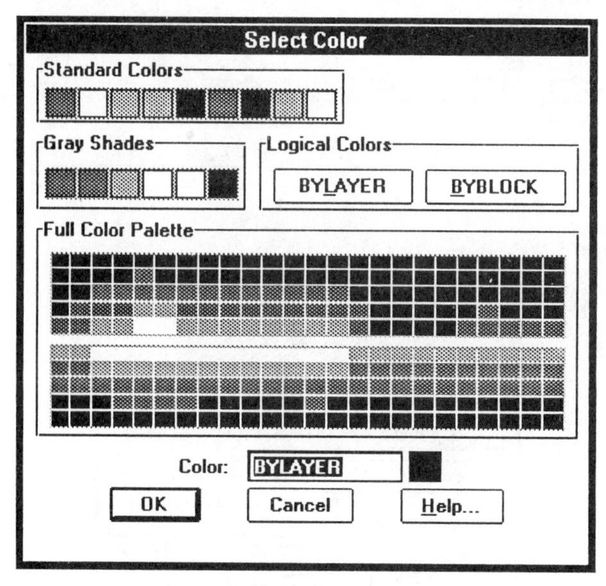

Linetype Sets the new working linetype; displays dialogue box.

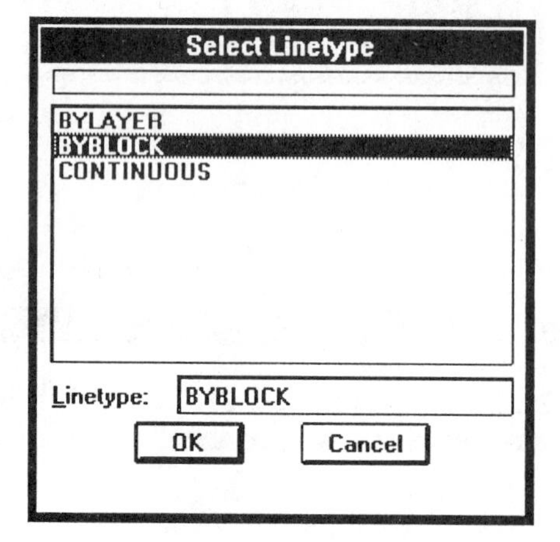

Text style Sets the new working textstyle; displays dialogue box.

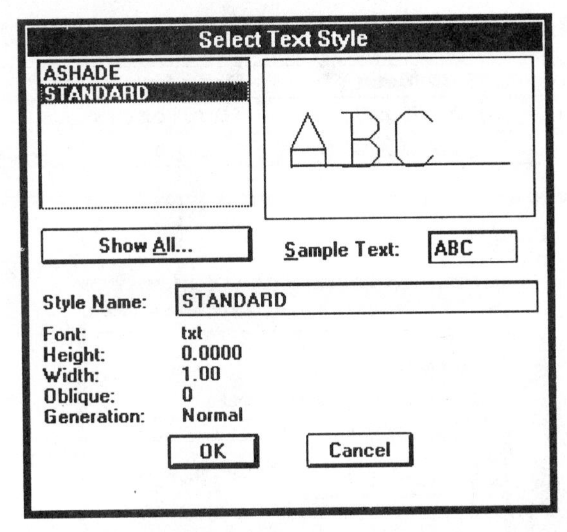

RELATED AUTOCAD COMMANDS

- **Change** Changes the color, layer, linetype, text elevation, and thickness.
- **ChProp** Changes the color, layer, linetype and thickness.
- **Color** Sets a new working color.
- **Colour** Sets a new working colour.
- **DdAttE** Edits attribute values.
- **DdChProp** Changes the color, layer, linetype, linetype scale, and thickness.
- **DdEdit** Changes text or attribute definition.
- **Elevation** Sets a new working elevation.
- **Layer** Sets a new working layer.
- **Linetype** Sets a new working linetype.
- **Style** Sets a new text style.

RELATED SYSTEM VARIABLES

- **CeColor** The current entity color.
- **CeLtype** The current entity linetype.
- **CLayer** The current layer name.
- **Elevation** The current elevation setting.
- **TextStyle** The current text style setting.
- **Thickness** The current thickness setting.

'DdGrips

Rel. 12

Turns object grips on and off; defines the size and color of grips (*an external command in DdGrips.lsp*).

Command	Alt+	Side Menu	Menu Bar	Tablet
'ddgrips	O,G	[OPTIONS] [DDgrips:]	[Options] [Grips]	...

Command: **ddgrips**

Displays dialogue box.

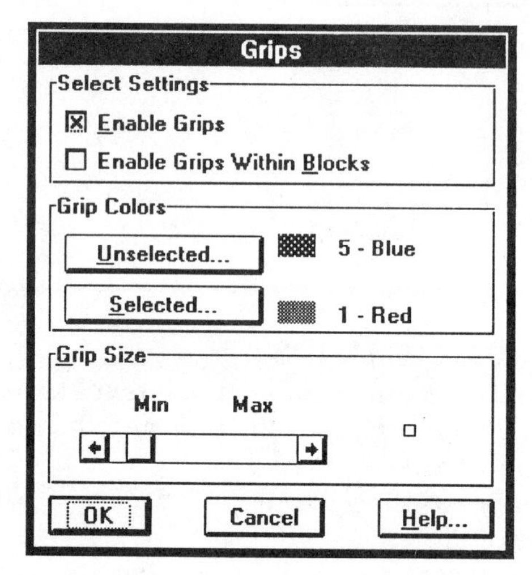

COMMAND OPTIONS

Enable grips Toggles the display of entity grips.

Enable grips within Blocks
Displays grips on entities within blocks.

Grip size Changes the size of the grip box.

Unselected Defines the color of unselected grips; displays dialogue box.
Selected Defines the color of selected grips; displays dialogue box.

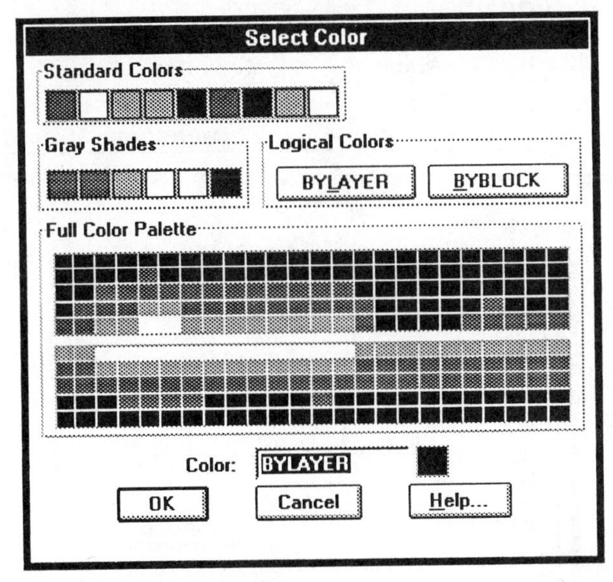

RELATED AUTOCAD COMMAND

- **Select** Creates a selection set of entities.

RELATED SYSTEM VARIABLES

- **Grips** Toggles use of grips:
  - **0** Disable grips.
  - **1** Enable grips (*default*).
- **GripBlock** Toggles display of grips inside blocks:
  - **0** Display grip only in block insertion point (*default*).
  - **1** Display grips on entities inside block.
- **GripColor** Color of unselected grips (*default=5, blue*).
- **GripHot** Color of selected grips (*default=1, red*).
- **GripSize** Size of grip, in pixels:
  - **1** Minimum size.
  - **3** Default size.
  - **255** Maximum size.

DDim

Rel. 12

Sets dimension styles and variables via a dialogue box (*short for Dialogue DIMension*).

Command	Alt+	Side Menu	Menu Bar	Tablet
ddim	D, D	[DATA]	[Data]	V 5-6
		[DDim:]	[Dimension Style]	

Command: **ddim**

Displays dialogue box.

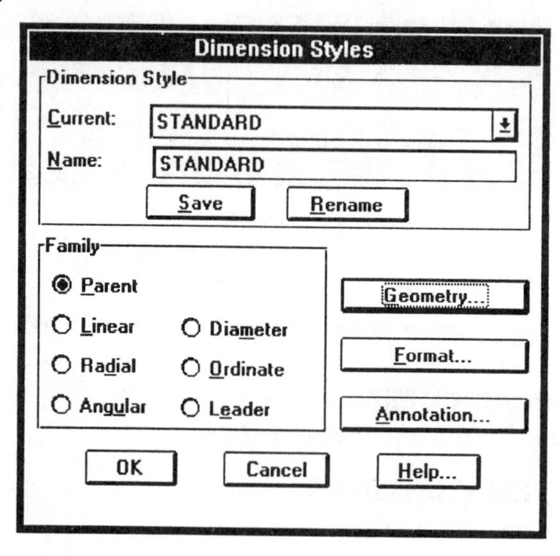

COMMAND OPTIONS

Dimension Style

Creates and select a dimension style (*or dimstyle, for short*):

Current Selects a dimstyle name.

Name Creates a new-named dimstyle.

Save Saves the dimstyle.

Rename Renames the dimstyle.

Geometry Specifies the dimension geometry variables: dimension, extension, and center lines; center marks; arrowheads; and overall scale; displays dialogue box.

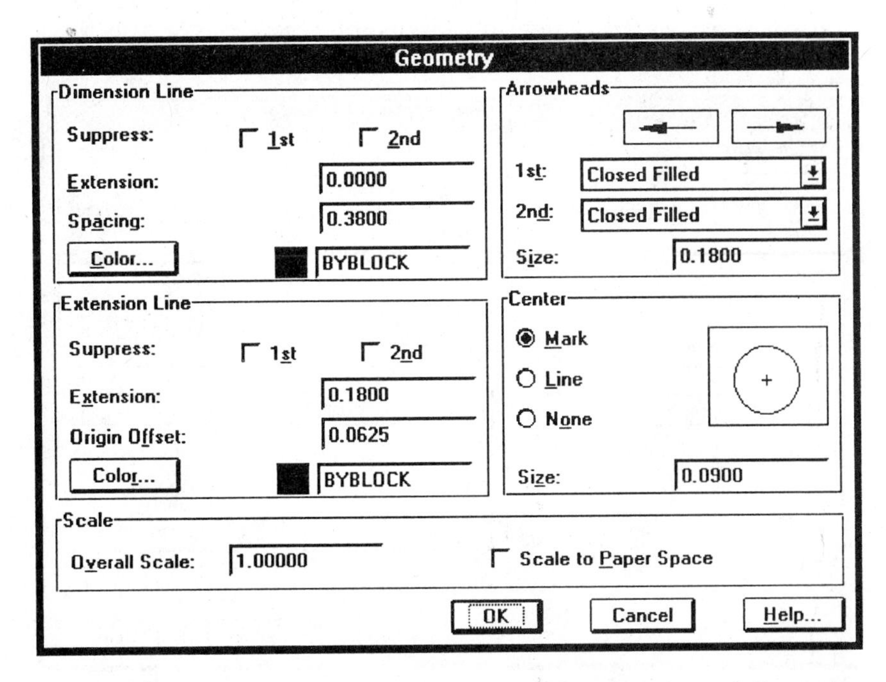

Format Specifies dimension format variables: location of dimension text; arrowheads; leader and dimension lines; displays dialogue box.

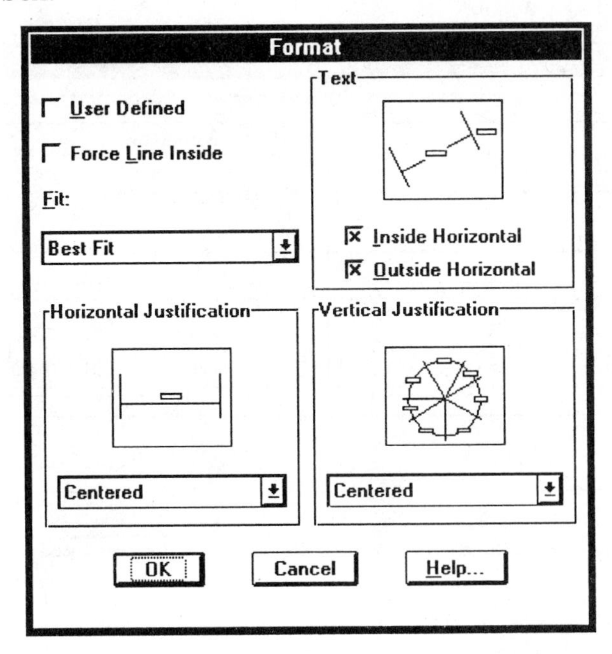

Annotation Specifies the dimension annotation variables: controls the look of dimension text; displays dialogue box.

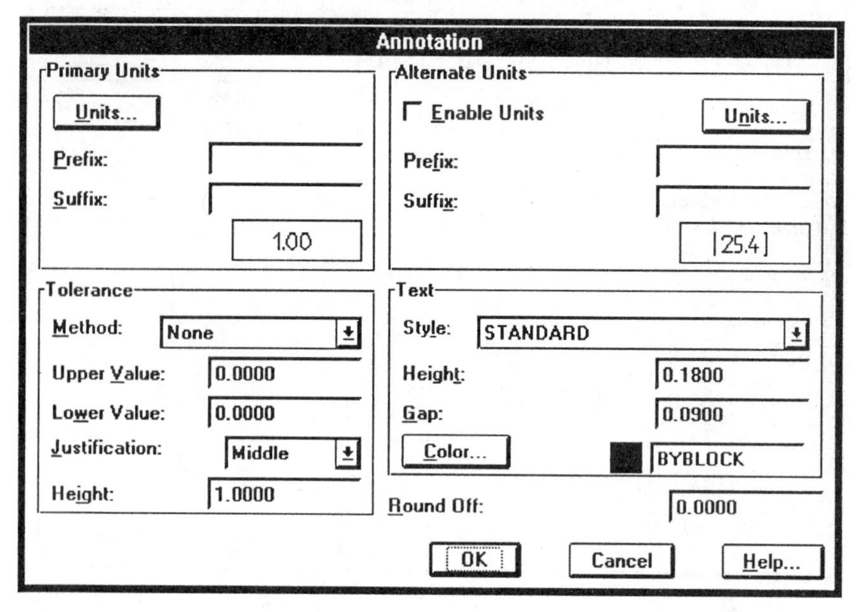

Primary Units

Specify the display of the primary dimensioning units; displays dialogue box.

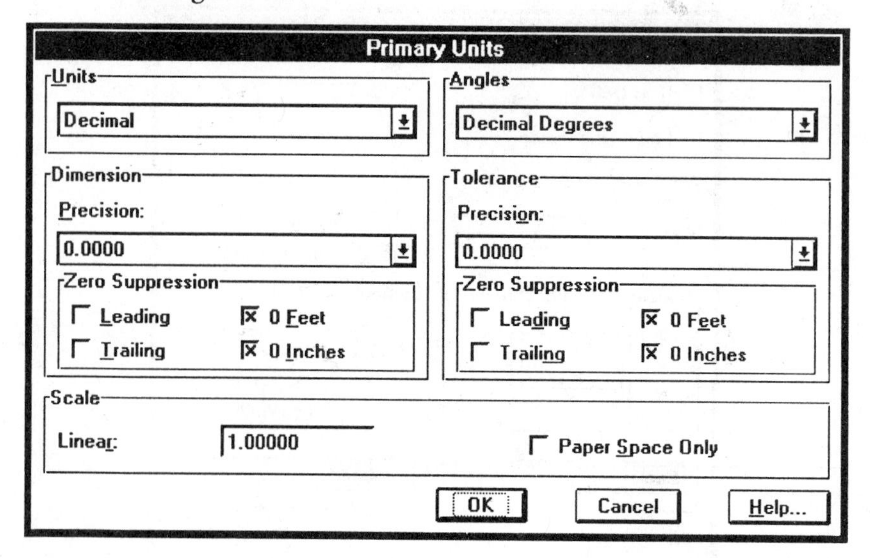

Alternate Units

Specifies the display of the alternate dimensioning units; displays dialogue box.

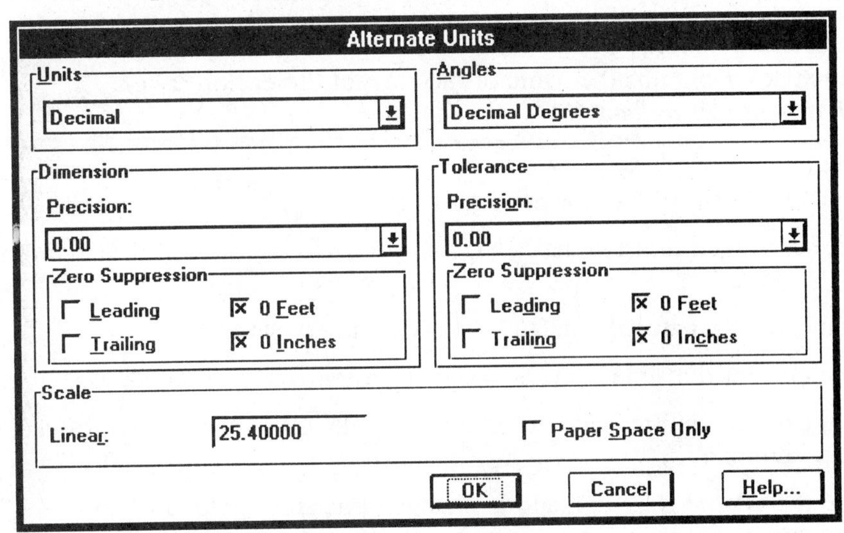

Color Specify the text color; displays dialogue box.

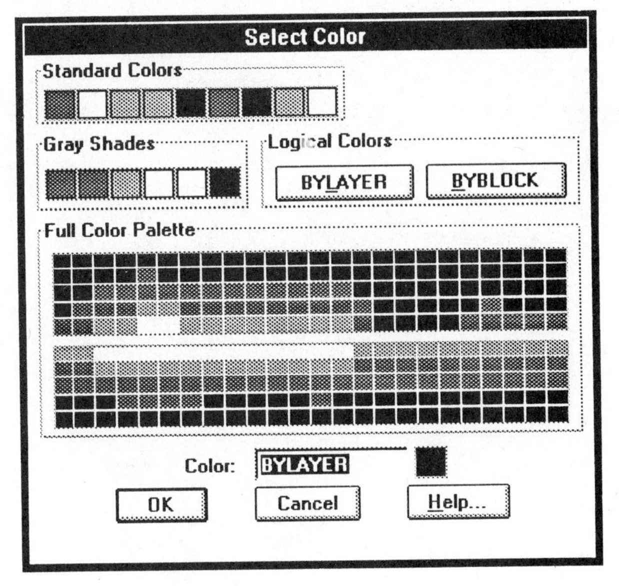

RELATED AUTOCAD COMMANDS

- **DimStyle** Changes dimension variables at the 'Command' prompt.
- *All commands prefixed with Dim.*

RELATED SYSTEM VARIABLES

- **DimStyle** Contains the name of the current dimension style.
- *All dimensioning variables.*

TIPS

- Dimstyle is shorthand for *dimension style.*

- An overridden dimstyle has a + (*plus*) prefix to the dimstyle name, as in '+STANDARD'.

- You cannot rename the default dimstyle named 'Standard.'

- The current dimstyle name is stored in system variable **DimStyle**.

- You can access dimstyles stored in externally-referenced drawings via the **XBind** command.

- Use the **DimStyle** command to change dimension styles at the 'Command' prompt.

DdInsert

Rel. 12

Insert blocks via dialogue box (*short for Dialogue Dynamic INSERT; an external command in DdInsert.Lsp*).

Command	Alias	Side Menu	Toolbar	Tablet
ddinsert	...	[DRAW 2] [DDinsert]	[Draw] [Block] [DdInsert]	W 10

Command: **ddinsert**

Displays dialogue box.

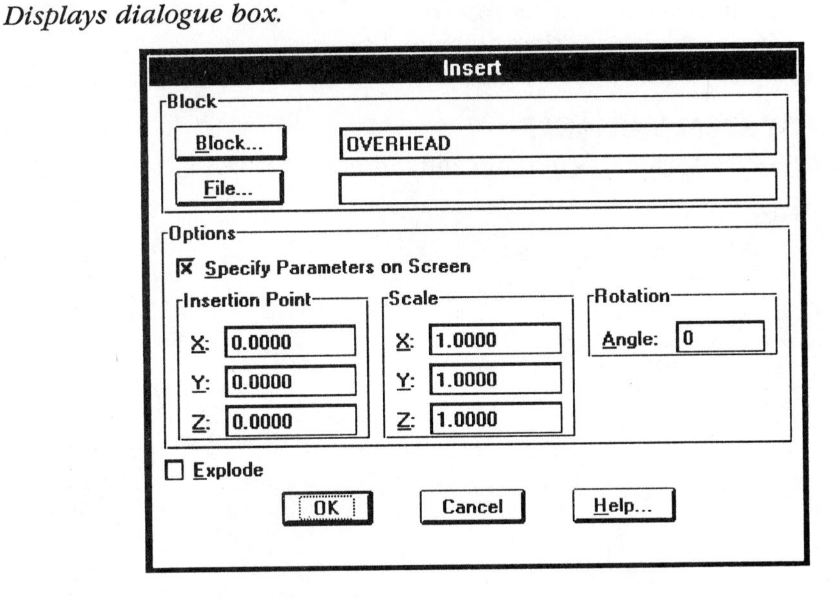

COMMAND OPTIONS

Block Selects the block name from a second dialogue box.

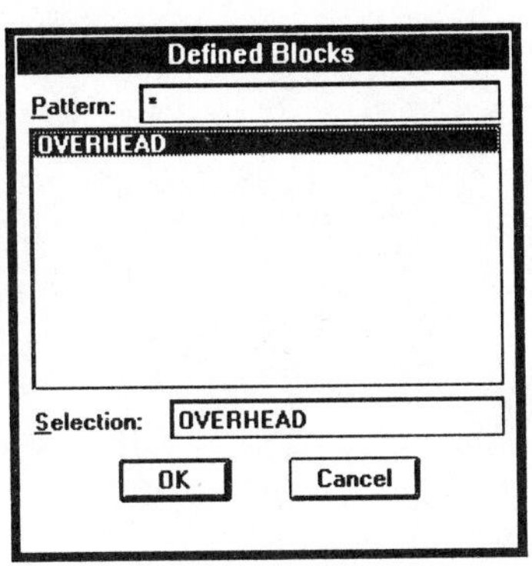

File Selects the drawing name from a second dialogue box.

Specify parameters on screen
Uses the cursor to position the block.

Insertion point Specifies the block's insertion point coordinates.

Scale Specifies the block's scale.

Rotation Specifies the block's angle of rotation.

Explode Inserts the block as individual entities.

RELATED AUTOCAD COMMANDS

- **Block** Creates a block from a group of objects.
- **Explode** Explodes a block after insertion.
- **Insert** Inserts blocks via the command line.
- **MInsert** Inserts an array as a block.

RELATED SYSTEM VARIABLE

- **InsName** Default name for most-recently inserted block.

'DdLModes

Rel. 9

Controls the layer settings in the drawing via a dialogue box (*short for Dynamic Dialogue Layer MODES*).

Command	Alt+	Side Menu	Menu Bar	Tablet
`'ddlmodes`	`D,L`	`[DATA]`	`[Data]`	`L 4-5`
		`[DDlmode:]`	`[Layers]`	

```
Command: ddlmodes
```

Displays dialogue box.

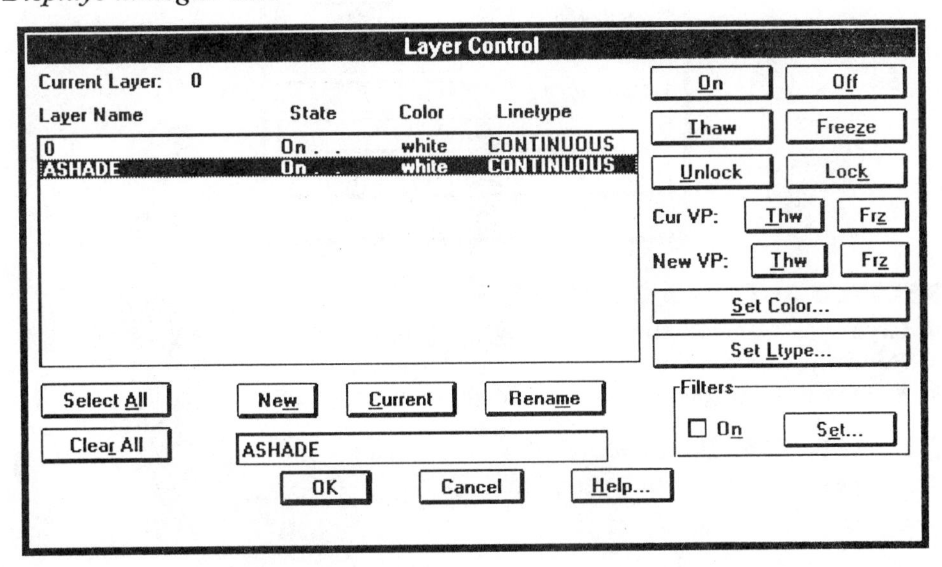

COMMAND OPTIONS

Clear all Clears all selected layers.
Current Sets the working (*current*) layer name.
Freeze Freezes the selected layers.
Lock Locks the selected layers.
New Creates a new layer name.
Off Turns off the selected layers.
On Turns on the selected layers.
Rename Renames a layer.
Select All Selects all layers.
Set Color Sets a new color for the selected layers.
Set Ltype Sets a new linetype for the selected layers.
Thaw Thaws the selected frozen layers.
Unlock Unlocks the selected locked layers.

Set filters Creates a filter set via dialogue box.

Set Layer Filters

On/Off: Both

Freeze/Thaw: Both

Lock/Unlock: Both

Current Vport: Both

New Vports: Both

Layer Names: *

Colors: *

Ltypes: *

Reset

OK Cancel Help...

RELATED AUTOCAD COMMANDS

- **Layer** Controls the current layer setting.
- **Rename** Renames a layer.
- **VpLayer** Controls the layer settings in paper space.

RELATED SYSTEM VARIABLE

- **CLayer** The current layer setting.

TIPS

- When a layer has been turned off, AutoCAD no longer displays nor plots entities on that layer.

- When a layer is frozen, AutoCAD no longer takes its entities into account during a regeneration; in addition, AutoCAD no longer displays nor plots entities on that layer.

- Freezing a layer is more efficient than turning the layer off.

- When a layer is locked, its entities are displayed and you can draw on the layer but you cannot edit the layer's entities.

- Locking layers is useful for redlining; you can make additions and notes but not change the drawing.

'DdLtype

Rel. 13

Loads linetype definitions, sets the working linetype and scale.

Command	Alt+	Side Menu	Menu Bar	Tablet
'ddltype	D, N	[DATA]	[Data]	Y 8
		[DDltype:]	[Linetype]	

Command: **ddltype**
Displays dialogue box:

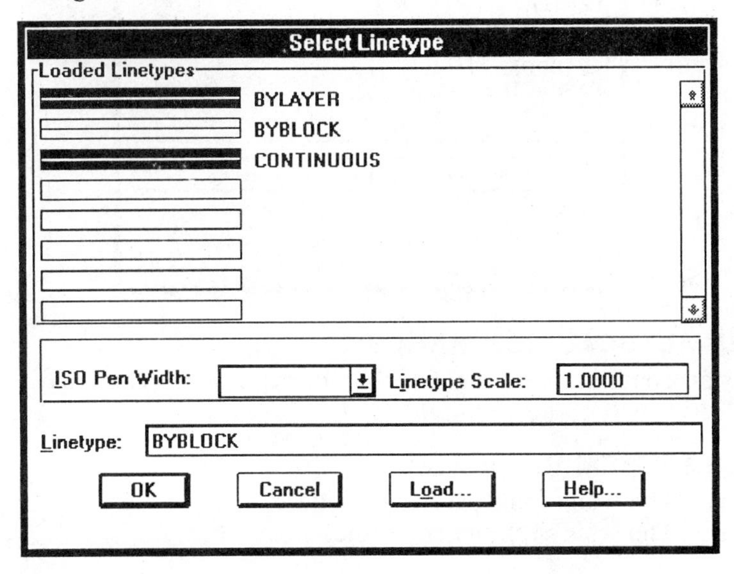

COMMAND OPTIONS

ISO Pen Width
Sets the pen width for ISO linetype (*measured in millimeters*).

Linetype Scale
Sets the working linetype scale.

Linetype Selects the name of working linetype pattern.

Load Loads linetype definitions from a LIN file:

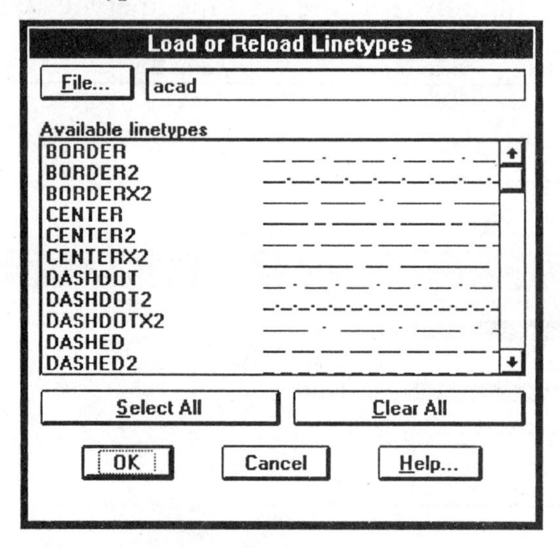

RELATED AUTOCAD COMMANDS

- **Linetype** Loads and sets the working linetype at the 'Command' prompt.
- **LtScale** Sets the linetype scale.

RELATED SYSTEM VARIABLES

- **CeLtype** The name of the current linetype.
- **LtScale** The scale of the current linetype.

RELATED FILES

- **Acad.Lin** Definitions of all linetypes used by AutoCAD.
- **TypeShp.Lin** Text shapes for linetypes.

TIPS

- See the **Linetype** command for the list of linetypes supplied with AutoCAD Release 13.

DdModify

Rel. 12

Views and edits properties of all entities via a dialogue box (*an external command in DdModify.Lsp*).

Command	Alt+	Side Menu	Menu Bar	Tablet
ddmodify	E, R	...	[Edit]	V 9-10
		...	[Properties]	

```
Command: ddmodify
Select object to list: [pick object]
```

A different dialogue box appears for type of object.

COMMAND OPTIONS

Color Changes the object's color via a dialogue box.
Layer name Moves the object to a different layer.
Linetype Changes the object's linetype via a dialogue box.
Linetype Scale
Changes the object's linetype scale.
Thickness Changes the object's thickness.

[pick arc] Displays the **Modify Arc** dialogue box.

Displays dialogue box.

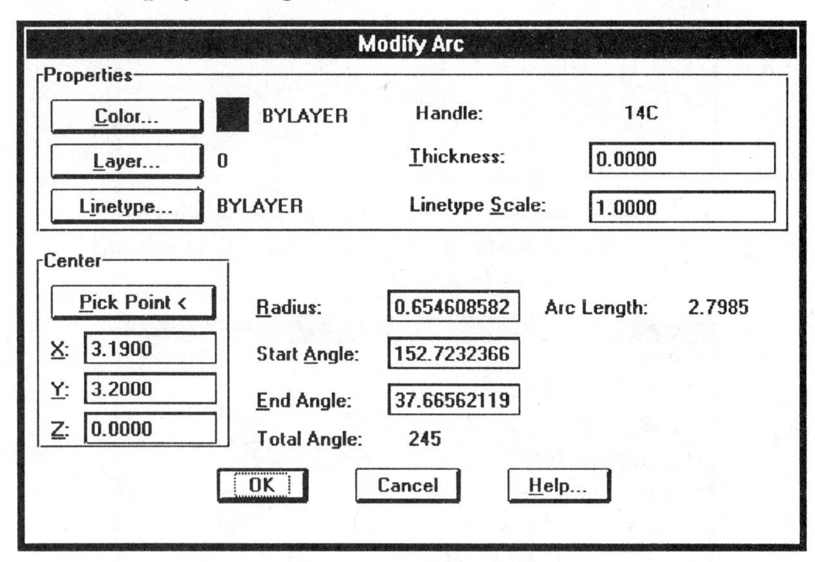

[pick hatch] Displays the **Modify Associative Hatch** dialogue box.
Displays dialogue box.

Modify Associative Hatch

Properties

Color...	BYLAYER	Handle:	192
Layer...	0	Thickness:	0.0000
Linetype...	BYLAYER	Linetype Scale:	1.0000

Block Name: *X6 - Associative Hatch

OK | Cancel | Hatch Edit... | Help...

[pick attribute] Displays the **Modify Attribute Definition** dialogue box.
Displays dialogue box.

Modify Attribute Definition

Properties

Color...	BYLAYER	Handle:	14C
Layer...	0	Thickness:	0.0000
Linetype...	BYLAYER	Linetype Scale:	1.0000

Tag: SAMPLE Prompt: Version Default: R13

Origin

Pick Point <	Height:	0.3000	Justify:	Left	
X: 3.1900	Rotation:	0	Style:	ROMANS	
Y: 3.2000	Width Factor:	0.8500	Upside Down		Constant
Z: 0.0000	Obliquing:	0	Backward		Verify
			Invisible		Preset

OK | Cancel | Help...

[pick block] Displays the **Modify Block Insertion** dialogue box.
Displays dialogue box.

Modify Block Insertion

Properties

Color...	BYLAYER	Handle:	137
Layer...	0	Thickness:	0.0000
Linetype...	BYLAYER	Linetype Scale:	1.0000

Block Name: OVERHEAD

At

Pick Point <	X-scale:	1.0000	Columns:	0
X: 8.4000	Y-scale:	1.0000	Rows:	0
Y: 5.3000	Z-scale:	1.0000	Col Spacing:	0.0000
Z: 0.0000	Rotation:	0	Row Spacing:	0.0000

OK | Cancel | Help...

[pick body] Displays the **Modify Body** dialogue box.
Displays dialogue box.

Modify Body
Properties
Color... BYLAYER Handle: 14C
Layer... 0 Thickness: 0.0000
Linetype... BYLAYER Linetype Scale: 1.0000
OK Cancel Help...

[pick circle] Displays the **Modify Circle** dialogue box.
Displays dialogue box.

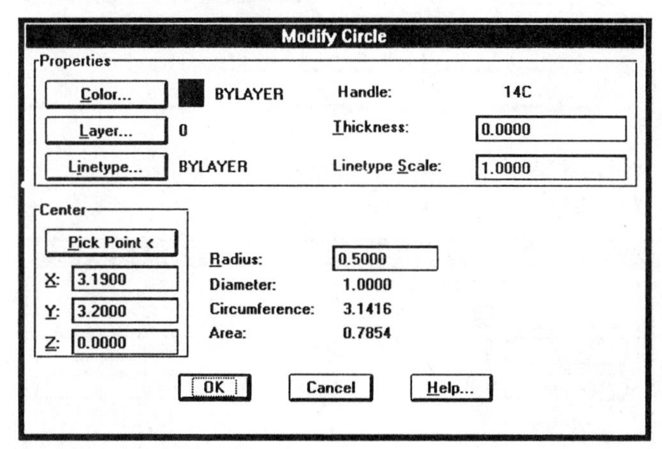

[pick dimension]
Displays the **Modify Dimension Entity** dialogue box.
Displays dialogue box.

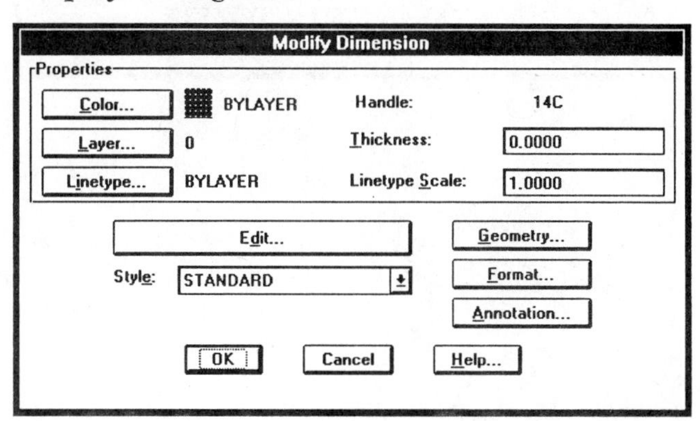

[pick ellipse]

Displays the **Modify Ellipse** dialogue box.
Displays dialogue box.

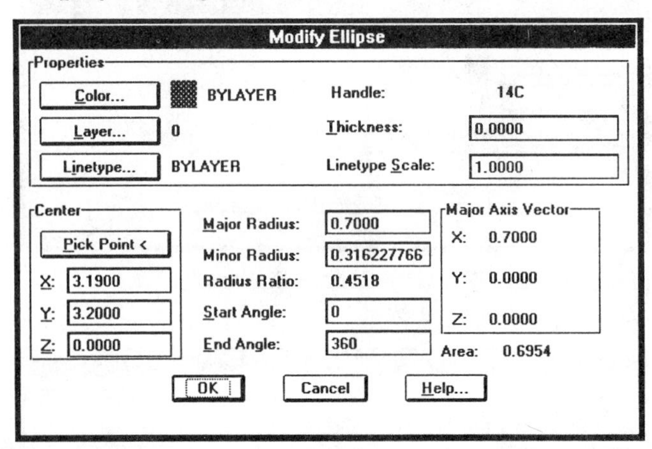

[pick leader]

Displays the **Modify Leader** dialogue box.
Displays dialogue box.

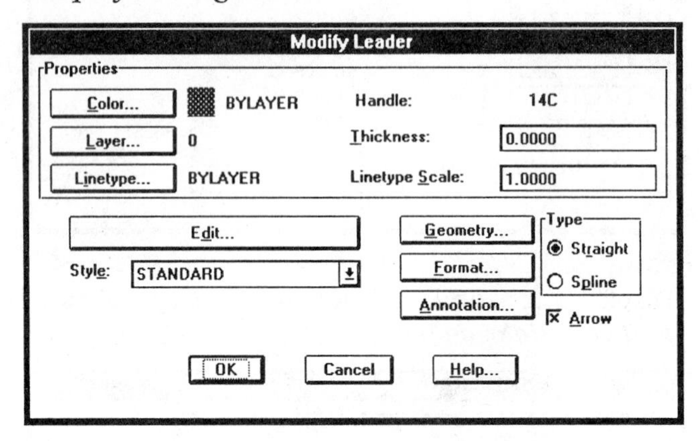

[pick line] Displays the **Modify Line** dialogue box.
Displays dialogue box.

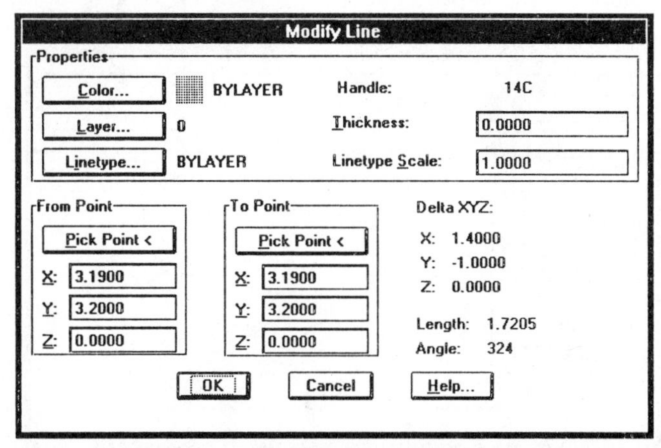

[pick multiline] Displays the **Modify MLine** dialogue box.
Displays dialogue box.

Modify MLine
Properties
Color... BYLAYER
Handle: 14C
Layer... 0
Thickness: 0.0000
Linetype... BYLAYER
Linetype Scale: 1.0000
MLine Style: STANDARD
OK Cancel Help...

[pick multi-line text] Displays the **Modify MText** dialogue box.
Displays dialogue box.

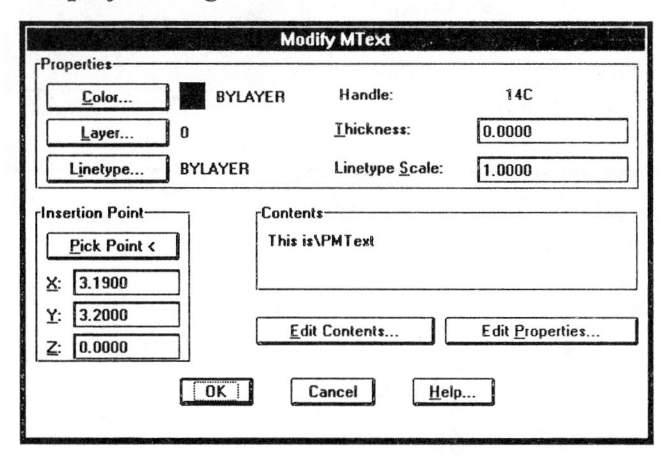

[pick point] Displays the **Modify Point** dialogue box.
Displays dialogue box.

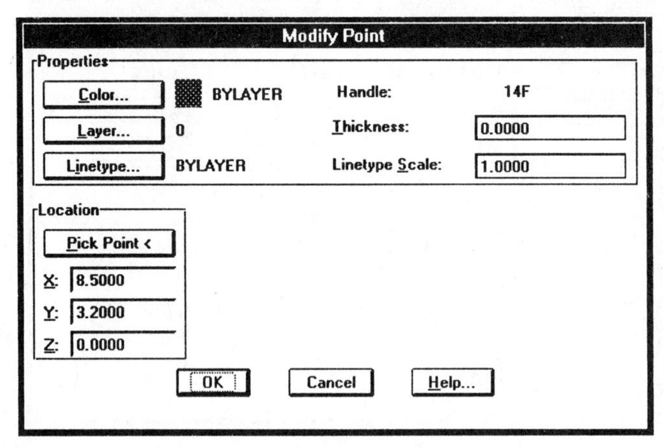

[pick polyline]

Displays the **Modify Polyline** dialogue box.
Displays dialogue box.

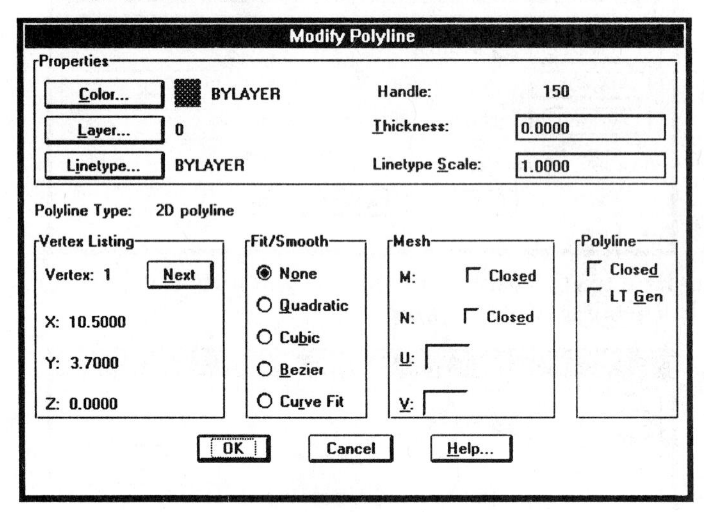

[pick ray] Displays the **Modify Ray** dialogue box.
Displays dialogue box.

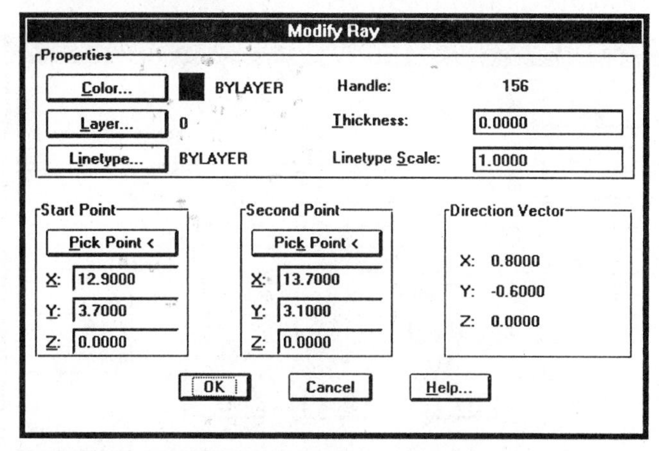

[pick region] Displays the **Modify Region** dialogue box.
Displays dialogue box.

Modify Region
Properties
Color... BYLAYER Handle: 15E
Layer... 0 Thickness: 0.0000
Linetype... BYLAYER Linetype Scale: 1.0000
OK Cancel Help...

[pick shape] Displays the **Modify Shape** dialogue box.
Displays dialogue box.

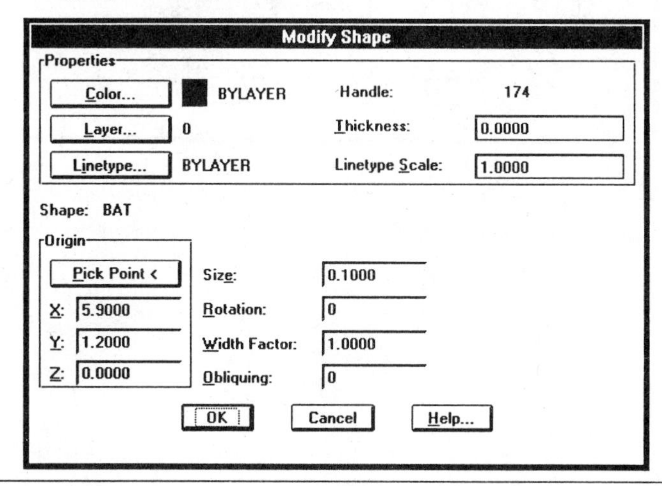

[pick solid] Displays the **Modify Solid** dialogue box.
Displays dialogue box.

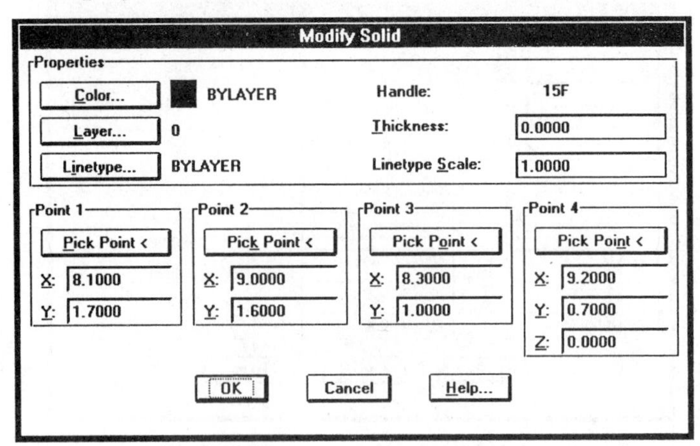

[pick spline]

Displays the **Modify Spline** dialogue box.
Displays dialogue box.

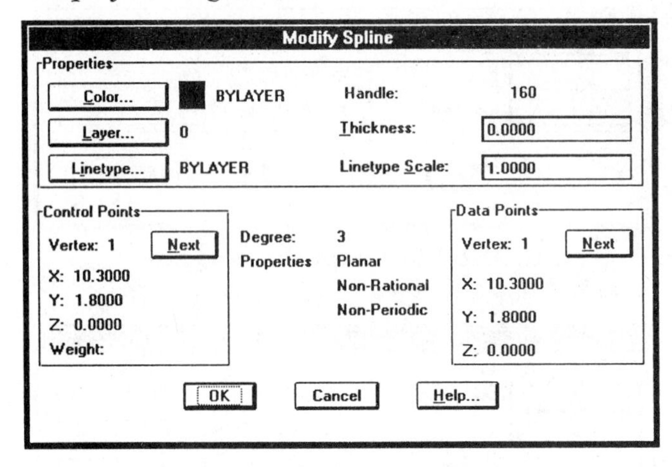

[pick text] Displays the **Modify Text** dialogue box.
Displays dialogue box.

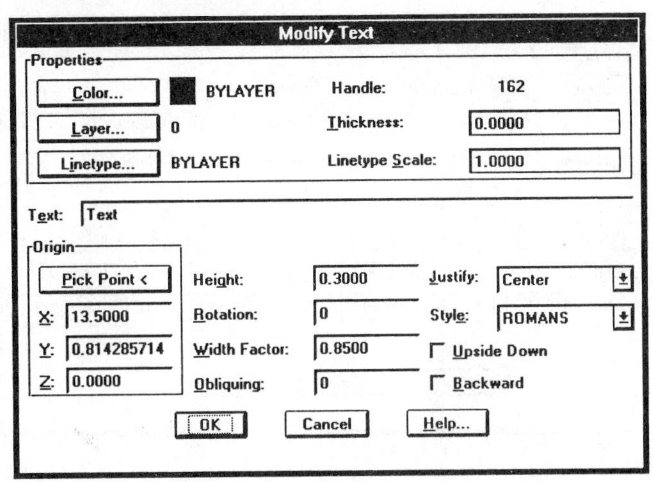

[pick tolerance]
Displays the **Modify Tolerance** dialogue box.
Displays dialogue box.

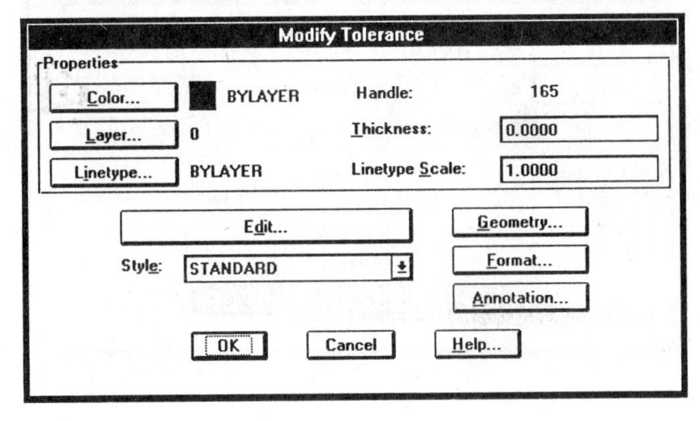

[pick trace] Displays the **Modify Trace** dialogue box.
Displays dialogue box.

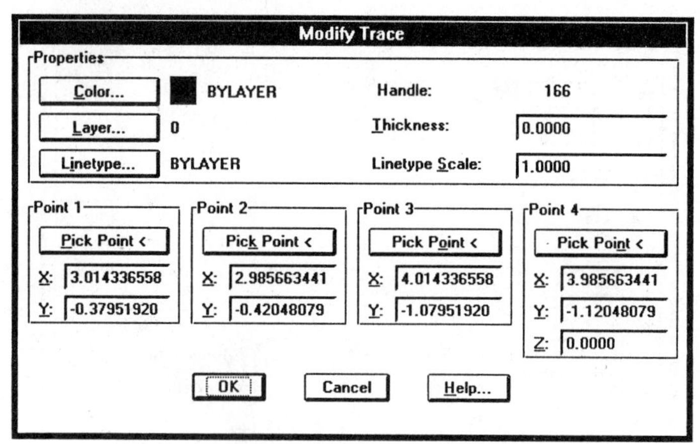

[pick viewport]

Displays the **Modify Viewport** dialogue box.
Displays dialogue box.

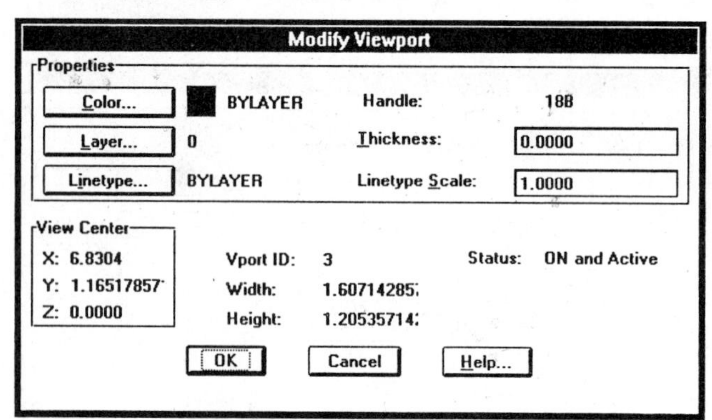

[pick xline] Displays the **Modify Xline** dialogue box.
Displays dialogue box.

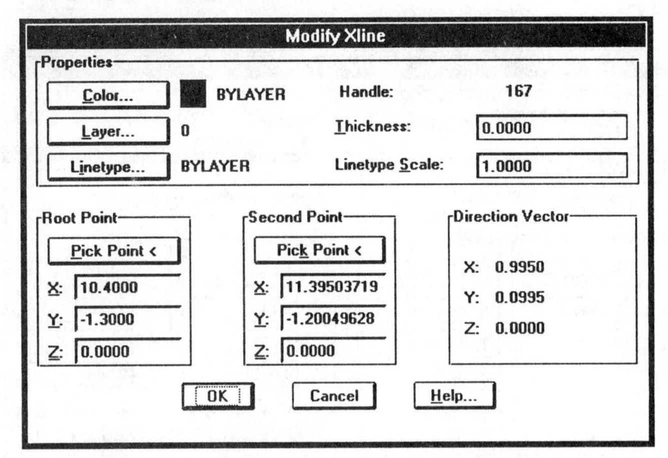

[pick xref] Displays the **Modify External Reference** dialogue box.
Displays dialogue box.

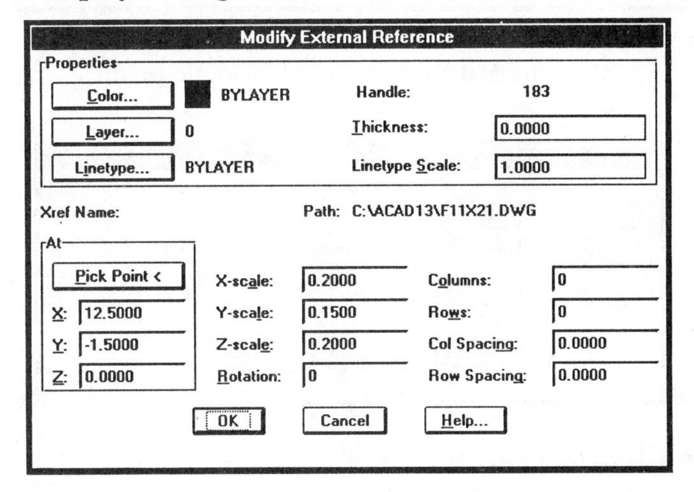

[pick 3D face]

Displays the **Modify 3D Face** dialogue box.
Displays dialogue box.

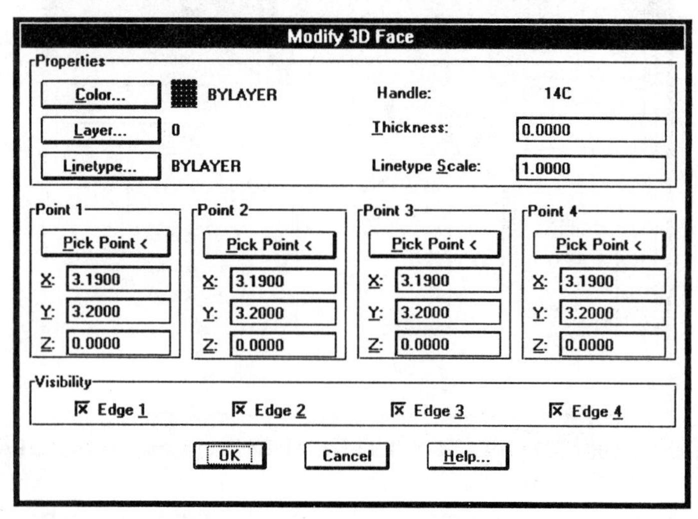

[pick 3D solid]

Displays the **Modify 3D Solid** dialogue box.
Displays dialogue box.

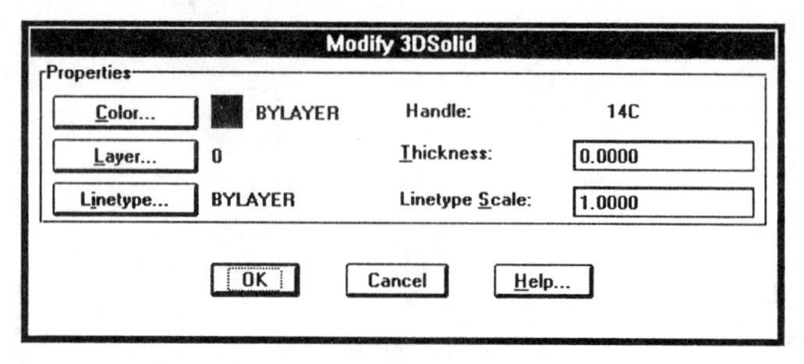

RELATED AUTOCAD COMMANDS

- **Change** Changes most properties of entities.
- **ChProp** Changes some properties of entities.
- **DdChProp** Edits properties via a dialogue box.
- **DdEdit** Edits text via a dialogue box.
- **PEdit** Edits polylines and meshes.

RELATED SYSTEM VARIABLES

- *Many system variables.*

'DdOsnap

Rel. 12

Sets object snap modes and aperture size via a dialogue box (*short for Dynamic Dialogue Object SNAP; an external command in DdOsnap.Lsp*).

Command	Button	Side Menu	Menu Bar	Tablet
'ddosnap	[#2]	[* * * *]	[Options] [Running Object Snap]	T12-U13

Command: **ddosnap**

Displays dialogue box.

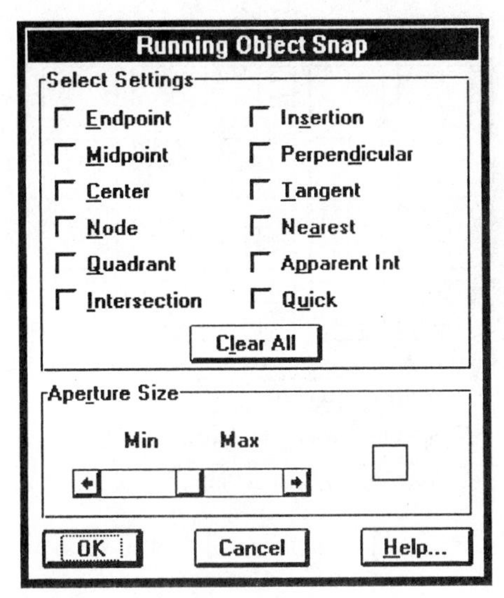

COMMAND OPTIONS

Select settings Selects the objects snaps.
Aperture size Changes the size of the aperture.

RELATED AUTOCAD COMMANDS

- **OSnap** Sets object snap modes via the command line.
- **<middle button>** Displays list of object snap modes.

RELATED SYSTEM VARIABLES

- **Aperture** Size of the object snap aperture.
- **OsMode** The current object snap modes.

'DdPtype

Rel. 12

Sets the type and size of points via a dialogue box (*short for Dynamic Dialogue Point TYPE; an external command in DdPtype.Lsp*).

Command	Alt+	Side Menu	Menu Bar	Tablet
`'ddptype`	`O,D,P`	`[OPTIONS]`	`[Options]`	...
		`[DISPLAY]`	`[Display]`	
		`[DDptype:]`	`[Point Style]`	

Command: **ddptype**

Displays dialogue box.

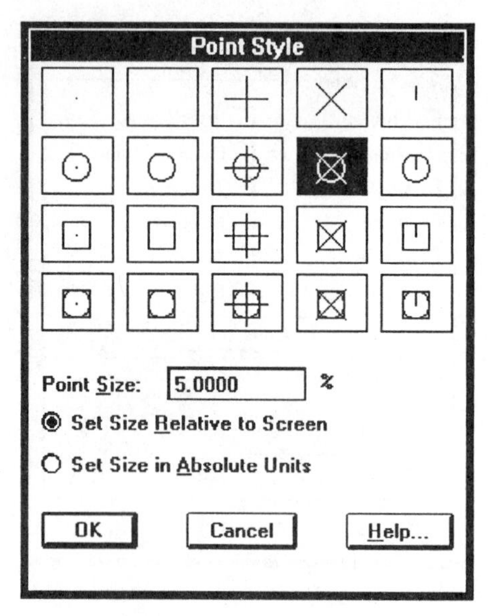

COMMAND OPTIONS

Point size Sets size in percent or pixels.

Set size Relative to screen
Sets size in percent.

Set size in Absolute units
Sets size in pixels.

RELATED AUTOCAD COMMAND

- **Point** Draws a point.

RELATED SYSTEM VARIABLES

- **PdMode** Determines the look of a point.
- **PdSize** Contains the size of the point.

DdRename

Rel. 12

Changes the names of blocks, dimension styles, layers, linetypes, text styles, UCS, views, and viewports via a dialogue box (*an external command in DdRename.Lsp*).

Command	Alt+	Side Menu	Menu Bar	Tablet
ddrename	D,R	[DATA] [Rename:]	[Data] [Rename]	...

`Command:` **`ddrename`**

Displays dialogue box.

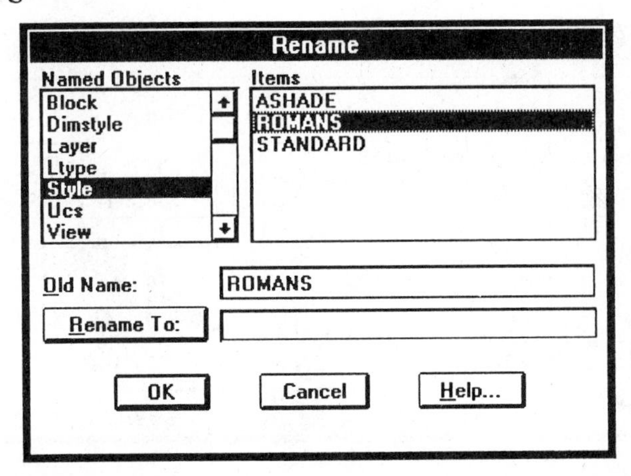

COMMAND OPTIONS

Old name Specifies the name (*or group of names*) to change.
Rename to Indicates the new name.

RELATED AUTOCAD COMMANDS

- **DdModify** Changes names (and all other attributes) of objects.
- **Rename** Changes the names of objects via the command line.

TIPS

- You cannot rename layer '0', dimstyle 'Standard', anonymous blocks, groups, and linetype 'Continuous.'

- To rename a group of similar names, use * (*the wildcard for "all"*) and ? (*the wildcard for a single character*).

- Names can be up to 31 characters in length, including the $, - and _ characters.

'DdRModes

Rel. 9

Controls the current settings of snap, snap angle, grid, axes, ortho, blips, and isometric modes (*short for Dynamic Dialogue dRawing MODES*).

Command	Alt+	Side Menu	Menu Bar	Tablet
'ddrmodes	O,D	[OPTIONS]	[Options]	Y 10
		[DDrmode:]	[Drawing Aids]	

Command: **ddrmodes**

Displays dialogue box.

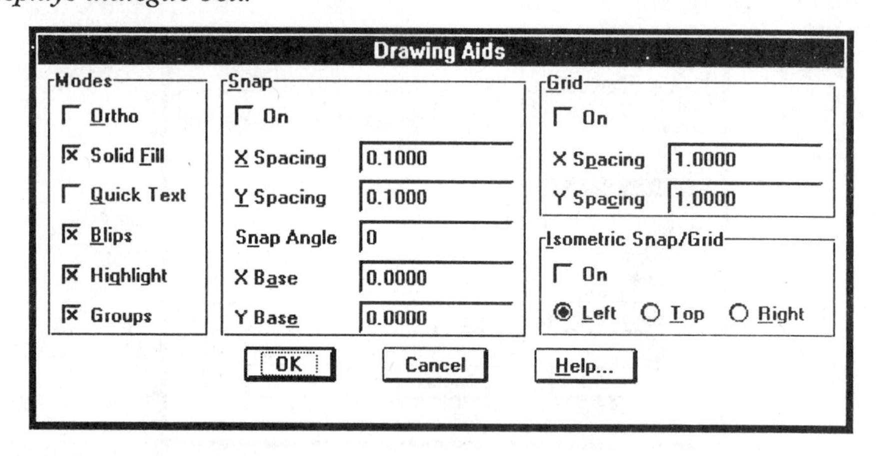

COMMAND OPTIONS

Ortho Turns orthographic mode on and off.
Solid Fill Turns solid fill on and off.
Quick Text Turns quick text on and off.
Blips Turns blipmarks on and off.
Highlight Turns object highlighting on and off.
Groups Turns automatic group selection on and off.
Snap Turns snap mode on and off:

Snap X spacing
Sets x-spacing for snap.

Snap Y spacing
Sets y-spacing for snap.

Snap angle Sets angle for snap and grid.
X Base Sets snap, grid hatch x-basepoint.
Y base Sets snap, grid hatch y-basepoint.

Grid Turns grid marks on and off:
- **X spacing** Sets grid x-spacing.
- **Y spacing** Sets grid y-spacing.

Isometric Turns isometric mode on and off:
- **Left** Switches to left isometric plane.
- **Top** Switchs to top isometric plane.
- **Right** Switch to right isometric plane.

RELATED AUTOCAD COMMANDS

- **Blipmode** Toggles visibility of blip markers.
- **Fill** Toggles fill mode.
- **Grid** Sets the grid spacing and toggles visibility.
- **Highlight** Toggles highlight mode.
- **Isoplane** Selects the working isometric plane.
- **Ortho** Toggles orthographic mode.
- **QText** Toggles quick text mode.
- **Snap** Sets the snap spacing and isometric mode.

RELATED SYSTEM VARIABLES

- **BlipMode** Current blip marker visibility:
  - 0 Off.
  - 1 On (*default*).
- **FillMode** Current fill mode:
  - 0 Off.
  - 1 On (*default*).
- **GridMode** Current grid visibility:
  - 0 Off (*default*).
  - 1 On.
- **GridUnit** Current grid spacing.
- **Highlight** Current highlight mode:
  - 0 Off.
  - 1 On (*default*).
- **OrthoMode** Current orthographic mode setting:
  - 0 Off (*default*).
  - 1 On.
- **QTextMode** Current quick text mode setting:
  - 0 Off (*default*).
  - 1 On.
- **PickStyle** Controls group selection:
  - 0 Groups and associative hatches not selected.
  - 1 Groups selected.
  - 2 Associative hatches selected.
  - 3 Both selected (*default*).

- **SnapAng** Current snap and grid rotation angle.
- **SnapBase** Base point of snap and grid rotation angle.
- **SnapIsoPair** Current isoplane:
  - 0 Left (*default*).
  - 1 Top.
  - 2 Right.
- **SnapMode** Current snap mode setting:
  - 0 Off (*default*).
  - 1 On.
- **SnapStyl** Snap style setting:
  - 0 Standard (*default*).
  - 1 Isometric.
- **SnapUnit** Current snap spacing.

TIPS

■ **DdRModes** is an alternative to the **SetVar** command for checking the status of the above 14 system variables.

■ Use the function key **F8** (*or control key [Ctrl]+O*) to change ortho mode during a command.

■ Use the function key **F9** (*or control key [Ctrl]+B*) to change snap mode during a command.

■ Use the function key **F7** (*or control key [Ctrl]+G*) to turn the grid on and off during a command.

■ Use the control key **[Ctrl]+A** to turn groups on and off during a command.

'DdSelect

Rel. 12

Defines the type of object selection mode and pickbox size via a dialogue box (*an external command in DdSelect.Lsp*).

Command	Alt+	Side Menu	Menu Bar	Tablet
'ddselect	O,S	[OPTIONS]	[Options]	...
		[DDselec:]	[Selection]	

```
Command: ddselect
```

Displays dialogue box.

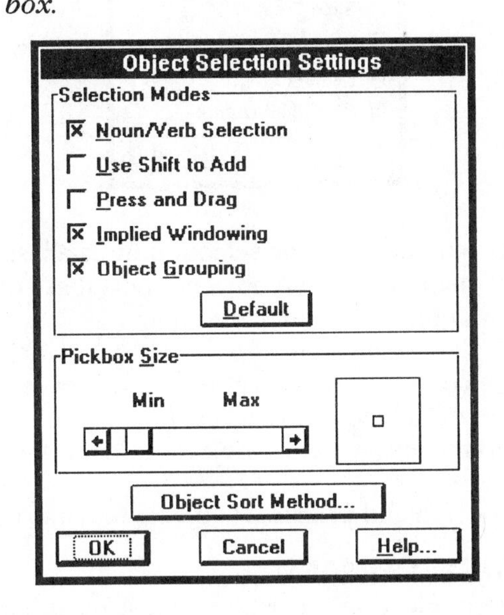

COMMAND OPTIONS

Selection modes
Specifies the style of object selection mode:

Noun/verb Selects objects first, then enter the command.

Use shift to add
The **[Shift]** key adds object selection set.

Press and drag
Creates windowed selection set by pressing mouse key and dragging window, rather than specifying two points.

Implied windowing
Automatically creates a windowed selection box.

Default selection mode
Resets modes to turn on Noun/verb and Implied windowing.

Pickbox size Interactively change the size of the pickbox.

Object sort method

Displays a second dialogue box to specify the commands that sort objects by drawing database order:

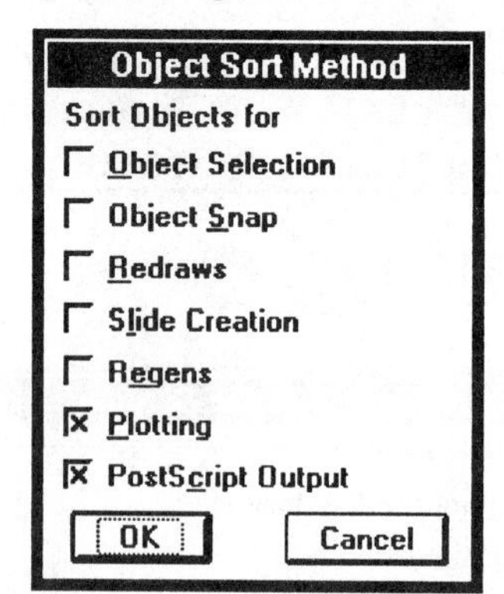

Object Selection
Objects are added to the selection in database order.

Object Snap Object snap modes find objects in database order.

Redraws Redraws objects in database order.

Slide Creation
MSlide command draws objects in database order.

Regens Regenerates objects in database order.

Plotting **Plot** command processes objects in database order.

PostScript Output
PsOut command processes objects in database order.

RELATED AUTOCAD COMMAND

Select Creates a selection set before executing an editing commands.

RELATED SYSTEM VARIABLES

- **PickAdd** Determines effect of **[Shift]** key on creating selection set:
  - 0 **[Shift]** key adds to selection set.
  - 1 **[Shift]** key removes from selection set (*default*).
- **PickAuto** Determines automatic windowing:
  - 0 Disabled.
  - 1 Enabled (*default*).
- **PickBox** Specifies the size of the pickbox.
- **PickDrag** Method of creating selection window:
  - 0 Click at both corners (*default*).
  - 1 Click one corner, drag to second corner.
- **PickFirst** Method of object selection:
  - 0 Enter command first.
  - 1 Select objects first (*default*).

■ **SortEnts** Objects are displayed in database order during:

- **0** Off.
- **1** Object selection.
- **2** Object snap.
- **4** Redraw.
- **8** Slide generation.
- **16** Regeneration.
- **32** Plots.
- **64** PostScript output.

TIPS

■ These commands work with noun-verb selection: **Array, Block, Change, ChProp, Copy, DdChProp, DView, Erase, Explode, Hatch, List, Mirror, Move, Rotate, Scale, Stretch,** and **WBlock**.

■ A larger pickbox makes it easier to select objects, but also makes it easier to inadvertently select objects.

■ Use **Object Sort Method** if the drawing requires that objects be processed in the order they appear in the drawing, such as for NC applications.

■ **Plotting** and **PostScript Output** are turned on, by default; setting more sort methods increases processing time.

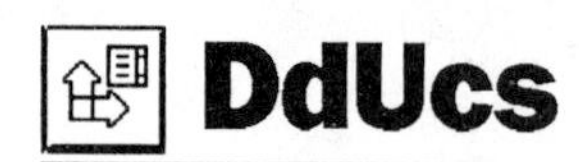

DdUcs

Rel. 10

Creates and controls UCS planes via a dialogue box (*short for Dynamic Dialogue User Coordinate System*).

Command	**Alt+**	**Side Menu**	**Menu Bar**	**Tablet**
`dducs`	`V,C`	`[VIEW]` `[DDucs:]`	`[View]` `[Named UCS]`	`J 5`

`Command:` **`dducs`**

Displays dialogue box.

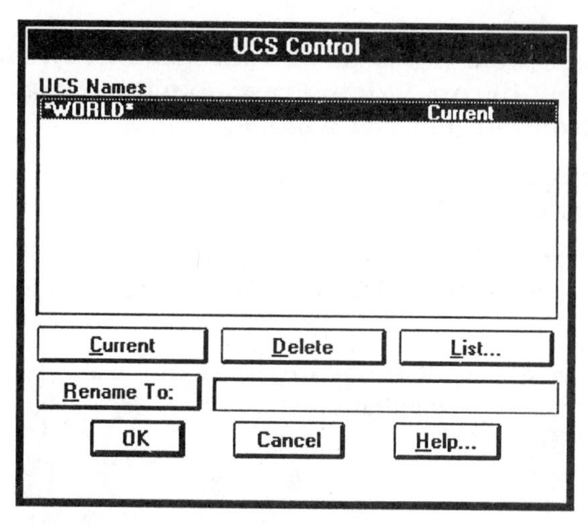

COMMAND OPTIONS

Current Makes the selected name the current UCS.
Delete Deletes a named UCS.
Rename to Renames a UCS.
List Lists information about the selected UCS in a dialogue box:

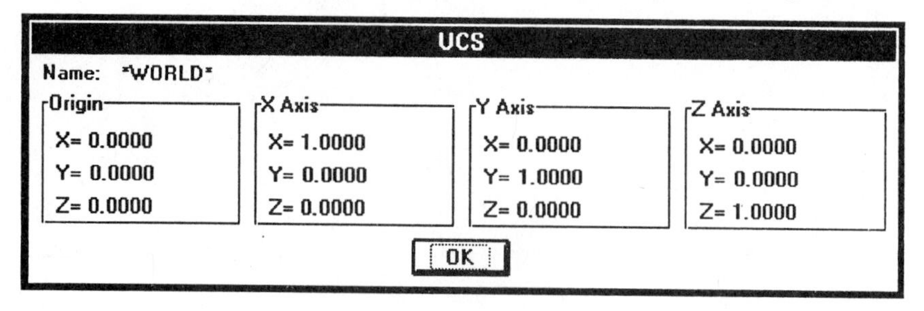

RELATED AUTOCAD COMMANDS

- **DdUcsP** Selects a predefined UCS from a dialogue box.
- **Ucs** Creates and saves user-defined coordinate systems.

RELATED SYSTEM VARIABLES

- **UcsFollow** New UCS is displayed in plan view.
- **UcsName** Current name of UCS.
- **UcsOrg** WCS origin of the current UCS.
- **UcsXdir** X-direction of the current UCS.
- **UcsYdir** Y-direction of the current UCS.
- **ViewMode** Current clipped viewing mode.
- **WorldUcs** UCS = WCS toggle.
- **WorldView** UCS or WCS for **DView** and **VPoint** commands.

DdUcsP

Rel. 12

Selects one of seven predefined user coordinate systems (*short for Dynamic Dialogue UCS Preset; an external command in DdUcsP.Lsp*).

Command	Alt+	Side Menu	Menu Bar	Tablet
ddducsp	V, U	[View]	[View]	K 5
		[DDucsp:]	[Preset UCS]	

Command: **dducsp**
Displays dialogue box.

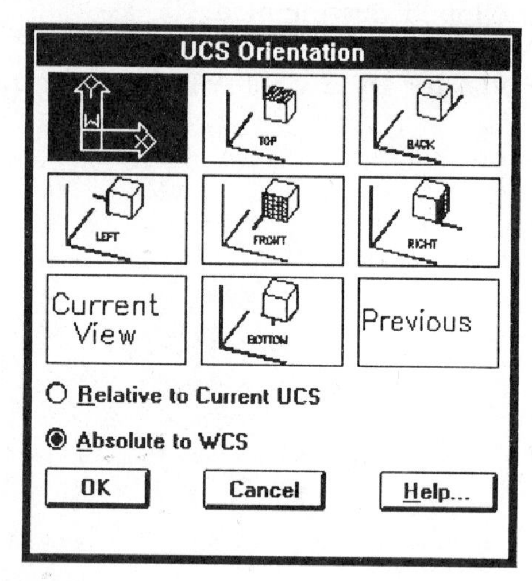

COMMAND OPTIONS

Relative to current UCS
Absolute to WCS

RELATED AUTOCAD COMMANDS

- **DdUcs** Creates and selects named UCS.
- **Ucs** Sets the current UCS via the command line.

RELATED SYSTEM VARIABLES

- **UcsFollow** New UCS is displayed in plan view.
- **UcsName** Current name of UCS.
- **UcsOrg** WCS origin of the current UCS.
- **UcsXdir** X-direction of the current UCS.
- **UcsYdir** Y-direction of the current UCS.
- **ViewMode** Current clipped viewing mode.
- **WorldUcs** UCS = WCS toggle.
- **WorldView** UCS or WCS for **DView** and **VPoint** commands.

'DdUnits

Rel. 12

Selects the display of units and angles via a dialogue box (*an external command in DdUnits.Lsp*).

Command	Alt+	Side Menu	Menu Bar	Tablet
'ddunits	D, U	[DATA] [Units:]	[Data] [Units]	Y 7

Command: **ddunits**
Displays dialogue box.

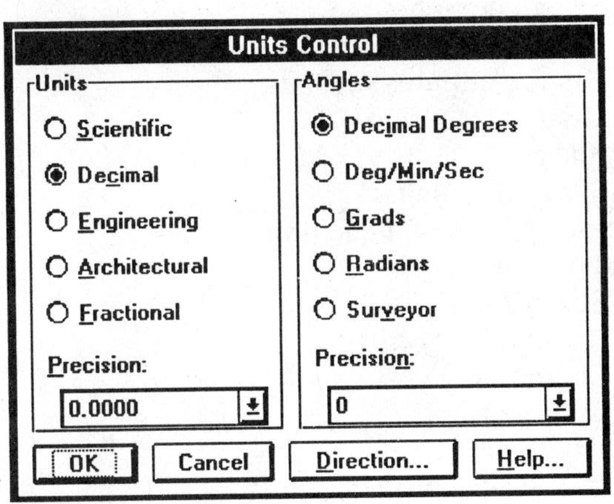

COMMAND OPTION

Direction Displays a second dialogue box to specify the direction of 0 degrees:

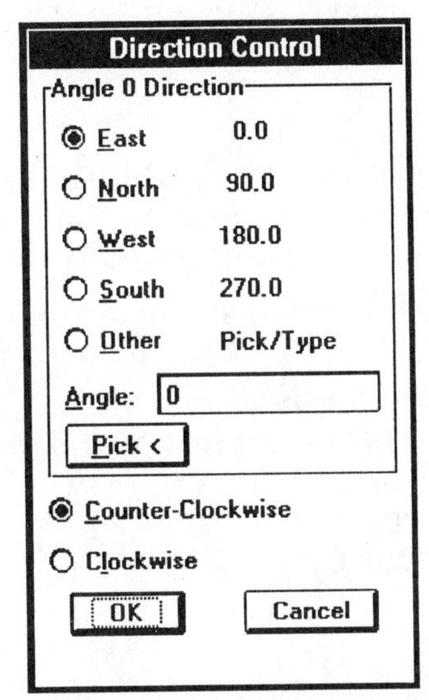

RELATED AUTOCAD COMMANDS

- **MvSetup** Sets up the drawing with border, title and units.
- **Units** Sets units and angles via the command line

RELATED SYSTEM VARIABLES

- **AngBase** Direction of zero degrees relative to the current UCS.
- **AngDir** Direction of angle measurement:
  0 Clockwise
  1 Counterclockwise (*default*).
- **AUnits** Style of angle units:
  0 Decimal degrees (*default*).
  1 Degree-minutes-seconds.
  2 Grad.
  3 Radian.
  4 Surveyor's units.
- **AuPrec** Decimal places of angle units
- **LUnits** Style of linear units:
  0 Scientific
  1 Decimal (*default*).
  2 Engineering.
  3 Architectural.
  4 Fractional.
- **LuPrec** Decimal places of linear units.
- **ModeMacro** Customizes the status line via the Diesel language.
- **UnitMode** Displays input units:
  0 As set by **DdUnits** or **Units** command (*default*).
  1 As input by the user.

TIPS

- Distance formats:
  - Decimal 0.0000 (*default*).
  - Architectural 0'-0/64" (*feet and fractional inches*)
  - Engineering 0'-0.0000" (*feet and decimal inches*)
  - Fractional 0 0/64 (*unitless fractional*)
  - Scientific 0.0000E+01

- Angular formats:
  - Decimal 0.0000 (*default*).
  - Deg-Min-Sec 0d0'0.0000" (*degrees, minutes, decimal seconds*)
  - Radian 0.0000r
  - Grad 0.0000g
  - SurveyorUnits N0d'0.0000"E

DdView

Rel. 12

Selects named views via a dialogue box (*external command in DdView.Lsp*).

Command	Alt+	Side Menu	Menu Bar	Tablet
ddview	V,N	[VIEW]	[View]	J 2
		[DDview:]	[Named Views]	

Command: **ddview**

Displays dialogue box.

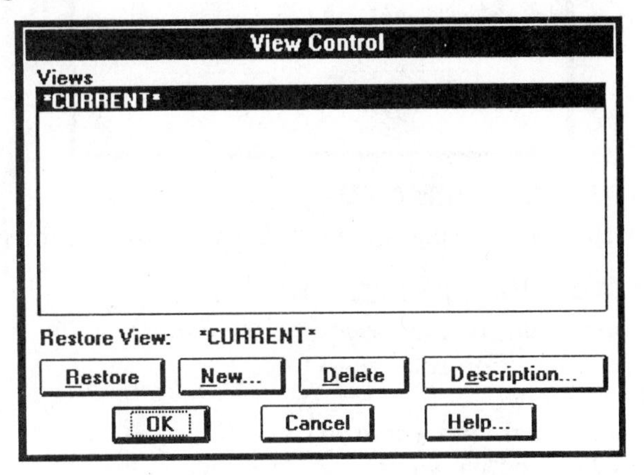

COMMAND OPTIONS

Views Lists the currently defined views.

Restore Restores a named view.

Delete Deletes a named view.

Description Lists the parameters of the selected view:

View Description

View Name: *CURRENT*

		Center (WCS)	Direction (WCS)
Width:	14.1831	X: 7.0915	X: 0.0000
Height:	9.0000	Y: 4.5000	Y: 0.0000
Twist:	0	Z: 0.0000	Z: 1.0000

Perspective:	OFF	Lens Length:	50.0000
Front Clipping:	OFF	Offset:	0.0000
Back Clipping:	OFF	Offset:	0.0000

OK

New Displays a second dialogue box to define a new view:

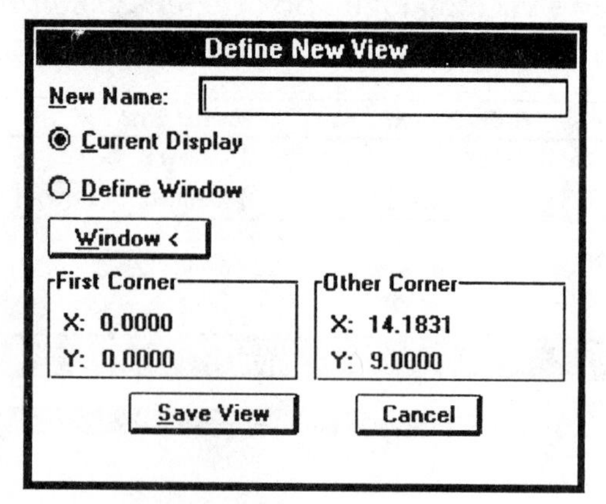

RELATED AUTOCAD COMMAND

- **View** Defines and displays named views via the command line.

RELATED SYSTEM VARIABLES

- **ViewCtr** Coordinates of the view's centerpoint.
- **ViewDir** View direction relative to UCS.
- **ViewMode** View Mode:
  - **0** Normal view (*default*)
  - **1** Perspective view.
  - **2** Front clipping.
  - **4** Back clipping.
  - **8** UCS-follow on.
  - **16** Front clip not at eye.
- **ViewSize** View height.
- **ViewTwist** Twist angle of current view.

DdVpoint

Rel. 12

Changes the viewpoint of drawings via a dialogue box (*short for Dynamic Dialogue ViewPOINT; an external command in DdVpoint.Lsp*).

Command	Alt+	Side Menu	Menu Bar	Tablet
ddvpoint	V,E,R	[VIEW]	[View]	J 1
		[Vpoint:]	[3D Viewpoint]	
		[DDvpoint]	[Rotate]	

Command: **ddvpoint**

Displays dialogue box.

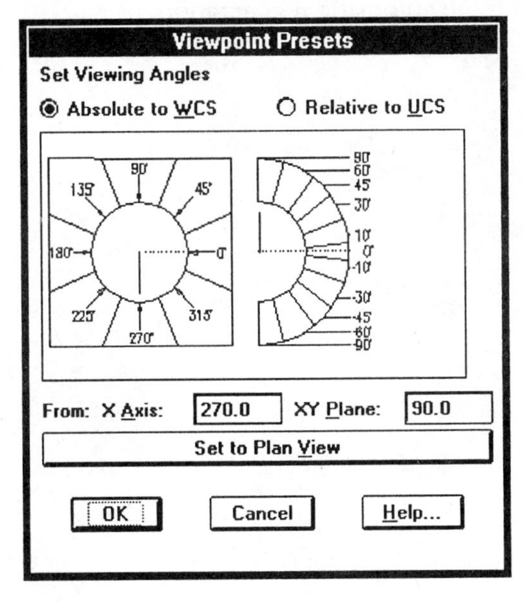

COMMAND OPTIONS

Set viewing angles
Selects view in:

Absolute to Wcs
WCS, the world coordinate system.

Relative to Ucs
UCS, the user-defined coordinate system.

From Measures viewpoint from:

X Axis X-axis.

Xy Plane X,y-plane

Set to Plan View
Changes view to plan view.

RELATED AUTOCAD COMMANDS

- **DView** Interactively changes the viewpoint.
- **VPoint** Adjusts the viewpoint from the command line.

RELATED SYSTEM VARIABLES

- **VpointX** X-coordinate of current 3D view.
- **VpointY** Y-coordinate of current 3D view.
- **VpointZ** Z-coordinate of current 3D view.
- **WorldView** Determines whether viewpoint coordinates are in WCS or UCS.

TIP

- After changing the viewpoint, AutoCAD performs an automatic **Zoom Extents**, along with the automatic regeneration.

'Delay

Ver. 1.4

Delays the next script command, in milliseconds.

Command	Alias	Side Menu	Menu Bar	Tablet
'delay	...	...	...	...

```
Command: delay
Delay time in milliseconds:
```

COMMAND OPTIONS

None

RELATED AUTOCAD COMMAND

- **Script** Initiates a script.

TIPS

- Use the **Delay** command to slow down the execution of a script file.
- The maximum delay is 32,767, just over 32 seconds.

Dim

Ver. 1.2

Changes the prompt from 'Command:' to 'Dim:' ; allows access to AutoCAD's old dimensioning commands (*short for DIMensions*).

Command	Alias	Side Menu	Menu Bar	Tablet
dim	...	...	...	...

```
Command: dim
Dim:
```

COMMAND OPTIONS

Aliases for the dimension commands are shown in uppercase (version or release introduced in brackets):

ALigned Draws linear dimension aligned with object (*ver. 2.0*)
ANgular Draws angular dimension that measures an angle (*ver. 2.0*).
Baseline Continues a dimension from a basepoint (*ver. 1.2*).
CEnter Draws a centermark on circle and arc centers (*ver. 2.0*).
COntinue Continues a dimension from the previous dimension's second extension line (*ver. 1.2*).
Diameter Draws diameter dimension on circles, arcs, and polyarcs (*ver. 2.0*).
Exit Returns to Command: prompt from Dim: prompt (*ver. 1.2*).
HOMetext Returns associative dimension text to its original position (*ver. 2.6*).
HORizontal Draws a horizontal dimension (*ver. 1.2*).
LEAder Draws a leader (*ver. 2.0*).
Newtext Edits text in associative dimensions (*ver. 2.6*).
OBlique Changes angle of extension lines in associative dimensions (*rel. 11*).
ORdinate Draws x- and y-coordinate dimensions (*rel. 11*).
OVerride Overrides the current set of dimension variables (*rel. 11*).
RAdius Draws radial dimension on circles, arcs, and polyline arcs (*ver. 2.0*).
REDraw Redraws the current viewport (*same as 'R; ver. 2.0*).
REStore Restores a dimension to the current dimstyle (*rel. 11*).
ROtated Draws a linear dimension at any angle (*ver. 2.0*).
SAve Saves the current setting of dimension styles as a dimstyle (*rel. 11*).
STAtus Lists the current settings of dimension variables (*ver. 2.0*).
STYle Sets a text style for the dimension text (*ver. 2.5*).
TEdit Changes location and orientation of text in associative dimensions (*rel. 11*).
TRotate Changes the rotation of text in associative dimensions (*rel. 11*)
Undo Undoes the last dimension action (*ver. 2.0*).
UPdate Updates selected associative dimensions to the current dimvar settings (*ver. 2.6*).
VAriables Lists values of variables associated with a dim style (*not dimvars; rel. 11*).
VErtical Draws vertical linear dimensions (*ver. 1.2*).

RELATED AUTOCAD COMMANDS

- **Ddim** Dialogue box for setting dimension variables.
- **Style** Determines the text style of the dimensioning text.
- **Units** Determines the angular and linear styles of dimensioning text.

RELATED DIM VARIABLES

- **DimAso** Determines whether dimensions are drawn associatively.
- **DimScale** Determines the dimension scale.

RELATED DIM BLOCK

- **Dot** Dim uses a dot in place of the arrowhead.

TIPS

- The dimension commands that operate at the 'Dim:' prompt are included for compatibility with Release 12.

- These dimension commands do not have dimension features new to Release 13. For example, the **Dim:Leader** command draws a leader from discrete objects (line, arrowhead, text); the new **Leader** command draws the leader using the new leader object and **MText**.

- Only transparent commands and dimension commands work at the 'Dim:' prompt. To use other commands, you must exit the 'Dim:' prompt back to the 'Command:' prompt with the **Exit** command.

- Most dimensions consist of four basic components: dimension line, extension lines, arrowheads, and text, as shown below.

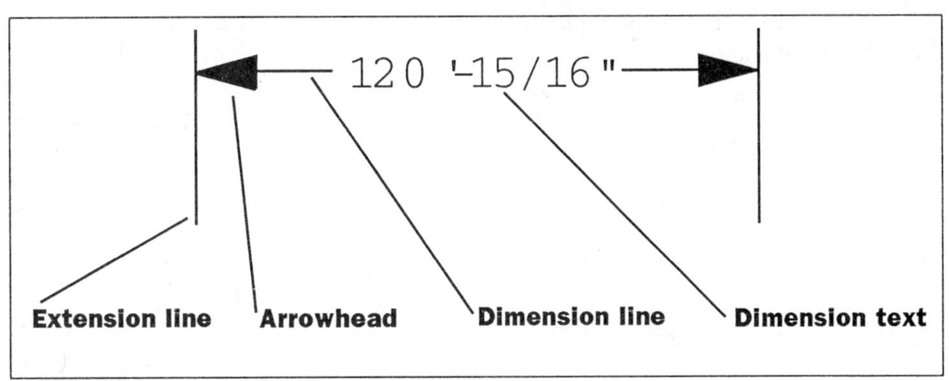

- All the components of an *associative dimension* are treated as a single object; components of a non-associative dimension are treated as individual objects.

Dim1

Ver. 2.5

Executes a single dimensioning command, then immediately returns to the 'Command:' prompt (*short for DIMension once; an undocumented command in Release 13*).

Command	Alias	Side Menu	Menu Bar	Tablet
`dim1`	...	...	...	...

```
Command: dim1
Dim:
```

COMMAND OPTIONS

*All "old" dimension commands; see **Dim** command for the complete list.*

RELATED AUTOCAD COMMANDS

- **DDim** Dialogue box for setting dimension variables.
- **Dim** Switches to "old" dimensioning mode and remains there.

RELATED DIM VARIABLES

- **DimAso** Determines whether dimensions are drawn associatively.
- **DimTxt** Determines the height of text.
- **DimScale** Determines the dimension scale.

RELATED DIM BLOCK

- **Dot** Dim uses a dot in place of the arrowhead.

TIP

- Use the **Dim1** command when you need to use just a single "old" dimension command.

DimAligned

Rel. 13

Draws linear dimensions aligned with an object (*formerly the Dim:ALigned command*).

Command	Alias	Side Menu	Toolbar	Tablet
dimaligned	dimali	[DRAW DIM] [Aligned:]	[Dimensioning] [DimAligned]	Y 5

```
Command: dimaligned
First extension line origin or RETURN to select: [Enter]
Select object to dimension: [pick]
Dimension line location (Text/Angle): T
Dimension text <>: [Enter]
```

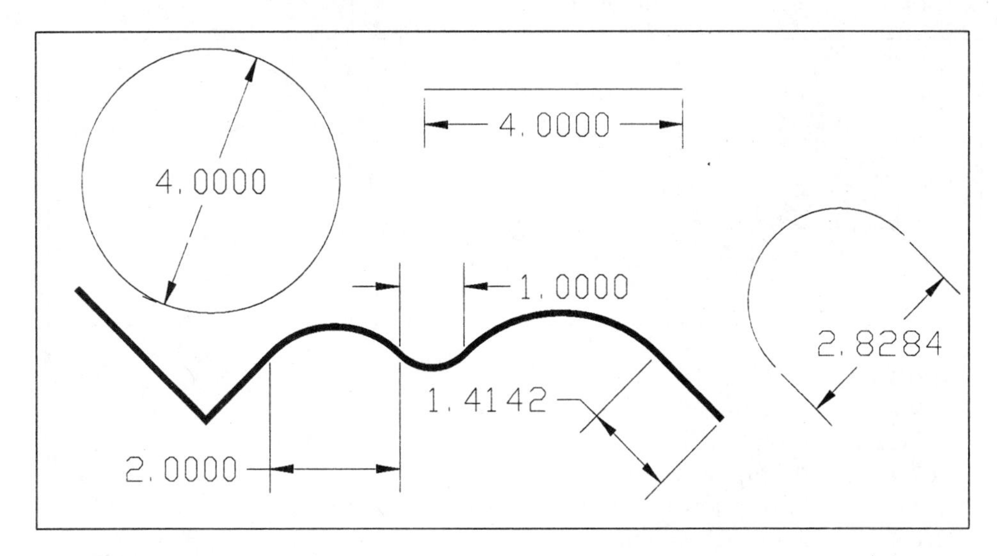

COMMAND OPTIONS

[Enter] Displays the submenu for selecting entities:
- **[pick line]** Dimensions line.
- **[pick arc]** Dimensions arc.
- **[pick circle]** Dimensions circle.
- **[pick pline]** Dimensions an individual segment.

Text Specifies dimension text.
Angle Indicates dimension angle.

RELATED DIM COMMAND

- **DimRotated** Draws angular dimension line with perpendicular extension line.

RELATED DIM VARIABLE

DimExo Dimension line offset distance.

DimAngular

Rel. 13

Draws a dimension that measures an angle (*formerly the* ***Dim:ANgular*** *command*).

Command	Alias	Side Menu	Toolbar	Tablet
dimangular	dimang	[DRAW DIM] [Angular:]	[Dimensioning] [DimAngular]	Y 3

Command: **dimangular**
Select arc, circle, line, or RETURN: **[pick]**
Second angle endpoint: **[pick]**
Dimension line arc location (Text/Angle): **T**
Dimension text <>: **[Enter]**

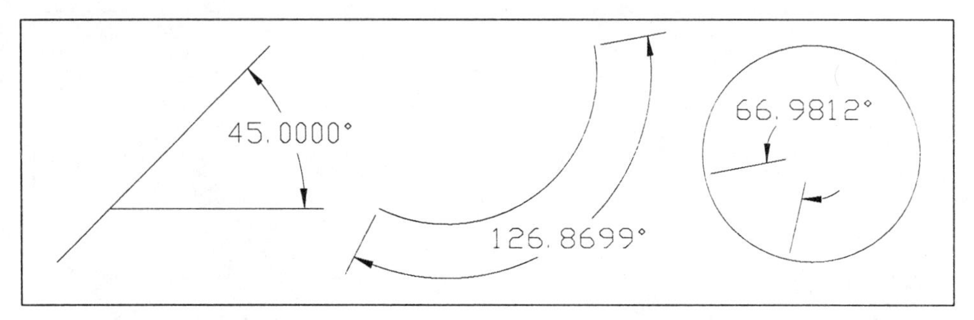

COMMAND OPTIONS

[pick arc] Measures the angle of the arc.
[pick circle] Prompts you to pick two points on the circle.
[pick line] Prompts you to pick two lines.
[Enter] Prompts you to pick points to make an angle.
Text Specifies dimension text.
Angle Indicates dimension angle.

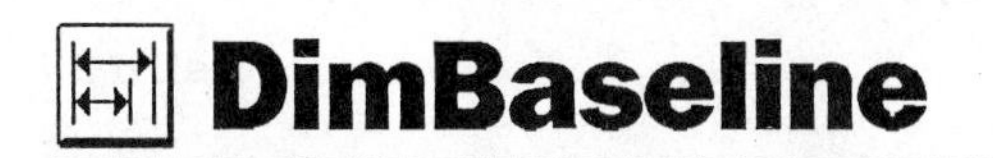

DimBaseline

Rel. 13

Draws linear dimension from the previous starting point (*formerly the **Dim:Baseline** command*).

Command	Alias	Side Menu	Toolbar	Tablet
dimbaseline	dimbase	[DRAW DIM] [Baselin:]	[Dimensioning] [DimBaseline]	X 2

```
Command: dimbaseline
Second extension line origin or RETURN to select: [pick]
Dimension text:
```

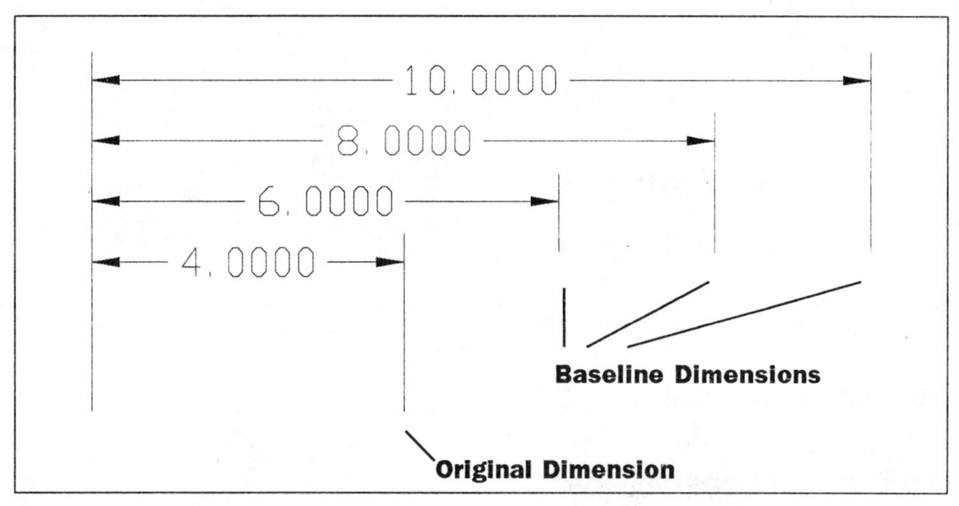

COMMAND OPTIONS

[Enter] Prompts to select original dimension.

RELATED DIM COMMAND

- **Continue** Continues linear dimensioning from last extension point.

RELATED DIM VARIABLES

- **DimDli** Specifies the distance between baseline dimension lines.
- **DimSe1** Suppress first extension line.

DimCenter

Rel. 13

Draws center lines on arcs and circles (*formerly the **Dim:CEnter** command*).

Command	Alias	Side Menu	Toolbar	Tablet
dimcenter	...	[DRAW DIM] [Center:]	[Dimensioning] [DimCenter]	W 1

```
Command: dimcenter
Select arc or circle: [pick]
```

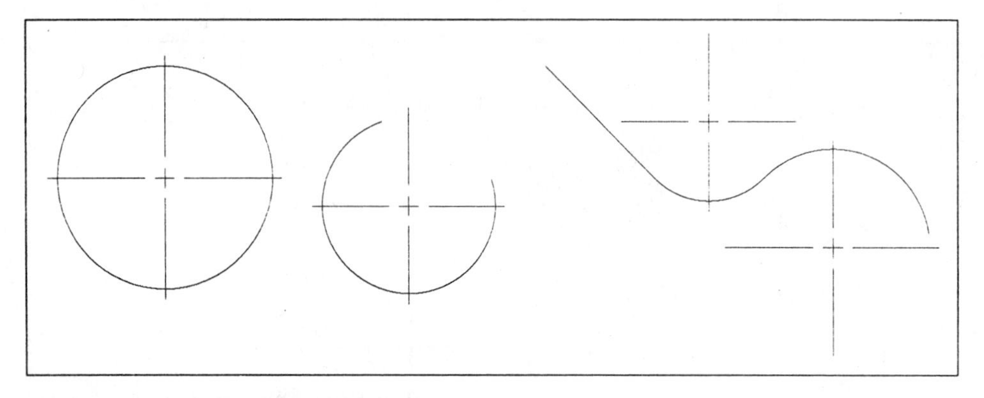

COMMAND OPTIONS

None

RELATED DIM COMMANDS

- **DimAligned** Dimensions arcs and circles.
- **DimDiameter** Dimensions arcs and circles by diameter value.
- **DimRadius** Dimensions arcs and circles by radius value.

RELATED DIM VARIABLE

- **DimCen** Size of the center mark.

DimContinue

Rel. 13

Continues a dimension from the second extension line of the previous dimension (*formerly the **Dim:COntinue** command*).

Command	Alias	Side Menu	Toolbar	Tablet
dimcontinue	dimcont	[DRAW DIM] [Continu:]	[Dimensioning] [DimContinue]	X 3

```
Command: dimcontinue
Select continued dimension: [pick]
Second extension line origin or RETURN to select: [pick]
Dimension text:
```

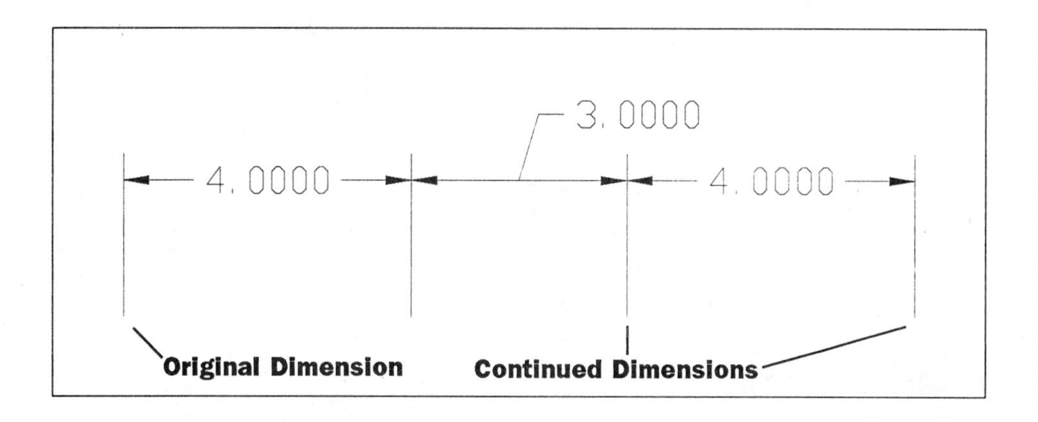

COMMAND OPTIONS

[Enter] Prompts to select originating dimension.

RELATED DIM COMMAND

- **DimBaseline** Continues dimensioning from first extension point.

RELATED DIM VARIABLES

- **DimDli** Distance between continuous dimension lines.
- **DimSe1** Suppress first extension line.

DimDiameter

Rel. 13

Draws a diameter dimension on arcs, circles, and polyline arcs *(formerly the **Dim:Diameter** command)*.

Command	Alias	Side Menu	Toolbar	Tablet
dimdiameter	dimdia	[DRAW DIM] [Diametr:]	[Dimensioning] [DimDiameter]	Y 2

```
Dim: diameter
Select arc or circle: [pick]
Dimension text: [Enter]
Enter leader length for text: [pick]
```

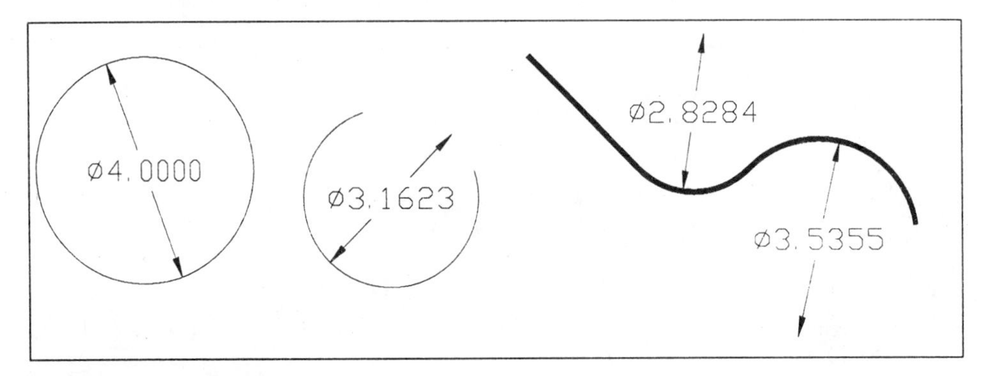

COMMAND OPTIONS

None

RELATED DIM COMMANDS

- **DimCenter** Marks the center point of arcs and circles.
- **DimRadius** Draws the radius dimension of arcs and circles.

RELATED DIM VARIABLE

- **DimCen** Controls the size of the center mark.

DimEdit

Rel. 13

Edits dimension text (*formerly the **Dim:HOMetext**, **Dim:Newtext**, **Dim:OBlique**, and **Dim:TRotate** commands*).

Command	Alias	Side Menu	Toolbar	Tablet
dimedit	dimed	[MOD DIM] [DimEdit:]	[Dimensioning] [DimEdit]	W 2

```
Command: dimedit
Dimension Edit (Home/New/Rotate/Oblique) <Home>:
```

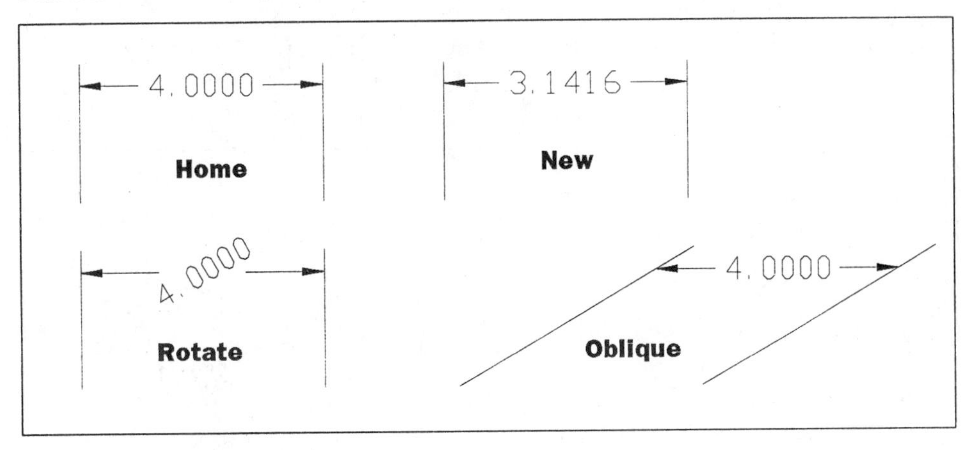

COMMAND OPTIONS

Angle Rotates the dimension text.
Home Returns dimension text to original position.
Left Moves dimension text to the left.
Right Moves dimension text to the right.

RELATED DIM COMMANDS

- *All*

RELATED DIM VARIABLES

- *Most*

TIPS

- When entering dimension text with **DimEdit**'s **New** option, AutoCAD recognizes the following metacharacters:

  [Square brackets] Alternate units format string.
  < Angle brackets > Prefix and suffix text format string.

- Use the **Oblique** option to angle dimension lines by 30 degrees, suitable for isometric drawings. Use the **Style** command to oblique text by 30 degrees.

DimLinear

Rel. 13

Draws horizontal dimensions (*formerly the **Dim:HORizontal**, **Dim:ROtated**, and **Dim:VErtical** commands*).

Command	Alias	Side Menu	Toolbar	Tablet
dimlinear	dimlin	[DRAW DIM] [Linear:]	[Dimensioning] [DimLinear]	X 4-5

```
Command: dimlinear
First extension line origin or RETURN to select: [Enter]
Select line, arc or circle: [pick]
Dimension line location(Text/Angle/Horizontal/Vertical/
Rotated): T
Dimension text <>: [Enter]
```

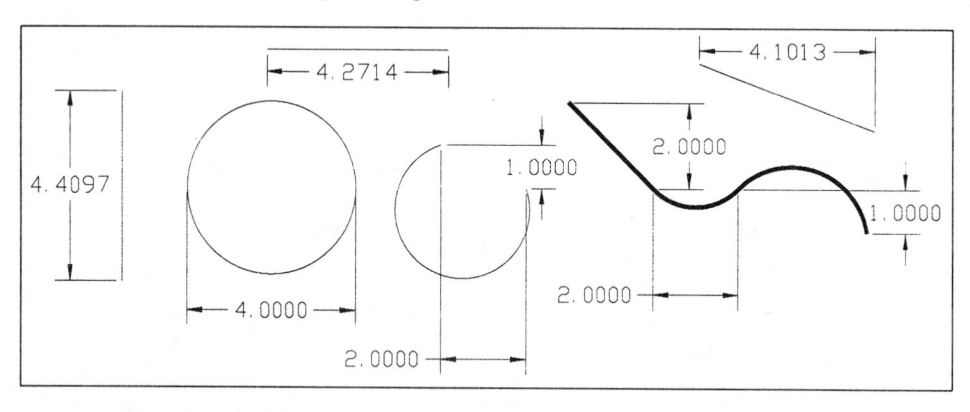

COMMAND OPTIONS

[Enter] Displays the submenu for dimensioning objects: **Select line, arc or circle**

Text Allows you to change the text of the dimension.

Angle Specify the angle of the text.

Horizontal Forces dimension to be horizontal.

Vertical Forces dimension to be vertical.

Rotated Forces dimension to be rotated.

RELATED DIM COMMAND

- **DimAligned**
  Draws linear dimensions aligned with objects.

DimOrdinate

Rel. 13

Draws an x- or y-ordinate dimension (*formerly the **Dim:ORdinate** command*).

Command	Alias	Side Menu	Toolbar	Tablet
dimordinate	dimord	[DRAW DIM] [Ordinat:]	[Dimensioning] [DimOrdinate]	Y 4

```
Command: dimordinate
Select Feature: [pick]
Leader endpoint (Xdatum/Ydatum/Text): [pick]
Leader endpoint: [pick]
```

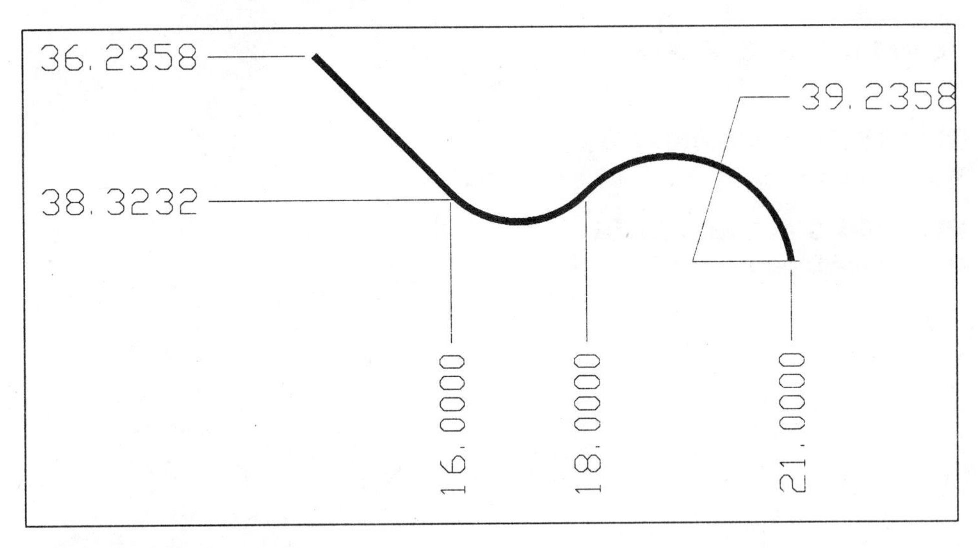

COMMAND OPTIONS

Text Places text, rather than dimension.

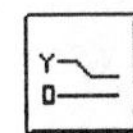 **Xdatum**

Forces x-ordinate dimension.

Ydatum

Forces y-ordinate dimension.

RELATED DIM COMMANDS

- **DimLinear** Draws regular horizontal and vertical dimensions.
- **Leader** Draws leader dimensions.

DimOverride

Rel. 13

Overrides the currently set dimension variables (*formerly the Dim:OVerride command*).

Command	Alias	Side Menu	Menu Bar	Tablet
dimoverride	dimover	[MOD DIM] [Overrid:]	...	...

```
Command: dimoverride
Dimension variable to override:
Current value xxx New Value:
Select objects: [pick]
```

COMMAND OPTIONS

None

RELATED DIM COMMAND

- **DimStyle** Creates and modifies dimension styles.

RELATED DIM VARIABLES

- *All dimension variables*

DimRadius

Rel. 13

Draws radial dimensions on circles, arcs, and polyline arcs (*formerly the **Dim:RAdius** command*).

Command	Alias	Side Menu	Toolbar	Tablet
dimradius	dimrad	[DRAW DIM] [Radius:]	[Dimensioning] [DimRadius]	Y 1

```
Command: dimradius
Select arc or circle: [pick]
Dimension line location (Text/Angle): T
Dimension text <>:
```

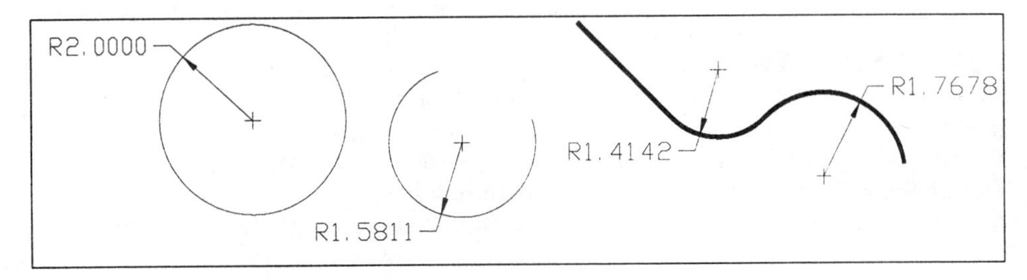

COMMAND OPTIONS

Text Places text.

Angle Changes angle of dimension text.

RELATED DIM COMMANDS

- **DimCenter**
  Draws center mark on arcs and circles.
- **DimDiameter**
  Draws diameter dimensions on arcs and circles.

RELATED DIM VARIABLE

- **DimCen** Determines the size of the center mark.

TIP

- To include the diameter symbol, use this **MText** control code: `\U+2205` .

DimStyle

Rel. 13

Creates and edits dimstyles (*short for dimension styles; formerly the **Dim:REStore, Dim:SAve, Dim:STAtus,** and **Dim:VAriables** commands*).

Command	Alias	Side Menu	Menu Bar	Tablet
dimstyle	dimsty	[MOD DIM] [DimStyl:]	[Dimensioning] [DimStyle]	V 2-3

```
Command: dimstyle
Dimension style STANDARD
Dimension style overrides: dimvar list
Dimension Style Edit (Save/Restore/STatus/Variables/Apply/?):
```

COMMAND OPTIONS

Save Saves current dimvar settings as a named dimstyle.
Restore Sets dimvar settings from a named dimstyle.
STatus Lists dimvars and current settings, then exits the **DimStyle** command.
Variables Lists dimvars and their current settings.
Apply Updates selected dimension objects with current dimstyle settings.
? Lists names of dimstyles stored in drawing.

INPUT OPTIONS

~dimvar (*Tilde prefix*) Lists differences between current and selected dimstyle.
[Enter] Lists dimvar settings for selected dimension object.

RELATED DIM COMMANDS

- **DDim** Changes dimvar settings.
- **DimScale** Determines the scale of dimension text.

RELATED DIM VARIABLES

- *All*
- **DimStyle** Name of the current dimstyle.

TIPS

- At the 'Dim:' prompt, the **Style** command sets the text style for dimension text and does *not* select a dimension style.

- Dimstyles cannot be stored to disk, except in a drawing.

- Read dimstyles from other drawings with the **XBind Dimstyle** command.

- Dimstyles stored in prototype drawings:
  - AutoCAD default — Acad.Dwg
  - American architectural — Us_Arch.Dwg
  - American mechanical — Us_Mech.Dwg
  - ISO — AcadIso.Dwg
  - JIS architectural — Jis_Arch.Dwg
  - JIS mechanical — Jis_Mech.Dwg

 DimTEdit

Rel. 13

Edits the location and orientation of text in associative dimensions (*formerly the **Dim:TEdit** command*).

Command	Alias	Side Menu	Toolbar	Tablet
dimtedit	dimted	[MOD DIM] [DimTEdt:]	[Dimensioning] [DimTEdit]	W 3-5

```
Command: dimtedit
Select dimension: [pick]
Enter text location (Left/Right/Home/Angle):
```

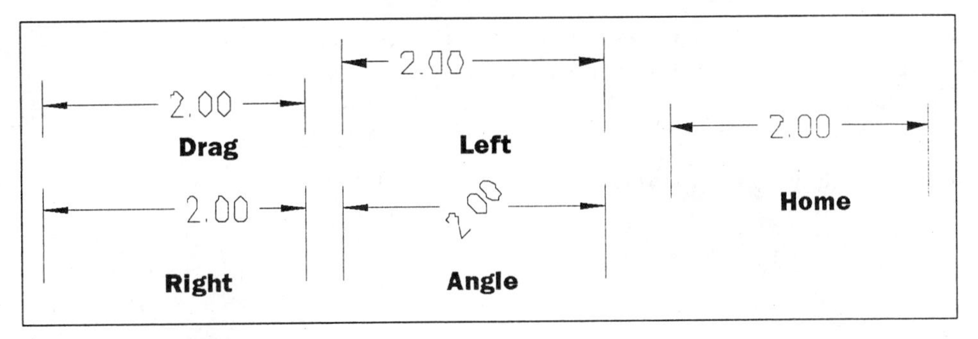

COMMAND OPTIONS

 **Angle** Rotates the dimension text.

 **Home** Returns dimension text to original position.

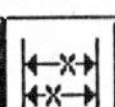 **Left** Moves dimension text to the left.

 **Right** Moves dimension text to the right.

RELATED DIM COMMANDS

- **DDim** Changes dimvar values.
- **DimEdit** Edits associative dimension text.
- **DimStyle** Creates and sets associative dimension styles.

RELATED DIM VARIABLES

DimSho Toggles whether dimension text dynamically updates while dragged.

DimTih Toggles whether dimension text is drawn horizontally or

RELATED DIM VARIABLES

DimSho	Toggles whether dimension text dynamically updates while dragged.
DimTih	Toggles whether dimension text is drawn horizontally or aligned with dimension line.
DimToh	Toggles whether dimension text is forces inside dimension lines.

TIPS

■ Use the **DdEdit** command to edit text in non-associative dimensions.

■ An angle of 0 returns dimension text to its default orientations.

■ When entering dimension text with **DimEdit**'s **New** option, AutoCAD recognizes the following metacharacters:

[Square brackets]	Alternate units format string.
< Angle brackets >	Prefix and suffix text format string.

■ Use the **Style** command to oblique text by 30 degrees.

'Dist

Ver. 1.0

Lists the 3D distance and angles between two points (*short for DISTance*).

Command	Alias	Side Menu	Menu Bar	Tablet
'dist	...	[ASSIST] [INQUIRY] [Dist:]	[Object Properties] [Inquiry] [Dist]	R 2

```
Command: dist
First point: [pick]
Second point: [pick]
```

Example result:

```
Distance=17.38, Angle in X-Y Plane=358, Angle from X-Y Plane=0
Delta X = 16.3000, Delta Y = -7.3000,  Delta Z = 0.0000
```

COMMAND OPTIONS

None

RELATED AUTOCAD COMMANDS

- **Area** Calculates the area and perimeter of objects.
- **Id** Lists the 3D coordinates of a point.
- **List** Lists information about selected entities.

RELATED AUTOLISP PROGRAMS

- *None*

RELATED SYSTEM VARIABLE

- **Distance** Last calculated distance.

TIP

- Use object snaps to precisely measure between two geometric features.

Divide

Ver. 2.5

Places points or blocks at an equally-divided distance along lines, arcs, and polylines.

Command	Alias	Side Menu	Menu Bar	Tablet
divide	...	[DRAW 2] [Divide:]	[Draw] [Point] [Divide]	W 22

```
Command: divide
Select object to divide: [pick]
<Number of segments>/Block: B
Block name to insert:
Align block with object? <Y>:
Number of segments: 10
```

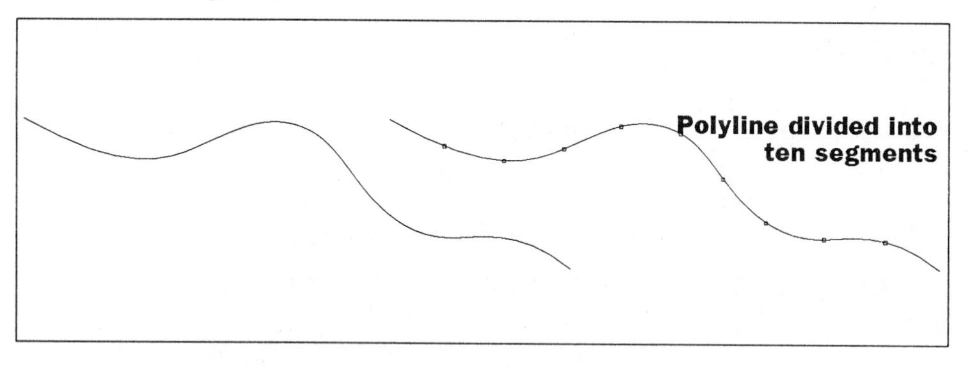

COMMAND OPTION

Block Specifies the name of the block to insert along the entity.

RELATED AUTOCAD COMMANDS

- **Block** Creates the block to use with the Divide command.
- **Insert** Places a single block in the drawing.
- **MInsert** Places an array of blocks in the drawing.
- **Measure** Divides an entity into measured distances.

RELATED SYSTEM VARIABLES

- **PdMode** Sets the style of point drawn.
- **PdSize** Sets the size of the point, in pixels.

TIPS

- Use **PdSize** and **PdMode** to make points visible along the object.
- Minimum number of segments is 2; maximum is 32,767.

O Donut *or* Doughnut

Ver. 2.5

Draws solid circles with a pair of wide polyline arcs.

Command	Alias	Side Menu	Menu Bar	Tablet
donut	...	[DRAW 1] [Dnut:]	[Draw] [Circle] [Donut]	O 9
doughnut				

```
Command: donut
Inside diameter <0.5000>:
Outside diameter <1.0000>:
Center of doughnut: [pick]
Center of doughnut: [Enter]
```

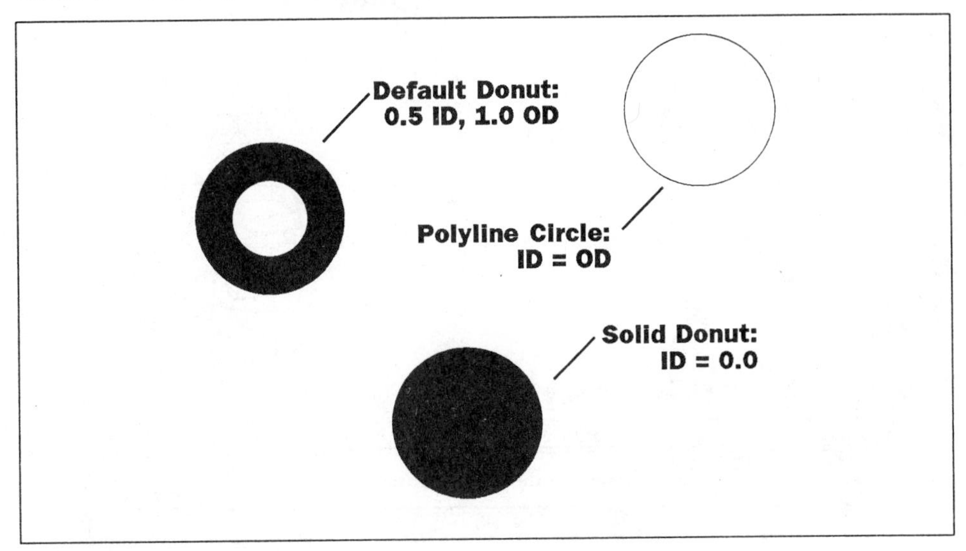

COMMAND OPTION

[Enter] Exits the **Donut** command.

RELATED AUTOCAD COMMAND

- **Circle** Draws a circle.

RELATED SYSTEM VARIABLES

- **DonutId** The current donut internal diameter.
- **DonutOd** The current donut outside diameter.

TIP

- Command automatically repeats itself until cancelled.

Dragmode

Ver. 2.0

Controls the display of an image during dragging operations.

Command	Alias	Side Menu	Menu Bar	Tablet
dragmode	...	...	...	...

```
Command: dragmode
ON/OFF/Auto <Auto>:
```

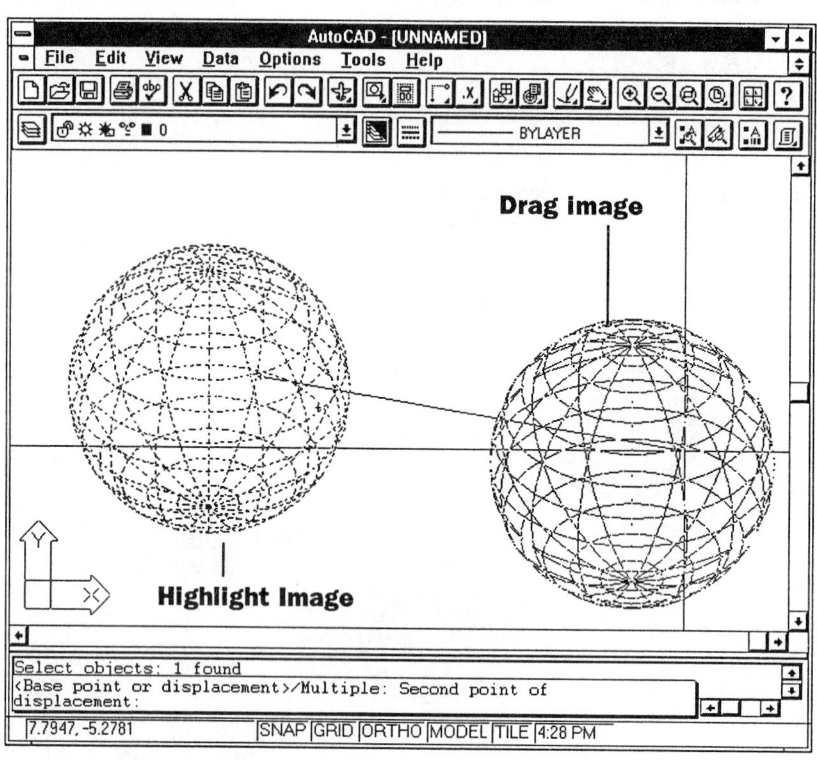

COMMAND OPTIONS

ON Enables dragging display only with the Drag option.
OFF Turns off all dragging displays.
Auto AutoCAD determines when to display dragging.

COMMAND MODIFIER

Drag Displays drag image when **Dragmode** = On.

RELATED SYSTEM VARIABLES

- **Dragmode** Current dragmode setting:
  - **0** No drag image.
  - **1** On if required.
  - **2** Automatic.
- **Drag1** Drag regeneration rate (*default=10*).
- **Drag2** Drag redraw rate (*default=25*).

'DsViewer

Rel. 13

Toggles the display of the bird's-eye view window; provides real-time pan and zoom (*short for "DiSplay Viewer"*).

Command	Alt+	Alias	Menu Bar	Tablet
'dsviewer	T,A	av	[Tools] [Aerial View]	...

Command: **dsviewer**

Displays Aerial View window.

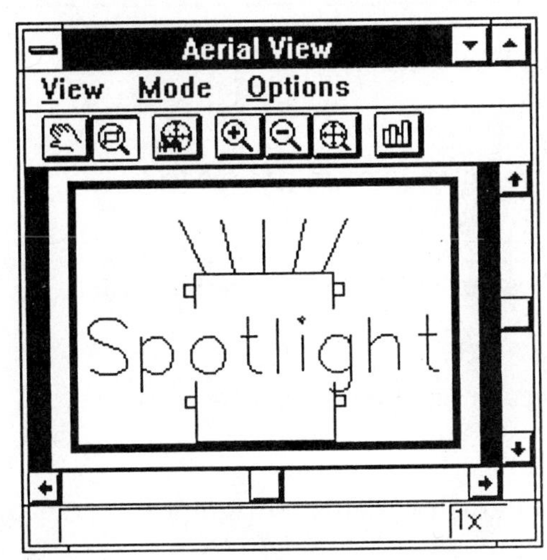

COMMAND OPTIONS

View Changes zoom level.

Mode Toggles between zoom and pan modes.

Options

Auto Viewport

Automatically displays the current model space viewport.

Dynamic Update

Automatically updates the Aerial View during editing.

Locator Magnification

Selects bird's-eye view magnification:

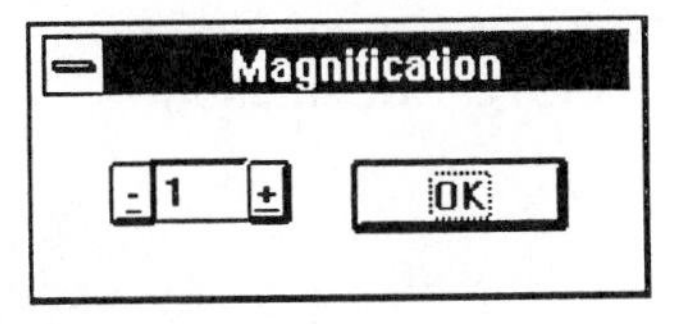

1 No magnification (*default*)

32 Maximum

Display Statistics

List the amount of memory consumed by the display-list:

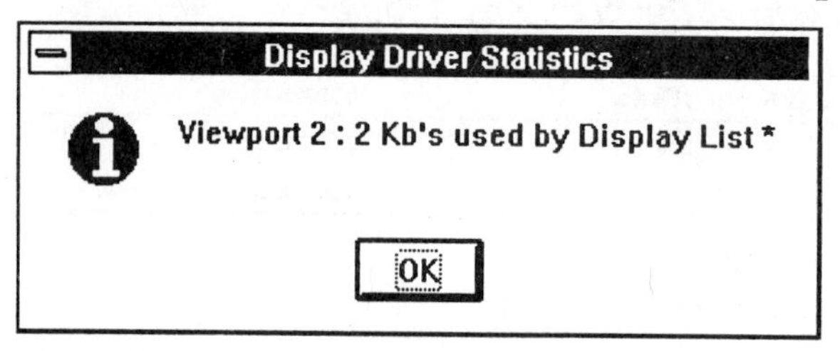

MENU BAR OPTIONS

 **Pan**

Switches Aerial View to pan mode.

 **Zoom**

Switches Aerial View to zoom mode (default).

 **Locator**

Switches Aerial View to bird's-eye view.

 **Zoom In**

Zooms in by user-defined magnification (*2x by default*).

 **Zoom Out**

Zooms out by user-defined magnification (*2x by default*).

 **Global**

Zooms extents to show entire drawing in Aerial View window.

 **Display List**

Displays amount of memory consumed by display list.

RELATED AUTOCAD COMMANDS

- **Pan** Moves the drawing view.
- **View** Creates and displays named views.
- **Zoom** Makes the view larger or smaller.

RELATED FILE

- **DsAi.Hlp** Help file for **DsViewer**.

DText

Ver. 2.5

Enters text in the drawing in a visual mode (*short for Dynamic TEXT*).

Command	Alias	Side Menu	Menu Bar	Tablet
dtext	...	[DRAW] [DText:]	[Draw] [Text] [DText]	T1 - U3

```
Command: dtext
Justify/Style/<Start point>: J
Align/Fit/Center/Middle/Right/TL/TC/TR/ML/MC/MR/BL/BC/BR:
Height:
Rotation Angle:
Text:
```

COMMAND OPTIONS

Justify Displays the justification submenu:

- **Align** Aligns the text between two points with adjusted text height.
- **Fit** Fits the text between two points with fixed text height.
- **Center** Centers the text along the baseline.
- **Middle** Center the text horizontally and vertically.

Right	Right-justifies the text.
TL	Top-left justification.
TC	Top-center justification.
TR	Top-right justification.
ML	Middle-left justification.
MC	Middle-center justification.
MR	Middle-right justification.
BL	Bottom-left justification.
BC	Bottom-center justification.
BR	Bottom-right justification.
<Start point>	Left-justifies the text (*default*).
Style	Displays the style submenu:
Style name	Indicates a different style name.
?	Lists the currently loaded styles.
[Enter]	Exit the **DText** command.

COMMAND MODIFIERS

%%c	Draws diameter symbol: Ø.
%%d	Draws degree symbol: °.
%%o	Starts and stops overlinning.
%%p	Draws the plus-minus symbol: ±.
%%u	Starts and stops underlinning.
%%%	Draw the percent symbol: %.

RELATED AUTOCAD COMMANDS

- **DdEdit** Edits the text.
- **Change** Changes the text height, rotation, style, and content.
- **Style** Creates new text styles.
- **Text** Adds new text to the drawing.
- **MText** Places paragraph text in drawings.

RELATED SYSTEM VARIABLES

- **TextSize** The current height of text.
- **TextStyle** The current style.
- **ShpName** The default shape name.

TIPS

- Use the **DText** command to easily place text in many locations.

- Be careful: the spacing between lines of text does not neccesarily match the current snap spacing.

- The popdown menus are not available during the **DText** command.

- Transparent commands do not work during the **DText** command.

- You can enter any justification modes at the '<Start point>:' prompt.

- The command automatically repeats until cancelled with **[Enter]**.

DView

Rel. 10

Dynamically zooms and pans 3D drawings, and turns on perspective mode (*short for Dynamic VIEW*).

Command	Alias	Side Menu	Menu Bar	Tablet
dview	dv	[VIEW] [Dview:]	[View] [3D Dynamic View]	J3 - K3

```
Command: dview
Select objects: [Enter]
CAmera/TArget/Distance/POints/PAn/Zoom/TWist/CLip/Hide/Off/
   Undo/<eXit>:
```

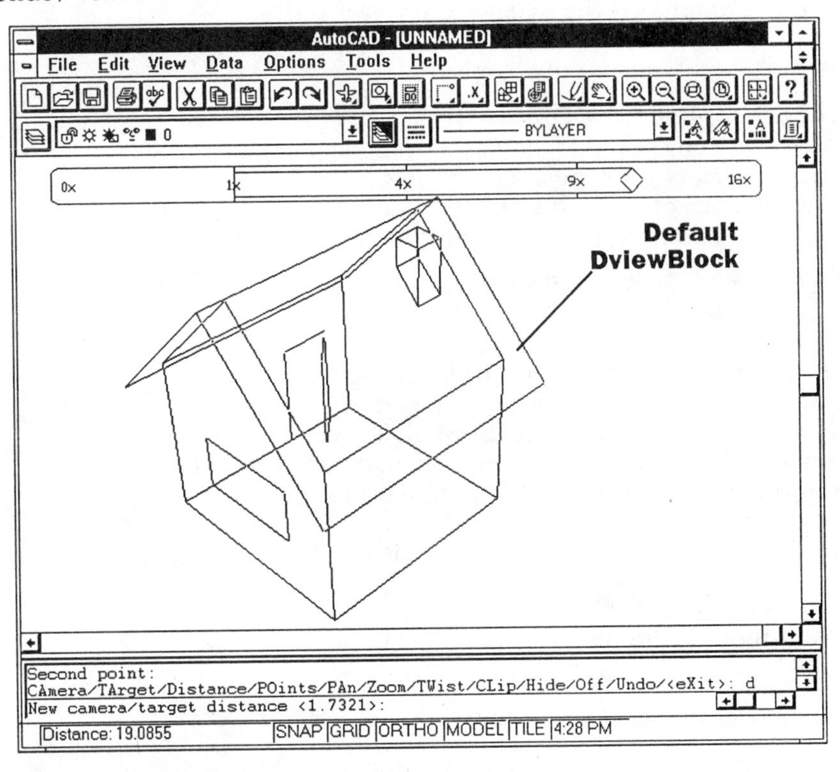

COMMAND OPTIONS

CAmera Indicates the camera angle relative to the target:

Toggle Switches between input angles.

TArget Indicates the target angle relative to the camera.

Distance Indicates camera-to-target distance; turns on perspective mode.

POints Indicates both the camera and target points.

PAn Dynamically pans the view.

Zoom Dynamically zooms the view.

TWist Rotates the camera.

CLip Displays the submenu for view clipping:
- **Back** Displays the submenu for back clipping:
  - **ON** Turns on the back clipping plane.
  - **OFF** Turns off the back clipping plane.
  - **<Distance from target>** Indicates the location of the back clipping plane.
- **Front** Displays the submenu for front clipping:
  - **Eye** Positions the front clipping plane at the camera.
  - **<Distance from target>** Indicates the location of the front clipping plane.
- **<Off>** Turns off view clipping.

Hide Performs hidden-line removal.
Off Turns off the perspective view.
Undo Undoes the most recent **DView** action.
<eXit> Exits the **DView** command.

RELATED AUTOCAD COMMANDS

- **Hide** Removes hidden-lines from a non-perspective view.
- **Pan** Pans a non-perspective view.
- **VPoint** Selects a non-perspective viewpoint of a 3D drawing.
- **Zoom** Zooms a non-perspective view.

RELATED SYSTEM VARIABLES

- **BackZ** Back clipping plane offset.
- **FrontZ** Front clipping plane offset.
- **LensLength** Perspective view lens length, in millimeters.
- **Target** UCS 3D coordinates of target point.
- **ViewCtr** 2D coordinates of current view center.
- **ViewDir** WCS 3D coordinates of camera offset from target.
- **ViewMode** Perspective and clipping settings.
- **ViewSize** Height of view.
- **ViewTwist** Rotation angle of current view.

RELATED SYSTEM BLOCK

- **DViewBlock** Alternate viewing object during DView.

TIPS

- The view direction is from 'camera' to 'target'.
- Press **[Enter]** at the 'Select objects:' prompt to display the house.
- To replace the house block with your own, redefining DVIewBlock.
- To view a 3D drawing in one-point perspective, use the **DView Zoom** command.
- Pull-down menus, zoom, and pan are not available during **DView**.
- Once the view is in perspective mode, you cannot use the **Sketch**, **Zoom**, and **Pan** commands.

Dxbin

Ver. 2.1

Imports a DXB-format file into the current drawing (*short for Drawing Exchange Binary INput*).

Command	Alt+	Side Menu	Menu Bar	Tablet
dxbin	F,I	[FILE]	[File]	...
		[IMPORT]	[Import]	
		[DXBin:]	[DXB]	

```
Command: dxbin
DXB file:
```

COMMAND OPTIONS

None

RELATED AUTOCAD COMMANDS

- **DxfIn** Reads DXF-format files.
- **Plot** Writes DXB-format files when configured for ADI plotter.

RELATED SYSTEM VARIABLES

- *None*

TIP

- Configure AutoCAD with the ADI plotter driver to produce a DXB file.

Dxfin

Ver. 2.0

Reads a DXF-format file into a drawing (*short for Drawing interchange Format INput*).

Command	Alt+	Side Menu	Menu Bar	Tablet
dxfin	F,I	[FILE]	[File]	...
		[IMPORT]	[Import]	
		[DXFin:]	[DXF]	

```
Command: dxfin
File name:
```

COMMAND OPTIONS

None

RELATED AUTOCAD COMMANDS

- **DxbIn** Reads a DXB-format file.
- **DxfOut** Writes a DXF-format file.

RELATED SYSTEM VARIABLES

- *None*

TIP

- The **IGESin** command was removed from Release 13.

DxfOut

Ver. 2.0

Writes a DXF-format file of part or all of the current drawing (*short for Drawing interchange Format OUTput*).

Command	Alt+	Side Menu	Menu Bar	Tablet
dxfout	F,E	[FILE]	[File]	...
		[EXPORT]	[Export]	
		[DXFout:]	[DXF]	

```
Command: dxfout
File name:
Enter decimal places of accuracy (0 to 16)/Objects/Binary
   <6>: e
Select objects: [pick]
Select objects: [Enter]
Enter decimal places of accuracy (0 to 16)/Binary <6>:
```

COMMAND OPTIONS

0 – 16 For ASCII binary files, indicates decimal places of accuracy.
Objects Selects objects to export in DXF format.
Binary Creates binary DXF file.

RELATED AUTOCAD COMMANDS

- **DxfIn** Reads a DXF-format file.
- **Save** Writes the drawing in DWG format.

RELATED SYSTEM VARIABLES

- *None*

TIPS

- Use the ASCII DXF format to exchange drawings with other CAD and graphics programs.

- A binary DXF file is much smaller and is created much faster than an ASCII binary file; however, few applications read a binary DXF file.

- There is no facility to create Release 12-compatible DXF files, as there is for DWG files (via the **SaveAsR12** command).

- The **IGESout** command was removed from Release 13.

Edge

Rel. 12

Toggles the visibility of 3D faces (*an external command in Edge.Lsp*).

Command	Alias	Side Menu	Toolbar	Tablet
edge	...	...	[Surfaces] [Edge]	...

```
Command: edge
Display/<Select edge>: D
Select/<All>: S
Select objects: [pick]
Display/<Select edge>: [Enter]
```

COMMAND OPTIONS

<Select edge> Selected edge is no longer visible.
Display Highlights invisible edges.
Select Regenerates selected hidden edges.
All Selects all hidden edges and regenerates them.

RELATED AUTOCAD COMMAND

- **3dFace** Creates 3D faces.

RELATED SYSTEM VARIABLE

- **SplFrame** Toggles visibility of 3D face edges.

TIPS

- Make edges invisible to make 3D objects look nicer.

- Command repeats itself until you press **[Enter]** at the 'Display/<Select Edge>:' prompt.

EdgeSurf

Rel. 10

Draws a 3D polygon mesh as a Coons surface patch between four boundaries (*short for EDGE-defined SURFace*).

Command	Alias	Side Menu	Toolbar	Tablet
edgesurf	...	[DRAW 2] [SURFACES] [Edgsurf:]	[Surfaces] [EdgeSurf]	O 8

```
Command: edgesurf
Select edge 1: [pick]
Select edge 2: [pick]
Select edge 3: [pick]
Select edge 4: [pick]
```

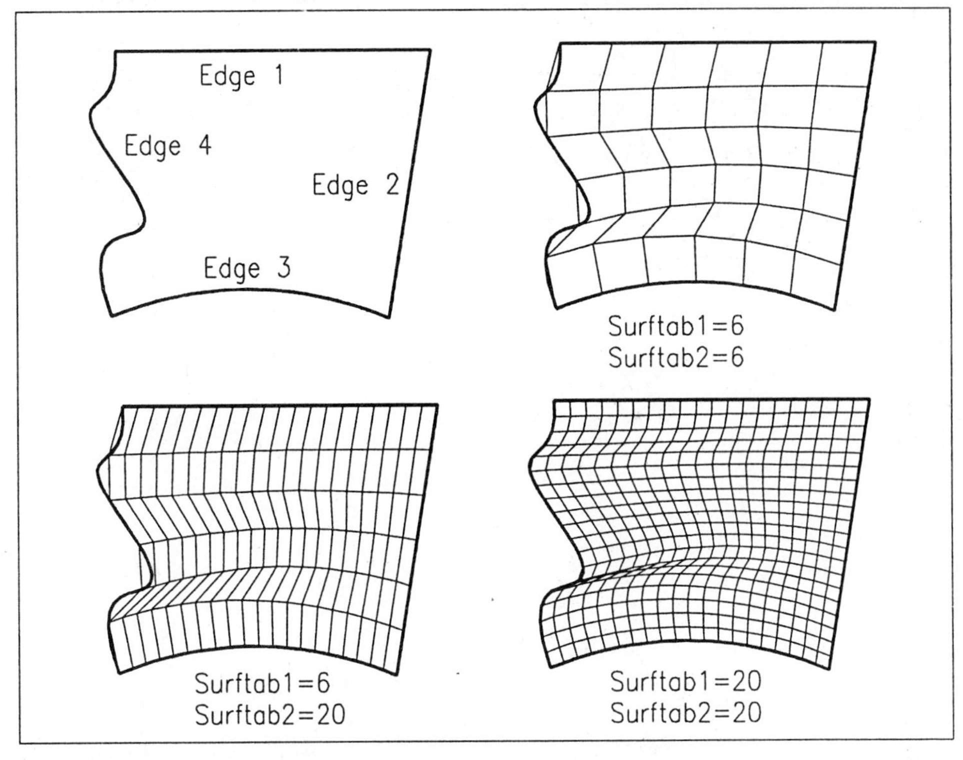

COMMAND OPTIONS

None

RELATED AUTOCAD COMMANDS

- **3Dmesh** Creates a 3D mesh by specifying every vertex.
- **3Dface** Creates a 3D mesh of irregular vertices.
- **Pedit** Edits the mesh created by **EdgeSurf**.
- **Tabsurf** Creates a tabulated 3D surface.
- **Rulesurf** Creates a ruled 3D surface.
- **Revsurf** Creates a 3D surface of revolution.

RELATED SYSTEM VARIABLES

- **Surftab1** The current M density of meshing.
- **Surftab2** The current N density of meshing.

TIPS

- The Coons surface created by the **EdgeSurf** command is an interpolated bi-cubic surface.

- The four boundary edges can be made from lines, arcs, and open 2D and 3D polylines; the edges must meet at their endpoints.

- The maximum mesh density is 32,767.

'Elev

Ver. 2.1

Sets elevation and thickness of extruded 3D objects (*short for ELEVation*).

Command	Alias	Side Menu	Menu Bar	Tablet
'elev	...	[SETTINGS] [ELEV:]	...	...

```
Command: elev
New current elevation <0.0000>:
New current thickness <0.0000>:
```

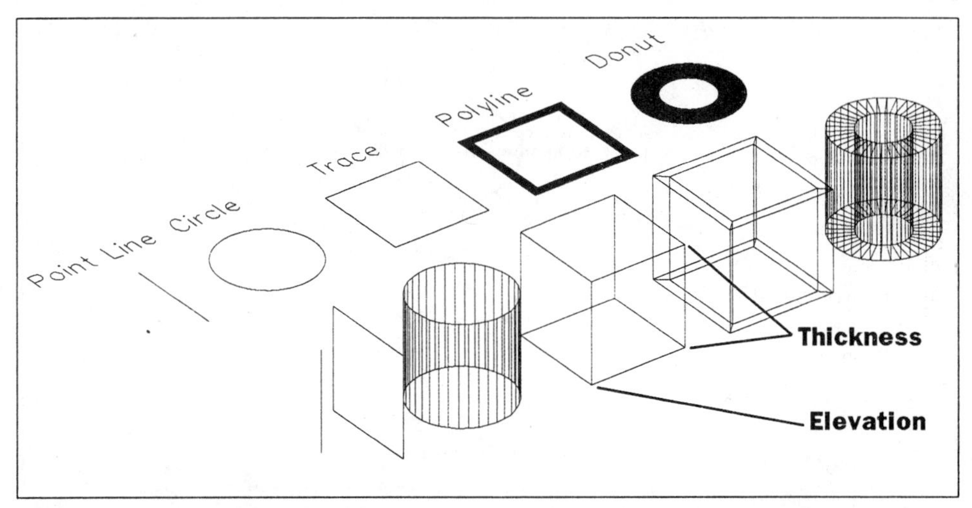

COMMAND OPTIONS

None

RELATED AUTOCAD COMMANDS

- **Change** Changes the thickness and elevation of objects.
- **Chprop** Changes the thickness of objects.
- **Move** Moves objects, including in the z-direction.

RELATED SYSTEM VARIABLES

- **Elevation** Stores the current elevation setting.
- **Thickness** Stores the current thickness setting.

TIPS

- The current value of elevation is used whenever a z-coordinate is not supplied.

- Thickness is measured up from the current elevation in the positive z-direction.

Ellipse

Ver. 25

Draws ellipses (*by four different methods*), elliptical arcs, and isometric circles.

Command	Alias	Side Menu	Toolbar	Tablet
ellipse	...	[DRAW 1] [Ellipse:]	[Draw] [Ellipse]	N 10

```
Command: ellipse
Arc/Center/Isocircle/<Axis endpoint 1>: [pick]
Axis endpoint 2: [pick]
<Other axis distance>/Rotation: [pick]
```

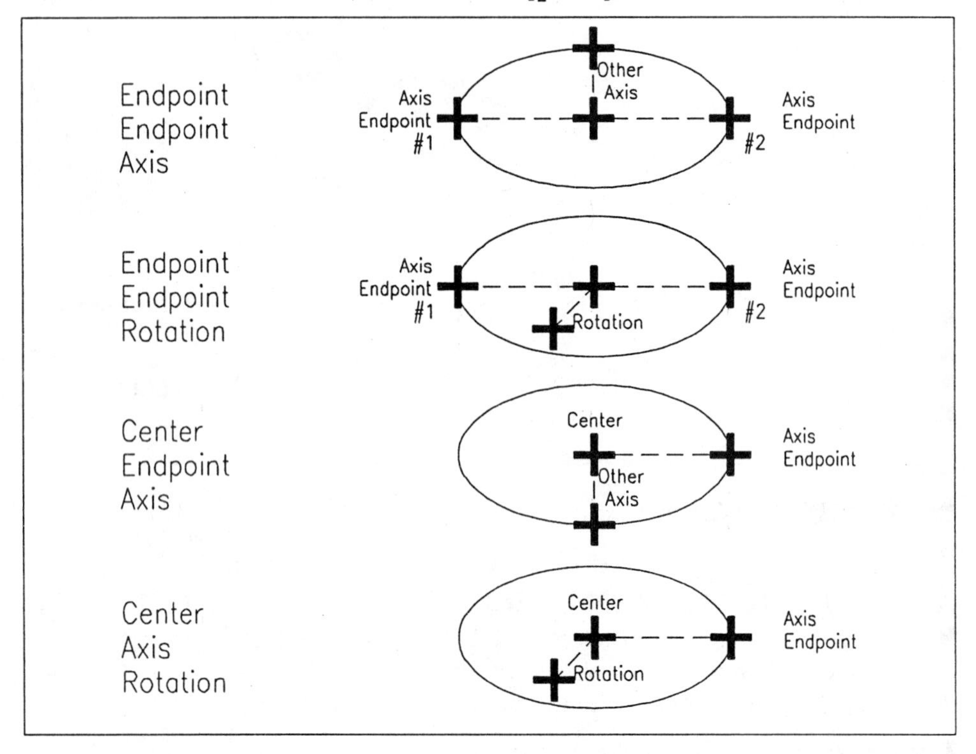

COMMAND OPTIONS

Axis endpoint 1

Indicates the first endpoint of the major axis.

Axis endpoint 2 Indicates the second endpoint of the major axis.

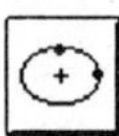

Center Indicates the center point of the ellipse.

Arc Draws an elliptical arc.

Rotation Indicates a rotation angle around the major axis.

Isocircle Draws isometric circles, appears only when **Snap** is set to isometric mode.

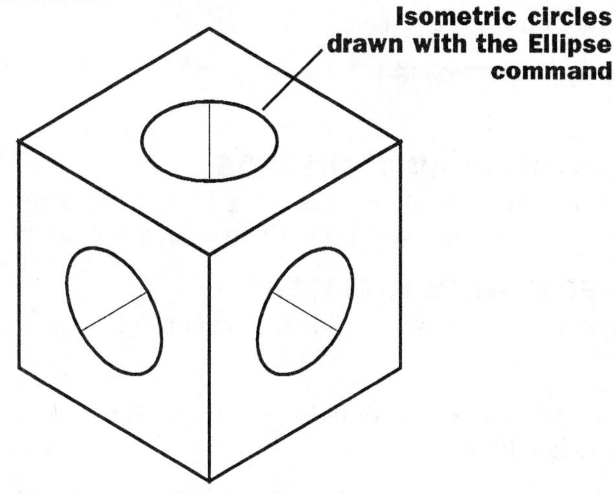

RELATED AUTOCAD COMMANDS

- **Isoplane** Sets the current isometric plane.
- **Pedit** Edits ellipses and other polyline entities.
- **Snap** Controls the setting of isometric mode.

RELATED SYSTEM VARIABLES

- **SnapIsoPair** Current isometric plane:
  - **0** Left.
  - **1** Top.
  - **2** Right.
- **SnapStyl** Regular or isometric drawing mode:
  - **0** Standard.
  - **1** Isometric.

TIPS

■ Previous to Release 13, the **Ellipse** command constructed the ellipse as a series of short polyline arcs; thus, ellipses translated by the **SaveAsR12** command are not mathematically exact.

■ When **PEllipse** = 1, the **Arc** option is not available.

■ Use ellipses to draw circles in isometric mode. When **Snap** is set to isometric mode, the **Ellipse** command's isocircle option projects a circle into the working isometric drawing plane. Use **[Ctrl]+E** to toggle isoplanes.

End

Ver. 1.0

Saves the drawing and exits AutoCAD to the operating system.

Command	Alias	Side Menu	Menu Bar	Tablet
end	...	...	...	...

Command: **end**

COMMAND OPTIONS

None

RELATED AUTOCAD COMMANDS

- **Saveas** Saves read-only drawings by another name.
- **Quit** Leaves AutoCAD without saving the drawing.

RELATED SYSTEM VARIABLES

- **Dbmod** Determines whether the drawing has been modified since loaded.

TIPS

- AutoCAD renames the drawing file to .BAK before saving the contents of the drawing editor.
- The **End** command does not work with drawings set to read-only; use the **SaveAs** command instead.

Erase

Ver. 1.0

Erases objects from the drawing.

Command	Alias	Side Menu	Toolbar	Tablet
erase	e	[MODIFY] [Erase:]	[Modify] [Erase]	V 7-8

```
Command: erase
Select objects: [pick]
```

COMMAND OPTIONS

None

RELATED AUTOCAD COMMANDS

- **Break** Removes a portion of a line, circle, arc, or polyline.
- **Change** Changes the length of a line.
- **Oops** Returns the most-recently erased objects to the drawing.
- **Trim** Reduces the length of a line, polyline, or arc.
- **Undo** Returns the erased objects to the drawing.

RELATED AUTOLISP PROGRAM

- **DelLayer.Lsp** Erases all objects on a layer.

RELATED SYSTEM VARIABLES

- *None*

TIPS

- The **Erase L** command combination erases the last-drawn item visible on the screen.

- The **Oops** command brings back the most-recently erased objects; use the **U** command to bring back other erased objects.

Explode

Ver. 2.5

Explodes a polyline, block, associative dimension, hatch, multiline, 3D solid, region, body, or polyface mesh into its constituent entities.

Command	Alias	Side Menu	Toolbar	Tablet
explode	...	[MODIFY] [Explode:]	[Modify] [Explode]	...

```
Command: explode
Select objects: [pick]
```

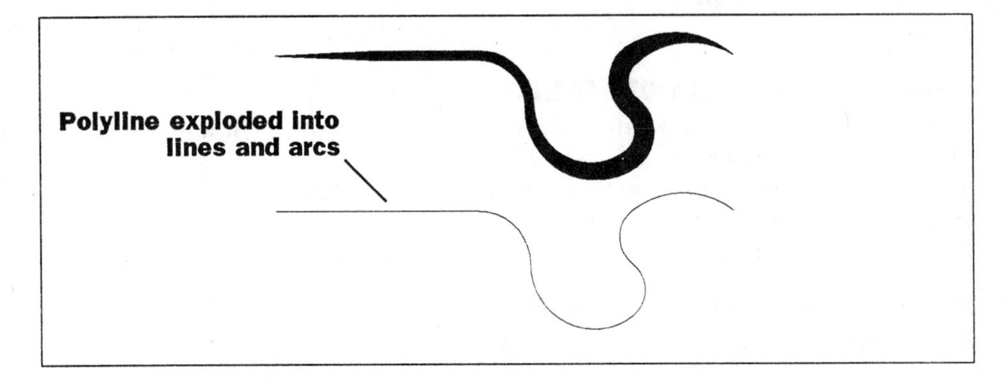

COMMAND OPTIONS

None

RELATED AUTOCAD COMMANDS

- **Block** Recreates a block after an explode.
- **PEdit** Converts a line into a polyline.
- **Region** Converts 2D objects into a region.
- **Undo** Reverses the effects of explode.
- **Xplode** Allows greater control over the exploding process.

RELATED SYSTEM VARIABLES

- **ExplMode** Toggles whether non-uniformly scaled blocks can be exploded:
  - **0** Does not explode (R12-compatible)
  - **1** Does explode (default)

TIPS

- The **Explode** command reduces:
  - Blocks into their constituent parts.
  - Associative dimension into lines, solids, and text.
  - 2D polylines into lines and arcs; width and tangency information is lost.

- 3D polylines into lines.
- Multilines into lines.
- Polygon meshes into 3D faces.
- Polyface meshes into 3D faces, lines, and points.
- 3D solids into regions and bodies.
- Regions into lines, arcs, ellipses, and splines.
- Bodies into single bodies, regions, and curves.

■ With Release 13, you can explode blocks inserted with unequal scale factors, mirrored blocks, and blocks created by the **MInsert** command.

■ You cannot explode xrefs and dependent blocks.

■ The parts making up exploded blocks and associative dimensions may change color and linetypes.

■ Resulting entities become the previous selection set.

■ A circle within a non-uniformly scaled block explodes into an ellipse; an arc into an elliptical arc.

Export

Rel. 13

Saves the drawing in formats other than DWG.

Command	Alt+	Side Menu	Menu Bar	Tablet
export	F,E	[FILE] [EXPORT]	[File] [Export]	...

```
Command: export
```

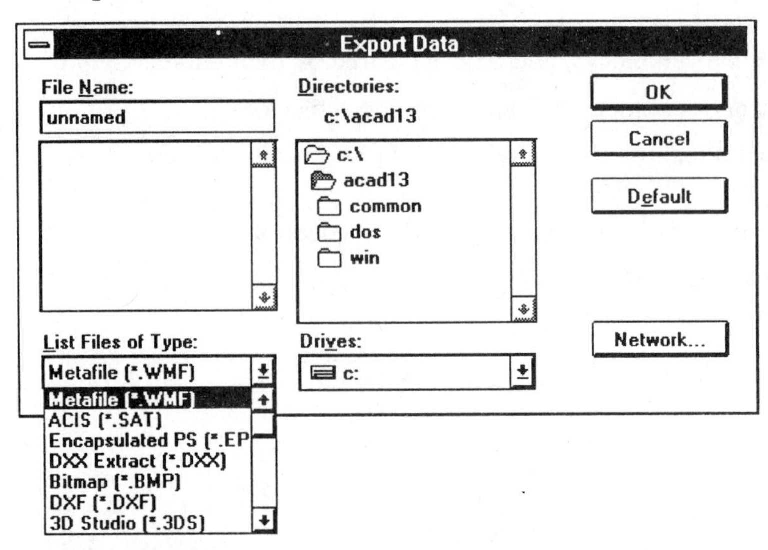

COMMAND OPTIONS

BMP	Windows bitmap format; executes the **BmpOut** command.
DWG	AutoCAD Release 13 format; executes the **SaveAs** command.
DXF	Drawing interchange format; executes the **DxfOut** command.
DXX	ASCII DXF attribute extract; executes the **AttExt** command.
EPS	Encapsulated PostScript; executes the **PsOut** command.
SAT	ASCII ACIS (*Save As Text*); executes the **AcisOut** command.
WMF	Windows metafile; executes the **WmfOut** command.
3DS	3D Studio format; executes the **3dsOut** command.

RELATED AUTOCAD COMMANDS

- **AcisOut** Exports ACIS solid objects in drawing as ASCII data.
- **AseExport** Exports external database data in drawing.
- **AttExt** Exports attribute data in drawing as CDF, SDF, or DXF formats.
- **BmpOut** Exports drawing as BMP format.
- **CopyClip** Exports drawing to Clipboard.
- **CopyHist** Exports text screen to Clipboard.
- **CopyLink** Exports drawing to Clipboard.

- **CutClip** Exports drawing to Clipboard.
- **DdAttExt** Exports attribute data in drawing as CDF, SDF, or DXF formats.
- **DxfOut** Exports drawing as ASCII or binary DXF format.
- **Import** Imports several vector and raster formats.
- **LogFileOn** Saves command line text to ASCII file Acad.Log.
- **MakePreview** Exports drawing as BMP file.
- **MassProp** Exports mass property data as ASCII text in MPR file.
- **MSlide** Exports current viewport as an SLD slide file.
- **Plot** Exports drawing in a couple of dozen vector and raster formats.
- **PsOut** Exports drawing as encapsulated PostScript file.
- **Save** Exports drawing as AutoCAD Release 13 DWG format.
- **SaveAsR12** Exports drawing as AutoCAD Release 12 DWG format.
- **SaveImg** Exports rendering as TIFF, Targa, or GIF format.
- **StlOut** Exports ACIS solid model in stereolithography format.
- **WBlock** Exports blocks in drawing as DWG files.
- **WmfOut** Exports drawing as WMF format.
- **3dsOut** Exports drawing as 3D Studio format.

TIP

- The **Export** command acts as a "shell" command; it launches other AutoCAD commands that perform the actual export function.

 Extend

Ver. 2.5

Extends the length of a line, open polyline, or arc to a boundary.

Command	Alias	Side Menu	Menu Bar	Tablet
extend	...	[MODIFY] [Extend:]	[Modify] [Trim] [Extend]	X 16

```
Command: extend
Select boundary edges:(Projmode=UCS,Edgemode=No extend)
Select objects: [pick]
Select objects: [Enter]
<Select object to extend>/Project/Edge/Undo: P
None/Ucs/View:
```

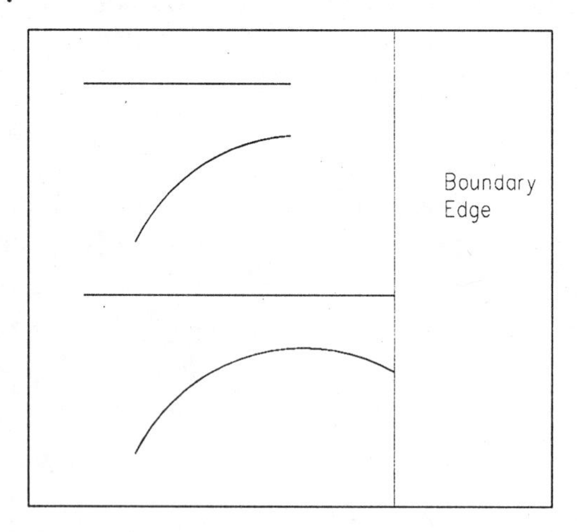

COMMAND OPTIONS

Project Specifies projection mode:
- **None** Extends objects to boundary (*Release 12 compatible*).
- **Ucs** Boundary is x,y-plane of current UCS.
- **View** Boundary is current view plane.

Edge Toggles actual or implied edge:
- **Extend** Extends to implied boundary.
- **No extend** Extends only to actual boundaries (*Release 12 compatible*).

Undo Undoes the most recent extend operation.

RELATED AUTOCAD COMMANDS

- **Change** Changes the length of lines.
- **Lengthen** Changes the length of open objects.
- **Stretch** Stretches objects wider or narrower.
- **Trim** Reduces the length of lines, polylines, and arcs.

RELATED SYSTEM VARIABLES

- **EdgeMode** Toggles boundary mode for **Extend** and **Trim** commands:
  - 0 Use actual edges (*Release 12 compatible; default*).
  - 1 Use implied edge.
- **ProjMode** Toggles projection mode for **Extend** and **Trim** commands:
  - 0 None (*Release 12 compatible*).
  - 1 Current UCS (*Default*).
  - 2 Current view plane.

TIPS

- The following objects can be used as edges:
  - Lines, rays, xlines, and multilines.
  - Arcs, circles, ellipses, and elliptical arcs.
  - 2D and 3D polylines, and splines.
  - Text and regions.

- Pick the object a second time to extend it to a second boundary line.

- Circles and other closed objects are valid edges: object is extended in direction nearest to the pick point.

- Extending a variable-width polyline widens it proportionately; extending a splined polyline adds a vertex.

Extrude

Rel. 11

Creates a 3D solid by extruding a 2D object with optional tapered sides *(formerly the **SolExt** command; an external command in Acis.Dll).*

Command	Alias	Side Menu	Toolbar	Tablet
extrude	...	[DRAW 2] [SOLIDS] [Extrude:]	[Solids] [Extrude]	K 8

```
Command: extrude
Select objects: [pick]
Path/<Height of extrusion>:
Extrusion taper angle <0>:
```

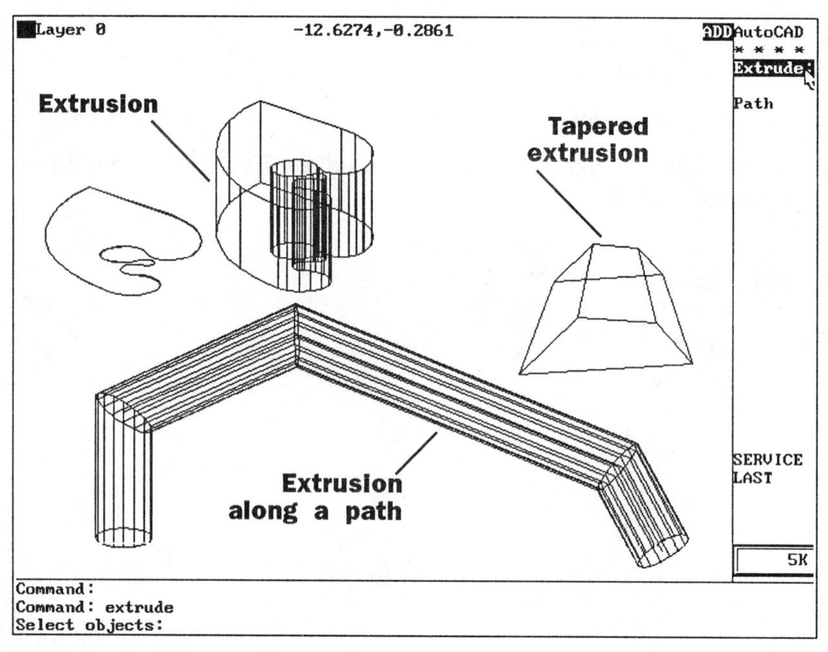

COMMAND OPTIONS

Path Selects an open object for the extrusion path.

Height Specifies the extrusion height.

Extrusion taper angle
Specifies the taper angle, -90 to +90 degrees.

RELATED AUTOCAD COMMAND

- **Revolve** Creates a 3D solid by revolving a 2D object.

TIPS

- You can extrude the following objects:
  - Circles, ellipses, and donuts.
  - Closed polylines, polygons, closed splines, and regions.
- You cannot extrude the following objects:
  - Objects within a block cannot be extruded; use the **Explode** command first.
  - Polylines with less than 3 or more than 500 vertices.
  - Crossing and self-intersecting polylines.
- The taper angle:
  - Must be between 0 (*default*) and 90 degrees.
  - Positive angle tapers in from base; negative angle taper out.
- The **Extrude** command does not work when a taper angle:
  - Is less than -90 degrees or more than +90 degrees.
  - If the combination of angle and height makes the object's extrusion walls intersect.
- You can use the following objects as extrusion paths:
  - Lines and polylines.
  - Arcs and elliptical arcs.
  - Circles and ellipses.
- Contrary to the official documentation, **Extrude** cannot use a spline as a path; the path cannot be in the plane of the profile.

FileOpen

Rel. 12

Opens a drawing file without displaying a dialogue box (*an undocumented command*).

Command	Alt+	Side Menu	Menu Bar	Tablet
`fileopen`	...	...	...	...

```
Command: fileopen
Name of drawing to open:
```

COMMAND OPTIONS

None

RELATED AUTOCAD COMMAND

Open Opens a drawing file via a dialogue box.

TIP

- Use the **FileOpen** command to open a drawing in menu macros.

Ver. 1.4

Displays the **File Utilities** dialogue box to allow list, copy, rename, delete, and unlock files.

Command	Alt+	Side Menu	Menu Bar	Tablet
'files	F,M,U	[FILE]	[File]	V 24
		[MANAGE]	[Management]	
		[Files:]	[Unlock File]	

Command: **files**

Displays dialogue box.

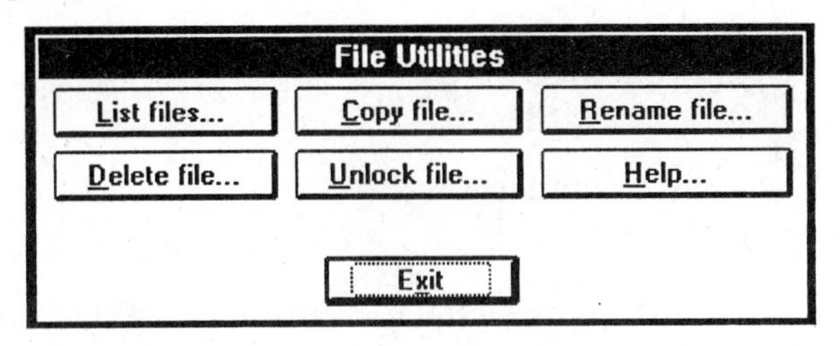

COMMAND OPTIONS

List files Lists files by extension.
Copy file Copies a file to another location.
Rename file Renames a file with a different name.
Delete file Deletes files.
Unlock file Erases *.??K lock files created by AutoCAD; sample dialogue box:

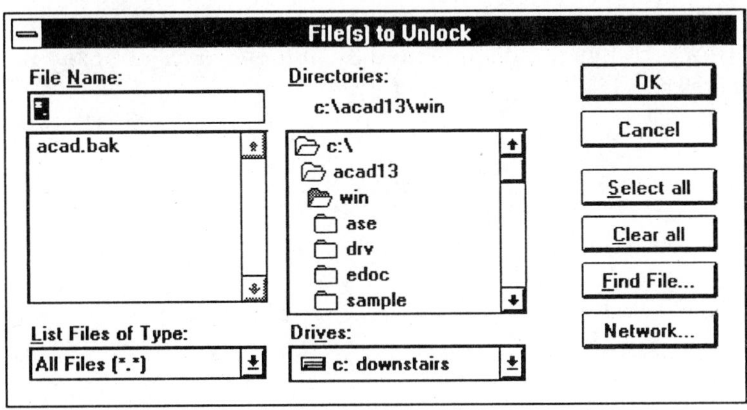

Exit Returns to 'Command:' prompt.

RELATED AUTOCAD COMMANDS

■ **Shell** Return temporarily to the operating system to perform file functions.

RELATED SYSTEM VARIABLE

■ **FileDia** Toggles display of dialogue box:

0 Functions display on text screen:

0. Exit File Utility menu.
1. List drawing files.
2. List user specified files
3. Delete files
4. Rename files
5. Copy files
6. Unlock files

1 Display dialogue box.

TIPS

■ Do not delete AutoCAD's temporary files while AutoCAD is running:

- *.AC$ and *.$A Temporary files.
- *.SWR Swap files.
- *.??K Lock files.

■ When AutoCAD crashes, these files (*listed above*) are often left behind and should be erased to recover disk space. Use the DOS **Erase** command.

■ Use the **Config** command to turn off file locking if your AutoCAD is not networked.

■ The **Unlock** command, documented by Autoesk, does not exist in AutoCAD Release 13.

'Fill

Ver. 1.0

Toggles wide objects (traces, multilines, solids, and polylines) to be displayed and plotted as solid-filled or as outlines.

Command	Alt+	Side Menu	Menu Bar	Tablet
'fill	O,D,O	[OPTIONS] [DISPLAY] [Fill:]	[Options] [Display] [Solid Fill]	...

Command: **fill**
ON/OFF <On>:

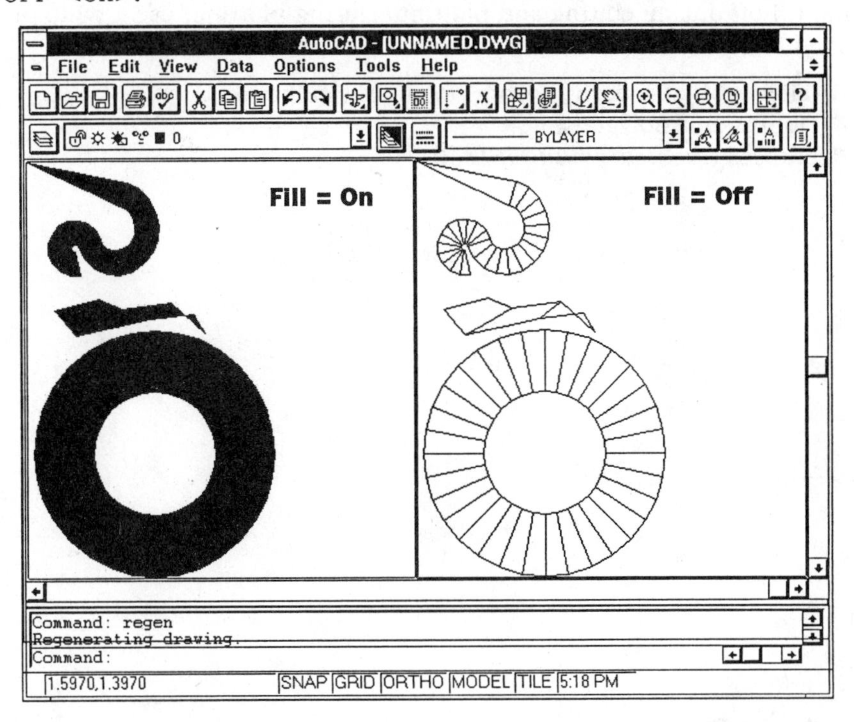

COMMAND OPTIONS

ON Turns fill on, after next regeneration.
OFF Turns fill off, after next regeneration.

RELATED SYSTEM VARIABLES

- **Fillmode** Current setting of fill status:
  - **0** Fill mode is off.
  - **1** Fill mode is on (*default*)
- **TextFill** Toggles whether PostScript and TrueType fonts are filled:
  - **0** Fonts not filled (*default*)
  - **1** Fonts filled.

RELATED AUTOCAD COMMAND

■ **Regen** Adjusts display to reflect fill-no fill status.

TIPS

■ The state of fill (*or no fill*) does not come into effect until the next regeneration.

■ Traces, solids, and polylines are not filled when the view is *not* in plan view, regardless of the setting of the **Fill** command.

■ Since filled objects take longer to regenerate, redraw, and plot, consider leaving fill off during editing and plotting. During plotting, use a wide pen for filled areas.

■ **Fill** affects objects derived from polylines, including:

- ■ Donuts and polygons.
- ■ Ellipses (*created with PEllipse = 1*).

Fillet

Ver. 1.4

Joins two intersecting lines, polylines, arcs, circles, or 3D solids with a radius.

Command	Alias	Side Menu	Toolbar	Tablet
fillet	...	[CONSTRCT] [Fillet:]	[Modify] [Feature] [Fillet]	X 19-20

```
Command: fillet
Polyline/Radius/Trim/<Select first object>: [pick]
Select second object: [pick]
```

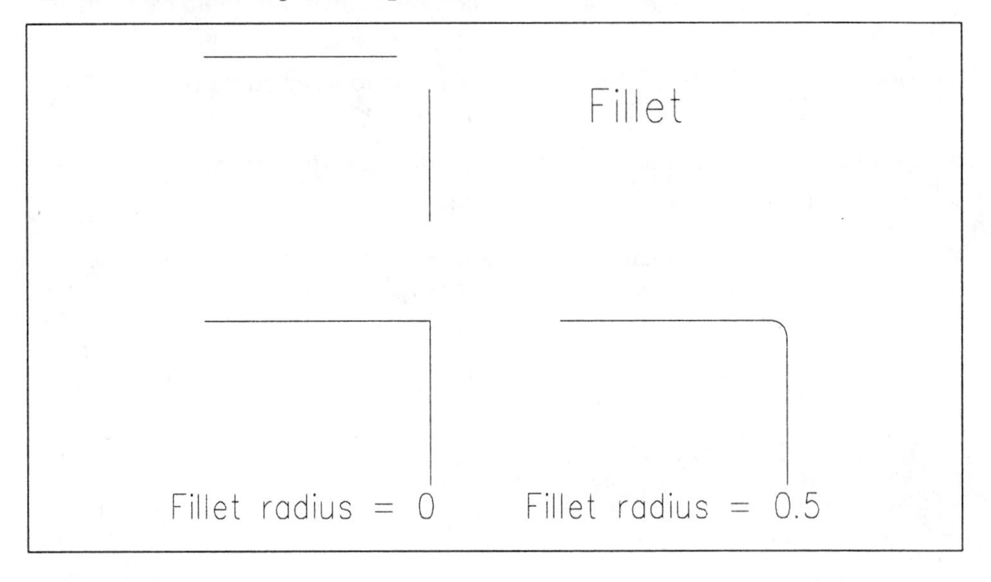

COMMAND OPTIONS

Polyline Fillets all vertices of a polyline.
Radius Indicates the filleting radius.
Trim Toggles whether objects are trimmed.

RELATED AUTOCAD COMMAND

- **Chamfer** Bevels intersecting lines or polyline vertices.

RELATED SYSTEM VARIABLES

- **FilletRad** The current filleting radius.
- **TrimMode** Toggles whether objects are trimmed.

TIPS

■ Pick the end of the entity you want filleted; the other end will remain untouched.

■ The lines, arcs, or circles need not touch.

■ As a faster substitute for the **Extend** and **Trim** commands, use the **Fillet** command with the radius of zero.

■ If the lines to be filleted are on two different layers, the fillet is drawn on the current layer.

■ The fillet radius must be smaller than the length of the lines. For example, if the lines to be filleted are 1.0m long, the fillet radius can be no more than 0.9999m.

■ Use the **Close** option of the **Polyline** command to ensure a polyline is filleted at all vertices.

■ You cannot fillet polyline segments from different polylines.

■ Filleting a pair of circles does not trim them.

■ As of Release 13, the **Fillet** command fillets a pair of parallel lines; the radius of fillet is automatically determined as half the offset distance.

REMOVED COMMAND: FilmRoll

The **FilmRoll** command, which imports AutoShade FLM filmroll files into AutoCAD, was removed from Release 13. There is no replacement.

'Filter

Rel. 12

Creates a filter list applied to selection sets (*an external command in Filter.Lsp*).

Command	Alias	Side Menu	Toolbar	Tablet
'filter	...	[ASSIST]	[Standard]	...
		[Filter:]	[Select Objects]	
			[Filter]	

```
Command: filter
```

Displays dialogue box.

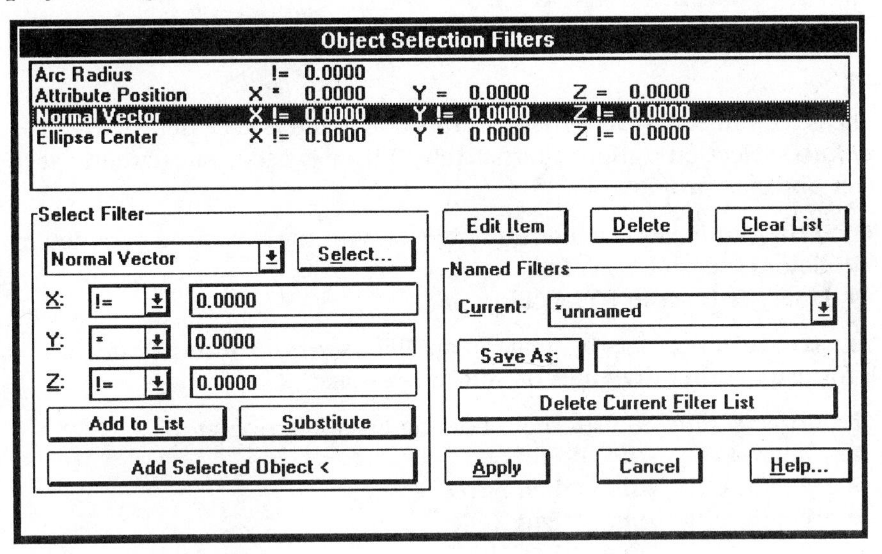

COMMAND OPTIONS

Select Filter Selects filter based on entity properties.
Select Displays all items of specified type in drawing.
X:, Y:, Z: Specifies object's coordinates.
Add to List Adds current select-filter option to filter list.
Substitute Replaces a highlighted filter with selected filter.
Add selected Object
Selects object from drawing.
Edit Item Lets you edit highlighted filter item.
Delete Deletes highlighted filter item.
Clear list Clears entire filter list.
Current named filter
Selects named filter from list.

Save as Saves filter list with name and .NFL extension.

Delete current Filter list

Deletes named filter.

Apply Applies filter operation.

RELATED AUTOCAD COMMANDS

■ *Any AutoCAD command with a 'Select objects:' prompt*

RELATED FILE

■ ***.NFL** Named filter list.

RELATED SYSTEM VARIABLES

■ *None*

TIPS

■ The selection set created by the **Filter** command is accessed via the 'P' (*previous*) selection option; alternatively, **'Filter** is used transparently at the 'Select objects:' prompt.

■ The **Filter** command cannot find objects with the following parameters:

- ■ Color is set to BYLAYER.
- ■ Linetype is set to BYLAYER.

■ You save selection sets by name to an NFL (*short for named filter*) file on disk for use in other drawings or editing sessions.

■ The **Filter** command uses the following grouping operators:

- ■ **Begin OR *and* **End OR
- ■ **Begin AND *and* **End AND
- ■ **Begin XOR *and* **End XOR
- ■ **Begin NOT *and* **End NOT

■ The **Filter** command uses the following relational operators:

- ■ < Less than.
- ■ <= Less than or equal to.
- ■ = Equal.
- ■ != Not equal to.
- ■ > Greater than.
- ■ >= Greater than or equal to.
- ■ * All values.

'Gifin

Rel. 12

Imports GIF raster files into the drawing as a block; converts pixels into solids (*an external file in RasterIn.Exp*).

Command	Alt+	Side Menu	Menu Bar	Tablet
'gifin	F,I	[FILE]	[File]	...
		[IMPORT]	[Import]	
		[GIFin:]	[GIF]	

```
Command: gifin
GIF file name:
Insertion point:
Scale factor:
```

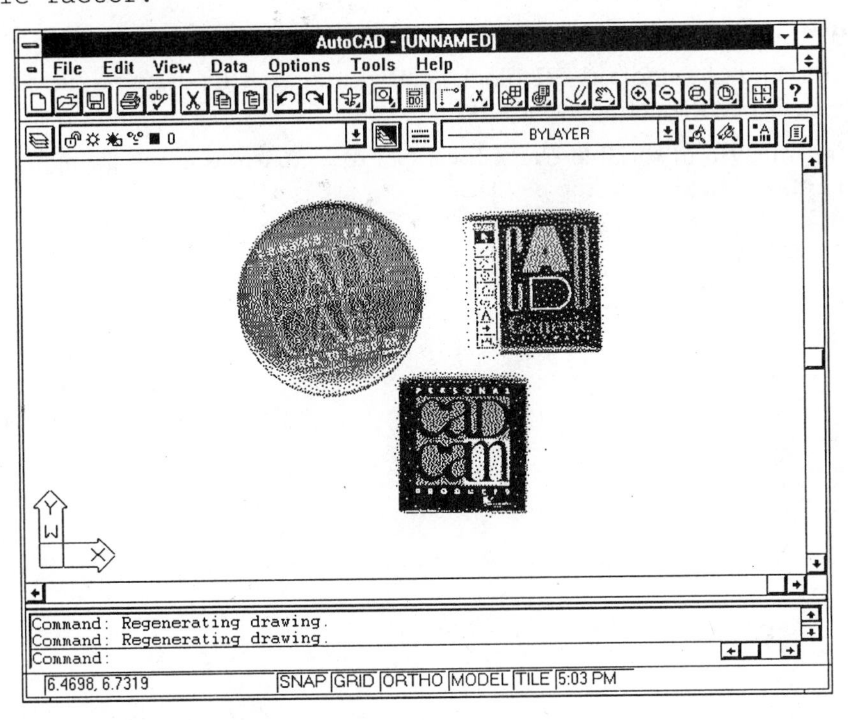

COMMAND OPTIONS

None

RELATED AUTOCAD COMMANDS

- **Pcxin** Imports PCX raster files.
- **PsIn** Imports EPS files.
- **Tiffin** Imports TIFF raster files.

RELATED SYSTEM VARIABLES

- **RiAspect** Adjust image's aspect ratio.
- **RiBackG** Change the image's background color.
- **RiEdge** Outline edges.
- **RiGamut** Specify number of colors.
- **RiGrey** Import as a grey scale image.
- **RiThresh** Control brightness threshold.

TIPS

- The GIF format (*short for Graphics Interchange Format*) was invented by CompuServe as a compressed graphics format that can be displayed on different hardware platforms.

- **GifIn** is limited to displaying a maximum of 255 colors.

- Exploding an imported GIF block greatly increases the drawing file size, since each pixel is defined as a solid.

- Turn off system variable **GripBlock** (set to 0) to avoid highlighting all the solid entities making up the block.

'GraphScr

Ver. 2.1

Switches the text screen back to the graphics screen in single-screen systems.

Command	Alias	Side Menu	Function Key	Tablet
'graphscr	...	...	[F2]	V 18

```
Command: graphscr
```

COMMAND OPTIONS

None

RELATED AUTOCAD COMMANDS

- **Script** Runs script files, which amke use of the **GraphScr** command.
- **TextScr** Switches from graphics screen to text screen.

RELATED RENDER COMMANDS

- **RendScr** Switches back to rendering display.

RELATED SYSTEM VARIABLES

- **Screenmode** Indicates whether current screen is text or graphics:
  - 0 Text screen.
  - 1 Graphics screen.
  - 2 Dual screen displaying both text and graphics.

TIP

- The **GraphScr** (and **TextScr**) commands do not work on a dual screen display.

’Grid

Ver. 1.0

Displays a grid of reference dots within the currently set limits.

Command	Ctrl+	Side Menu	Status bar	Function	Tablet
'grid	G	[ASSIST] [Grid:]	GRID	[F7]	V 20

```
Command: grid
Grid spacing(X) or ON/OFF/Snap/Aspect <1.0000>:
```

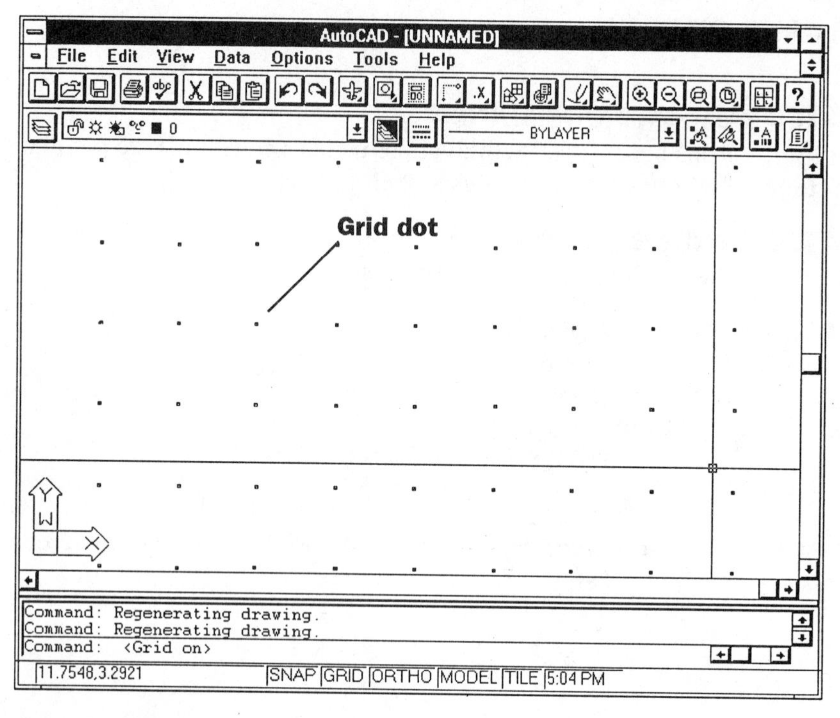

COMMAND OPTIONS

Aspect Indicates different spacing for the x- and y-direction.

Grid spacing(X)
Sets the x- and y-direction spacing; an *X* following the value sets the grid spacing a multiple of the current snap setting.

OFF Turns grid markings off.

ON Turns grid markings on.

Snap Makes the grid spacing the same as the snap spacing.

RELATED AUTOCAD COMMANDS

- **Ddrmodes** Sets the grid via a dialogue box.
- **Limits** Sets the limits of the grid in WCS.
- **Snap** Sets the snap spacing.

RELATED SYSTEM VARIABLES

- **Gridmode** Current grid visibility:
  - 0 Grid is off.
  - 1 Grid is on.
- **Gridunit** Current grid x,y-spacing.

TIPS

- The grid is most useful when set to the snap spacing, or to a multiple of the snap spacing.
- You can set a different grid spacing in each viewport; and a different grid spacing in the x- and y-directions.
- Rotate the grid with the **Snap** command's **Rotate** option; **Snap**'s **Isometric** option creates an isometric grid.
- If a very dense grid spacing is selected, the grid will take a long time to display; press **[Esc]** to cancel the display.
- AutoCAD does not display a too-dense grid and gives the message, "Grid too dense to display."
- Grid markings are not plotted; to create a plotted grid, use the **Array** command to place an array of points.

Rel. 13

Creates a named selection set of objects.

Command	Ctrl+	Side Menu	Toolbar	Tablet
group	A	[ASSIST] [Group:]	[Standard] [Group]	V 12

```
Command: group
```

Displays dialogue box.

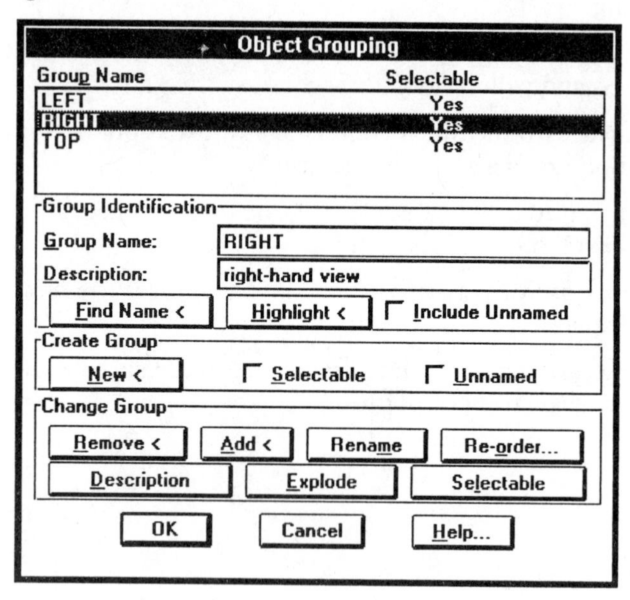


```
Command: -group
?/Order/Add/Remove/Explode/REName/Selectable/<Create>:
```

COMMAND OPTIONS

?	Lists names and descriptions of currently-defined groups.
Add	Adds objects to group.
Remove	Removes objects from group.
Explode	Removes group definition from drawing.
REName	Renames the group.
Selectable	Toggles whether group is selectable.
Create	Creates a new named group from objects selected.

Order Changes the order of objects within the group:

Order Group

Group Name

LEFT
RIGHT
TOP

Description

right-hand view

Remove from position (0 - 0):

Replace at position (0 - 0):

Number of objects (1 - 1):

Re-Order | Highlight | Reverse Order

OK | Cancel | Help...

RELATED AUTOCAD COMMANDS

- **Block** Creates a named symbol from a group of objects.
- **Select** Creates a selection set.

RELATED SYSTEM VARIABLE

- **PickStyle** Determines how groups are selected:
  - **0** Groups and associative hatches are not selected.
  - **1** Groups are included in selection sets.
  - **2** Associate hatches included in selection sets.
  - **3** Both are selected (*default*).

TIPS

- Toggle groups on and off with the **[Ctrl]+A** key.

- Consider a group as a named selection set; unlike a regular selection set, a group is not "lost" when the next group is created.

- Group names are up to 31 characters long, including $, _ , and - .

- Group descriptions are up to 64 characters long.

- Anonymous groups are unnamed; AutoCAD refers to them as *A*n*.

REMOVED COMMAND: Handles

The **Handles** command has been removed from Release 13; the system variable **Handles** is always turned on.

Hatch

Ver. 1.4

Draws a non-associative cross-hatch pattern within a closed boundary.

Command	Alias	Side Menu	Menu Bar	Tablet
hatch	...	...	...	...

```
Command: hatch
Pattern (? or name/U,style):
Scale for pattern <1.0000>:
Angle for pattern <0>:
Select hatch boundaries or RETURN for direct hatch option,
Select objects: [Enter]
Retain polyline <N>:
From point: [pick]
Arc/Close/Length/Undo/<Next Point>: [pick]
From point or RETURN to apply hatch: [Enter]
```

COMMAND OPTIONS

Pattern Indicates name of hatch pattern.
? Lists the hatch pattern names.
U Creates a user-defined hatch pattern.
style Displays the sub-menu of hatching styles:
- **N** Hatches alternate boundaries (*short for Normal*).
- **O** Hatchs only outermost boundaries (*short for Outermost*).
- **I** Hatch everything within boundary (*short for Ignore*).

Direct hatch options:

Retain polyline
Yes: leaves boundary in place after hatch is complete.
No: erases boundary after hatching.

From point Begins drawing the hatch boundary.
Next point Draws straight line.
Arc Draws an arc hatch boundary.
Close Closes the hatch boundary.
Length Continues the boundary by specified distance.
Undo Undoes last-drawn segment.

RELATED AUTOCAD COMMANDS

- **BHatch** Automatic, associative hatching.
- **Boundary** Automatically creates polyline or region boundary.
- **Explode** Reduces hatch pattern to its constituent lines.
- **PsFill** Fills closed polyline with a PostScript pattern.
- **Snap** Changes the hatch pattern's origin.

RELATED SYSTEM VARIABLES

- **HpAng** Hatch pattern angle.
- **HpDouble** Doubled hatch pattern.
- **HpName** Hatch pattern name.
- **HpScale** Hatch pattern scale.
- **HpSpace** Hatch pattern spacing.
- **SnapBase** Controls the origin of the hatch pattern.
- **SnapAng** Controls the angle of the hatch pattern.

RELATED FILE

- **Acad.Pat** Hatch pattern definition file.

TIPS

- The **Hatch** command draws non-associative hatch patterns; the pattern remains in place when its boundary is edited.

- For complex hatch areas, you may find it easier to outline the area with a polyline (*using object snap*) or to use the **BHatch** command.

- By default, the hatch is created as a block; the name begins with the letter "X" followed by a consecutive number. To create the hatch as line entities, precede the pattern name with * (*asterisk*).

- AutoCAD includes the following hatch patterns in the file Acad.Pat:

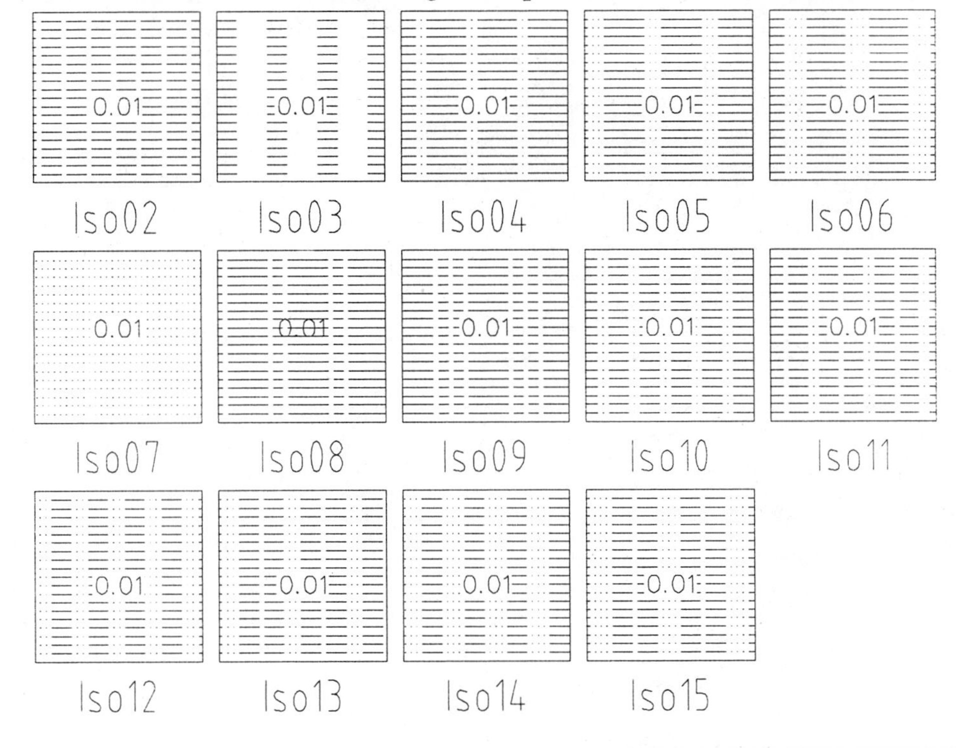

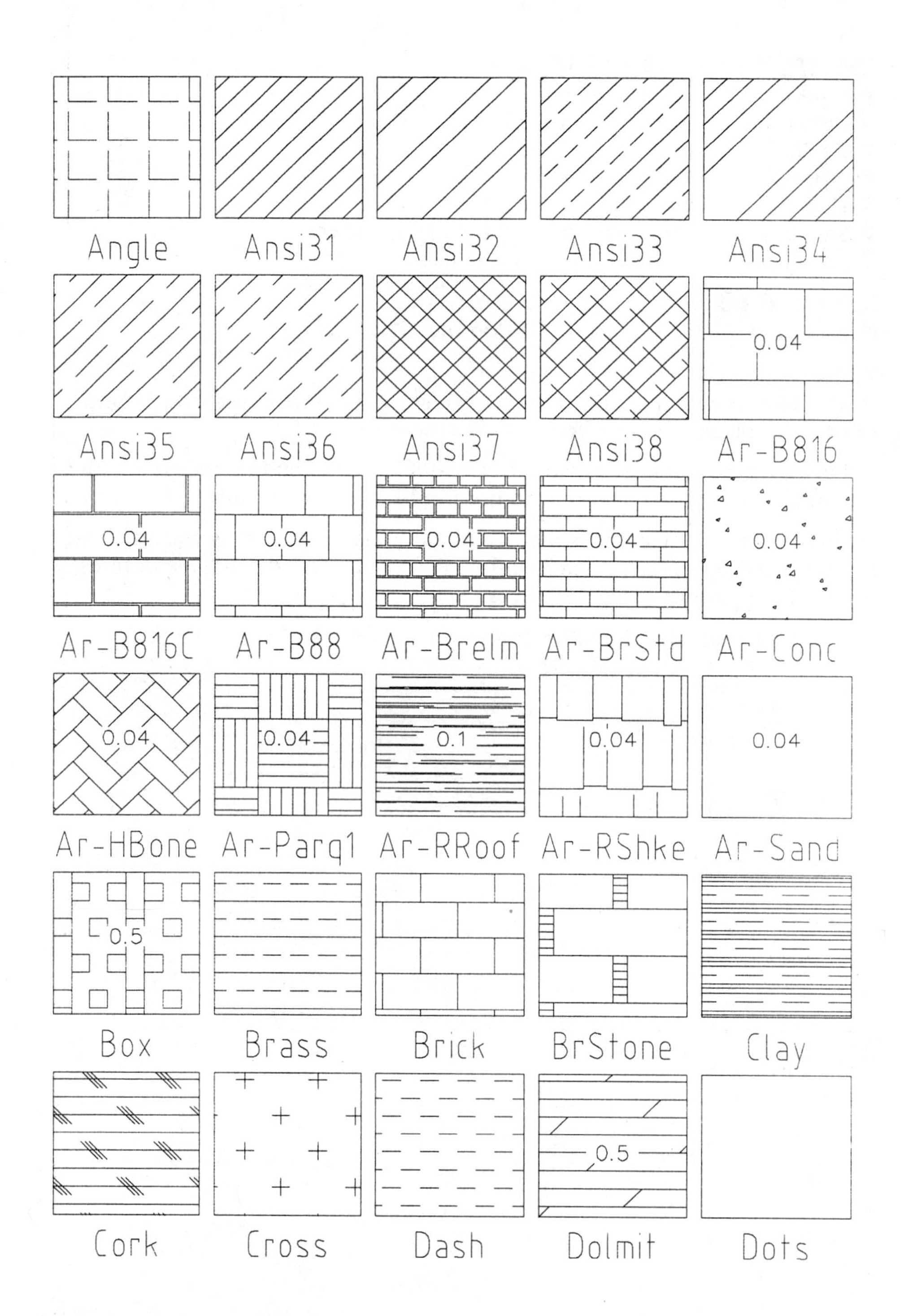
Angle
Ansi31
Ansi32
Ansi33
Ansi34
Ansi35
Ansi36
Ansi37
Ansi38
0.04
Ar-B816
0.04
0.04
0.04
0.04
0.04
Ar-B816C
Ar-B88
Ar-Brelm
Ar-BrStd
Ar-Conc
0.04
0.04
0.1
0.04
0.04
Ar-HBone
Ar-Parq1
Ar-RRoof
Ar-RShke
Ar-Sand
0.5
Box
Brass
Brick
BrStone
Clay
0.5
Cork
Cross
Dash
Dolmit
Dots

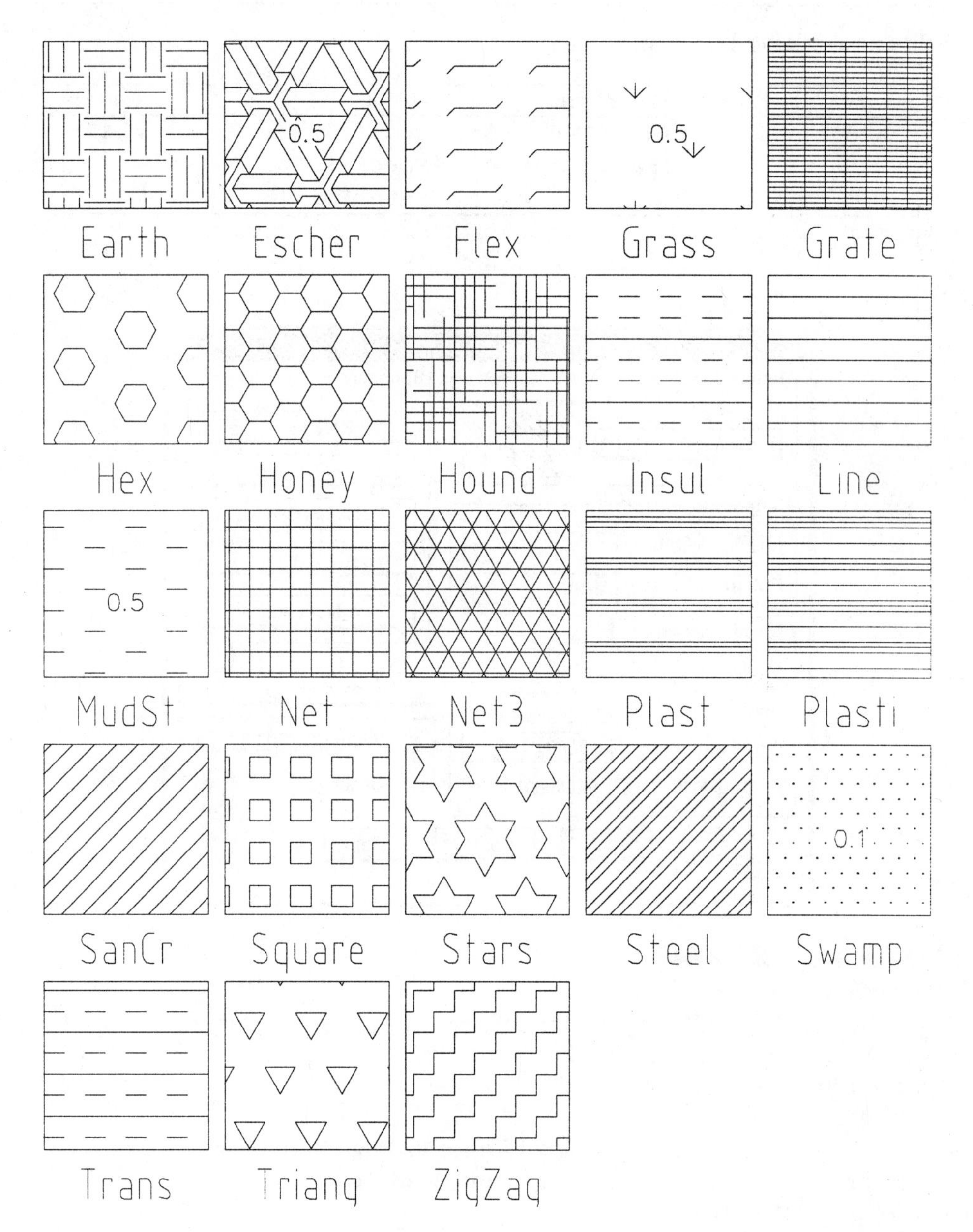

■ Patterns are drawn at scale factor 1.0; those marked with a scale number (*0.01, 0.5, etc.*) are shown at a smaller scale.

HatchEdit

Rel. 13

Edits associative hatch patterns.

Command	Alias	Side Menu	Toolbar	Tablet
`hatchedit`	...	[MODIFY] [HatchEd:]	[Modify] [Special Edit] [HatchEdit]	W 17

Command: **hatchedit**

Displays dialogue box.

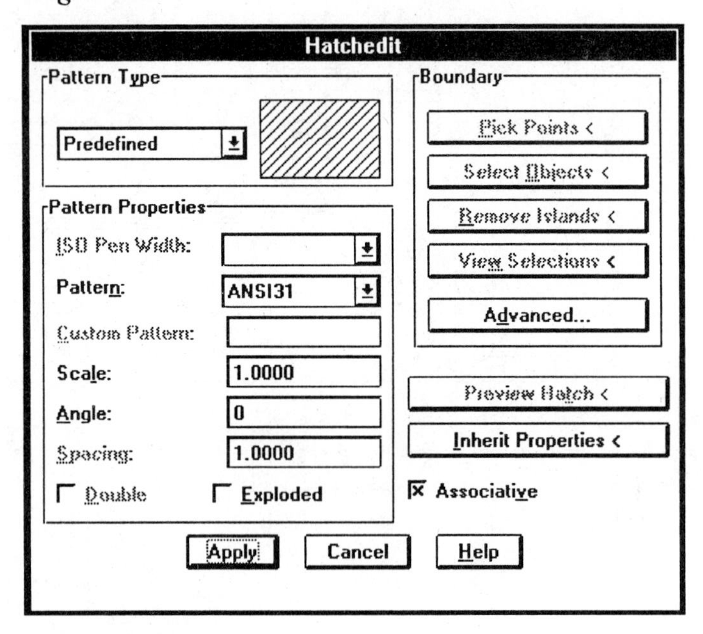

COMMAND OPTIONS

Pattern Type
Toggles between 'Predefined' in Acad.Pat, 'User-defined' on the fly, and 'Custom' in another PAT file.

Pattern Properties:

ISO Pen Width
Specifies plotting width for ISO hatch patterns.

Pattern Names of hatch pattern in Acad.Pat.

Custom Pattern
Names of custom hatch pattern.

Scale Hatch pattern scale.

Angle Global hatch pattern angle.

Spacing Spacing between pattern lines.

Double Double hatches (*applied at 90 degrees*).
Exploded Places hatch pattern as lines, rather than as a block.
Inherit Properties
Select another hatch pattern to match parameters.
Associative Toggle associativity of hatch pattern.

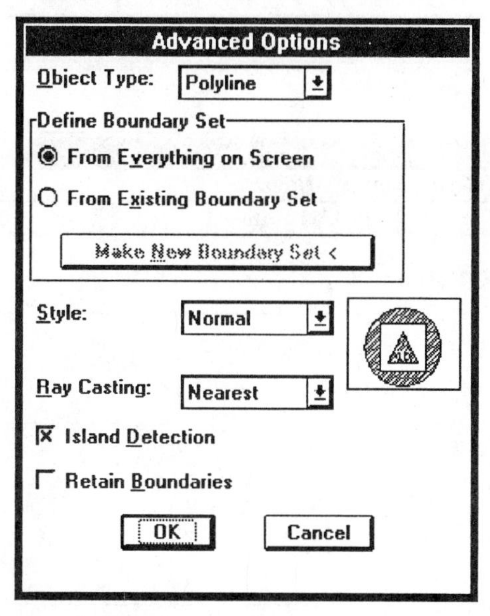

RELATED AUTOCAD COMMANDS

- **BHatch** Applies associative hatch pattern.
- **Hatch** Applies non-associative hatch pattern.
- **Explode** Explodes a hatch pattern block into lines.

RELATED SYSTEM VARIABLES

- **HpAng** Hatch pattern angle.
- **HpDouble** Doubled hatch pattern.
- **HpName** Hatch pattern name.
- **HpScale** Hatch pattern scale.
- **HpSpace** Hatch pattern spacing.

'Help *or* '?

Ver. 1.0

Lists text screens of information for using AutoCAD's commands.

Command	Alias	Side Menu	Menu Bar	Funtion key	Tablet
`'help`	`'?`	`[INQUIRY]`	`[Help]`	`F1`	`T 7-8`
		`[HELP:]`	`[Help]`		

`Command:` **`help`**

Displays dialogue box.

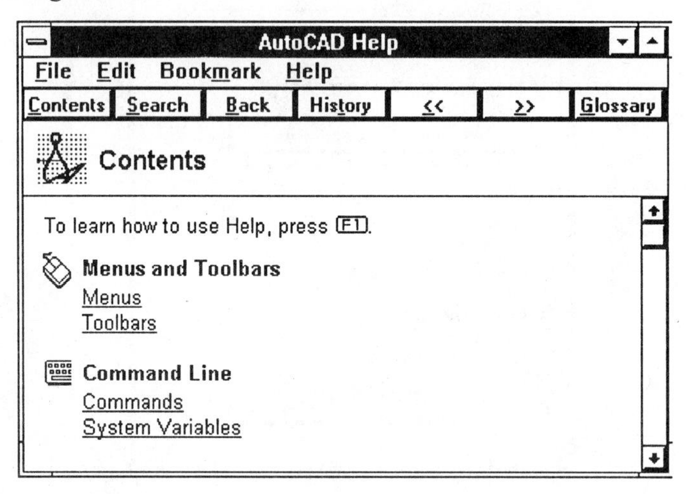

COMMAND OPTIONS

File	Selects HLP help file; prints help topic; exits **Help**.
Edit	Copies help topic to the Windows Clipboard.
Bookmark	Marks the help topic with a bookmark.
Help	Helps using the Windows help system.
Contents	Displays the table of contents for the help screens.
Search	Searches for key words in the help file.
Back	Moves back to the previous help screen.
History	Displays list of most-recently accessed help topics.
<<	Goes back to previous topic.
>>	Accesses next help topic.

Glossary Goes to glossary page:

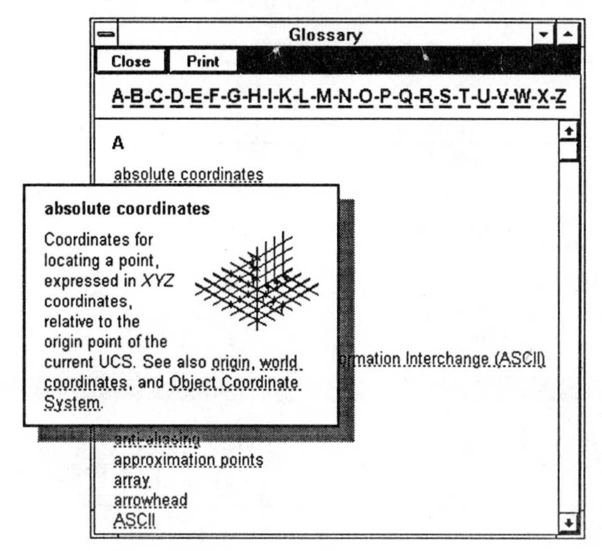

RELATED AUTOCAD COMMANDS

All

RELATED SYSTEM VARIABLES

All

RELATED FILES

- **Acad.Hlp** The AutoCAD help file, located in \Acad13\Win\Support.
- **DsAdi.Hlp** The accelerated display driver help file, in \Acad13\Win.

TIPS

- Since **Help**, **?**, and **[F1]** are transparent commands, you can use them during another command to get help on the command's options.

- The text of the Windows help file is stored in the file Acad.Hlp – this file is not ASCII and cannot be edited; it can be customized with the **Edit | Annotate** command.

- The **DlxHelp** command does not exist in the Windows version of AutoCAD Release 13; instead, load the DsAdi.Hlp file into the Windows Help program for help on the Windows accelerated display driver.

Hide

Ver. 2.1

Removes hidden lines from 3D drawings.

Command	Alias	Side Menu	Toolbar	Tablet
help	...	[TOOLS] [Hide:]	[Render] [Hide]	N1
		[TOOLS] [SHADE] [Hidden]		

```
Command: hide
Regenerating drawing.
Removing hidden lines: 25
```

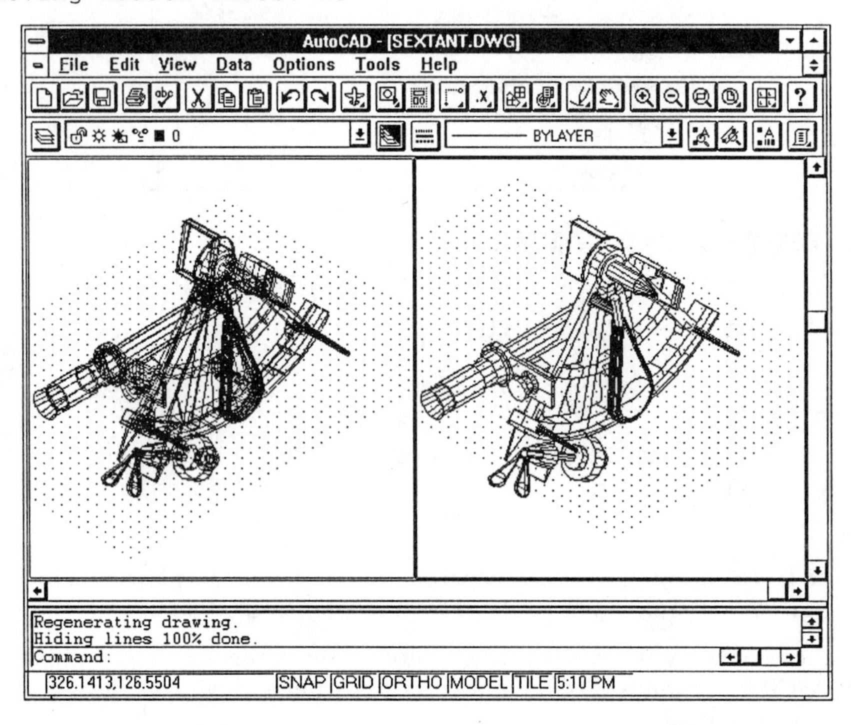

COMMAND OPTIONS

None

RELATED AUTOCAD COMMANDS

- **DView** Removes hidden lines of perspective 3D views.
- **MView** Removes hidden lines during plots of paper space drawings.
- **Plot** Removes hidden lines during plotting of 3D drawings.
- **Regen** Returns the view to wireframe.
- **Render** Performs realistic renderings of 3D models.

- **Shade** Performs quick renderings and quick hides of 3D models.
- **VPoint** Selects the 3D viewpoint.

RELATED SYSTEM VARIABLES

None

TIPS

- The **Hide** command considers the following entities as opaque:
  - Circle.
  - Solid, trace, and wide polyline.
  - 3D face, polygon mesh, and
  - The extrusion of any object with thickness.

- Use the **MSlide** (or **SaveImg**) to save the hidden-line view as an SLD (or TIFF, TGA, or GIF) file. View the saved image with the **VSlide** (or **Replay**) command.

- The **Shade** command simulates hidden-line removal when **ShadEdge** system variable is set to 2.

- Freezing layers speeds up the hide process since **Hide** ignores those objects.

- **Hide** does not consider the visibility of text and attributes.

- To create a hidden-line view when plotting in paper space, select the **HidePlot** option of the **MView** command.

- Use the **Regen** command to return to the wireframe view.

- As of Release 11, AutoCAD no longer supports the "Release 11" hidden-line algorithm.

HpMPlot

Rel. 12

Allows HPGL/2 plotters to plot one rendered viewport in paper space (*short for Hewlett-Packard Mixed PLOT; an external command in HpMPlot.Exp*).

Command	alias	Side Menu	Menu Bar	Tablet
hpmplot	...	...	...	...

Command: **hpmplot**
Displays dialogue box.

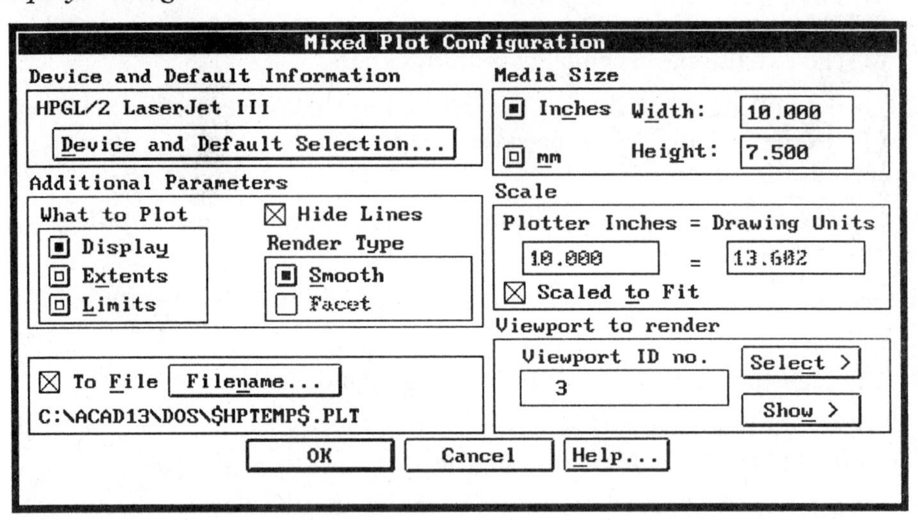

COMMAND OPTIONS

Device and Default Information
Selects HPGL/2 output device.

Additional Parameters
Selects the view, type of rendering, and whether hidden lines are removed from wireframe viewports.

To File Option to plot to an HPGL/2-format plot file.

Media Size Specifies size of media.

Scale Specifies plot scale.

Viewport to Render
Selects the one viewport to be rendered.

RELATED AUTOCAD COMMANDS

- **Config** Configures AutoCAD for output devices.
- **Plot** Plots non-rendered paper space and model space drawings.
- **RConfig** Configures AutoCAD for rendering plots and displays.
- **Render** Displays renderings to screen or common file formats.

TIPS

■ **HpMPlot** is useful for plotting a rendering and regular wireframe on a single sheet.

■ When you attempt to use a non-HPGL/2 device with **HpMPlot**, it complains, "Only HPGL/2 devices are valid for mixed plots."

QUICK START: Setting Up HpMPlot

To set up AutoCAD for using **HpMPlot**, follow these steps:

1. Configure AutoCAD for an HPGL/2 plotter with the **Config** command.

2. Configure **Render** for an HPGL/2 hardcopy rendering device with the **RConfig** command:

   ■ When AutoCAD complains about a missing rendering driver, supply 'RHpRtl.Exp' as the driver name.

   ■ Select from one of the following HPGL/2 plotters:
      ■ HP DesignJet 650C, 600, 200
      ■ HP PaintJet XL300
      ■ HP 7600 Series Monochrome and Color
      ■ HP LaserJet III and 4

3. Use the **HpConfig** command to configure the output parameters; displays a dialogue box:

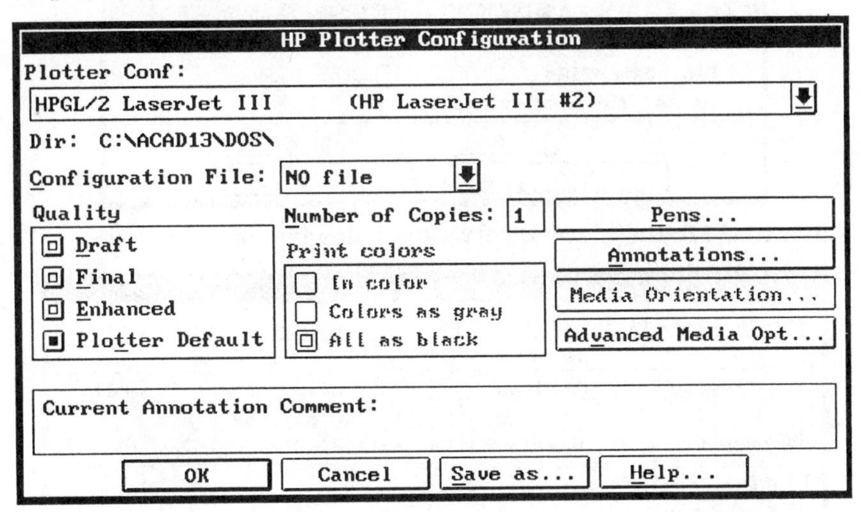

The **Pens** button displays the following dialogue box:

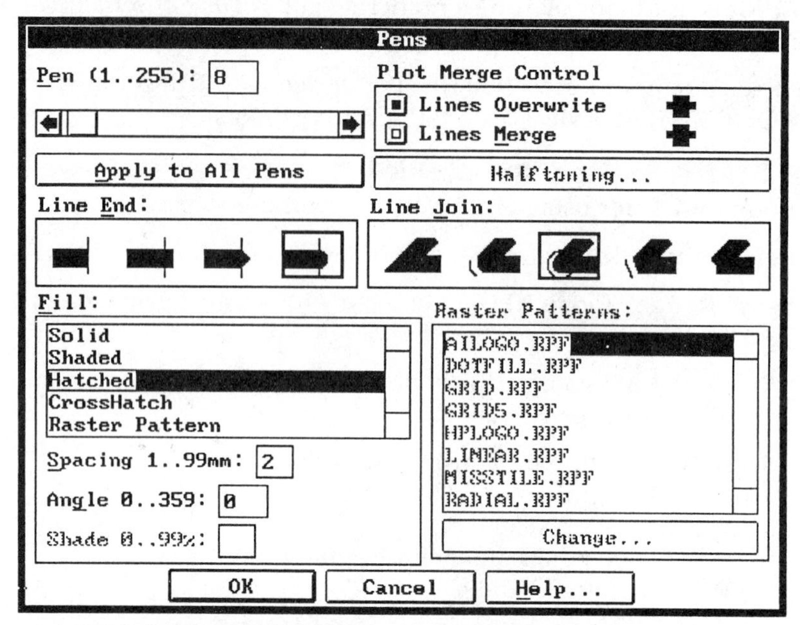

The **Annotations** button displays the following dialogue box:

Annotations

Drawing Filename
Driver Info
Plot Date/Time
Crop Marks
Comments: (50 Characters Max)

OK | Cancel | Help...

The **Advanced Media** button displays the following dialogue box:

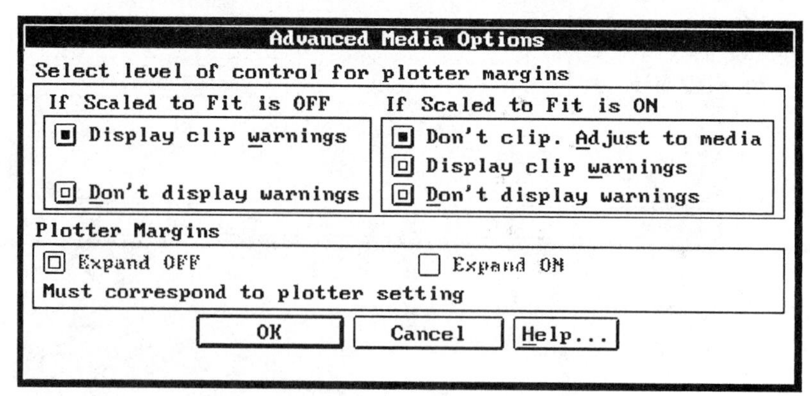

4. Save the output parameters to an HPC file with the **Device and Default Selection** option.

5. Set **TileMode** = 0.

6. Use the **MView** command to create a pair paperspace viewports.

7. Start the **HpMPlot** command. Select one viewport to be the rendered viewport. Create the mixed plot.

- The **HpRender** command renders a single viewport to the output device; displays dialogue box:

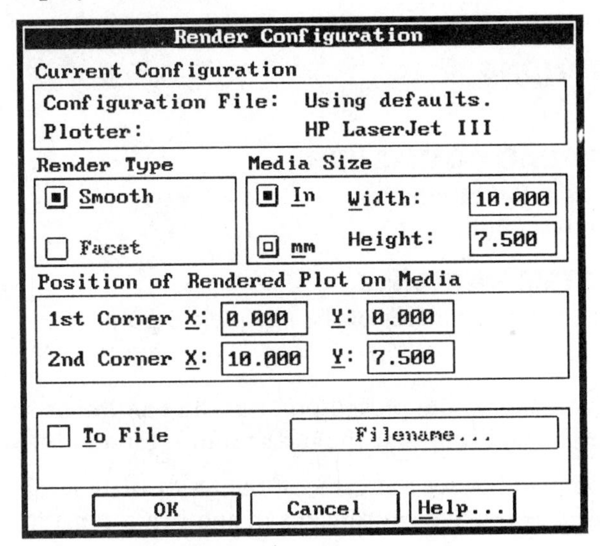

RELATED FILES

- ***.HPC** Stores configuration parameters of the **HpConfig** command.
- ***. PLT** HPGL/2 plot files.
- ***.RPF** Raster pattern files used by **HpMPlot**, found in subdirectory \Acad13\Dos\Support:
  AiLogo, DotFill, Grid, Grid5, HpLogo, Linear, MissTile, Radial, Rivrston, Sediment, Shingle, and SwampGrs.

'Id

Ver. 1.0

Identifies the 3D coordinates of a point (*short for IDentify*).

Command	Alias	Side Menu	Toolbar	Tablet
'id	...	[ASSIST]	[Object Properties]	Q 1
		[INQUIRY]	[Inquiry]	
		[Id:]	[Id]	

```
Command: id
Point: [pick]
```

Example output:

```
X = 1278.0018        Y = 1541.5993        Z = 0.0000
```

COMMAND OPTIONS

None

RELATED AUTOCAD COMMANDS

- **List** Lists information about a picked entity.
- **Point** Draws a point.

RELATED SYSTEM VARIABLE

- **LastPoint** The 3D coordinates of the last picked point.

TIPS

- The **Id** command stores the picked point in the **LastPoint** system variable. Access that value by entering @ at the next prompt for a point value.

- Use the **Id** command to set the value of the **Lastpoint** system variable, which can be used as relative coordinates in another command.

- The z-coordinate displayed by the **Id** command is the current elevation setting; if you use the **Id** command with an object snap, then the z-coordinate is the osnap'ed value.

- The **Id** command used in a menu macro or AutoLISP routine can quickly label points on the screen.

REMOVED COMMANDS: IgesIn and IgesOut

The **IgesIn** and **IgesOut** commands were removed from Release 13. The work-around is to purchase Autodesk's extra-cost IGES Translator package.

Import

Rel. 13

Import vector and raster files into the current drawing.

Command	Alt+	Side Menu	Menu Bar	Tablet
export	F,E	[FILES]	[File]	...
		[IMPORT]	[Import]	

Command: **export**

Displays dialogue box.

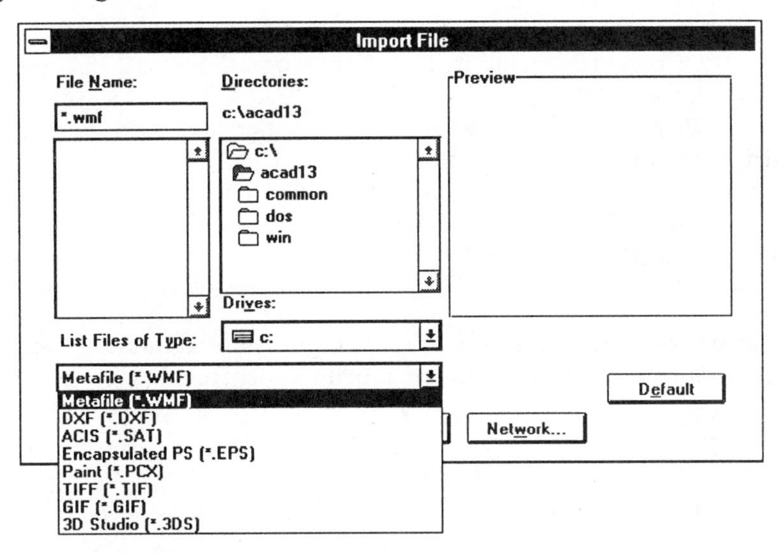

COMMAND OPTIONS

DXF Drawing interchange format; executes the **DxfIn** command.
EPS Encapsulated PostScript; executes the **PsIn** command.
GIF Graphic Interchange Format; executes the **GifIn** command
PCX PCX format; executes the **PcxIn** command.
SAT ASCII ACIS (Save As Text); executes the **AcisIn** command.
TIF Tagged image file format; executes the **TiffIn** command.
WMF Windows metafile; executes the **WmfIn** command.
3DS 3D Studio format; executes the **3dsIn** command.

RELATED AUTOCAD COMMANDS

- **AppLoad** Loads AutoLISP, ADS, and ARx routines.
- **AcisIn** Imports ACIS solid objects into drawing from SAT file.
- **DxbIn** Imports DXB file.
- **DxfIn** Imports ASCII or binary DXF format file.
- **Export** Exports drawing in several vector and raster formats.
- **Load** Imports SHX shape objects.

- **Insert** Places another drawing in the current drawing.
- **InsertObj** Places OLE object in drawing via Clipboard.
- **MatLib** Imports rendering material definitions.
- **Menu** Loads menu file into AutoCAD.
- **GifIn** Imports GIF format file.
- **Open** Opens AutoCAD (*any version*) DWG file.
- **PasteClip** Pastes object from the Clipboard.
- **PasteSpec** Pastes or links object from the Clipboard.
- **PcxIn** Imports PCX file.
- **PsIn** Imports encapsulated PostScript file.
- **Replay** Displays rendering in TIFF, Targa, or GIF format.
- **TiffIn** Imports TIFF file.
- **VSlide** Displays an SLD slide file.
- **WmfIn** Imports WMF format file.
- **XBind** Imports named objects from another DWG file.
- **XRef** Displays another DWG file in the current drawing.
- **3dsIn** Imports 3D Studio format file.

TIP

- The **Import** command acts as a "shell" command; it launches other AutoCAD commands that perform the actual export function.

Insert

Ver. 1.0

Inserts a previously-defined block into the drawing.

Command	Alias	Side Menu	Menu Bar	Tablet
insert	...	...	...	...

```
Command: insert
Block name (or ?):
Insertion point: [pick]
X scale factor <1> / Corner / XYZ: [Enter]
Y scale factor (default=X): [Enter]
Rotation angle <0>: [Enter]
```

COMMAND OPTIONS

Block name Indicates the name of the block to be inserted.
? Lists the names of blocks stored in the drawing.
X scale factor <1>
Indicates the x-scale factor.
Corner Indicates the x- and y-scale factors by pointing on the screen.
XYZ Displays the x-, y- and z-scale submenu.
P Supplies predefined block name, scale, and rotation values.

INPUT OPTIONS

- In response to the 'Block Name:' prompt, you can enter:
  - ~ Displays a dialogue box of blocks stored on disk.
  - = Redefines existing block with a new block, as in:
    `Block name: oldname=newname`
- In response to the 'Insertion point:' prompt, you can enter:
  - **Scale** Specifies x-, y-, z-scale factors.
  - **PScale** Presets the x-, y-, and z-scale factors.
  - **Xscale** Specifies x-scale factor.
  - **PsScale** Presets x-scale factor.
  - **Yscale** Specifies y-scale factor.
  - **PyScale** Presets y-scale factor.
  - **Zscale** Specifies z-scale factor.
  - **PzScale** Presets the z-scale factor.
  - **Rotate** Specifies the rotation angle.
  - **PRotate** Presets the rotation angle.

RELATED AUTOCAD COMMANDS

- **Block** Creates a block of a group of entities.
- **DdInsert** Dialogue box for inserting blocks.
- **Explode** Reduces an inserted block to its constituent entities.
- **MInsert** Inserts blocks as a blocked rectangular array.

- **Rename** Renames blocks.
- **WBlock** Writes blocks to disk.
- **XRef** Displays drawings stored on disk in the drawing.

RELATED SYSTEM VARIABLES

- **ExplMode** Toggles whether non-uniformly scaled blocks can be exploded:
  0 Cannot explode (*Release 12 compatible*).
  1 Can be exploded (*default*).
- **InsBase** Name of most-recently inserted block.

TIPS

- You can insert any other AutoCAD drawing into the current drawing.

- A "preset" scale factor or rotation means the dragged image is shown at that factor.

- Drawings are normally inserted as a block; prefix the filename with an * (asterisk) to insert the drawing as separate entities.

- Redefine an existing block by adding the suffix = (equal) after its name at the 'Block name:' prompt.

- Insert a mirrored block by supplying a negative x- or y-scale factor (*such as 'X scale factor: -1'*); AutoCAD converts negative z-scale factors into their absolute value (makes them always positive).

- As of Release 13, you can explode a mirrored block and a block inserted with different scale factors when system variable **ExplMode** is turned on.

InsertObj

Rel. 13

Places an OLE object as a linked or embedded object (*short for INSERT OBJect*).

Command	Alt+	Side Menu	Menu Bar	Tablet
insertobj	E,O	...	[Edit] [Insert Object]	...

Command: **insertobj**

Displays dialogue box.

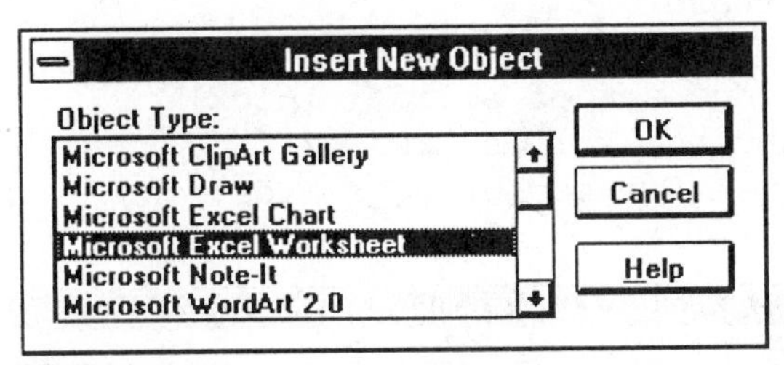

COMMAND OPTIONS

Object Type Selects an object type from the list; the related application automatically launches.

RELATED AUTOCAD COMMAND

- **PasteSpec** Places an object from the Clipboard in the drawing as a linked object.

RELATED WINDOWS COMMANDS

- **Edit | Copy**
  Copies an object to the Clipboard in another Windows application.
- **File | Update**
  Updates an OLE object.

Interfere

Rel. 11

Determines the interference of two or more 3D solid objects; creates a 3D solid body of the volumes in common (*formerly the **SolInterf** command; an external command in Acis.Dll*).

Command	Alias	Side Menu	Toolbar	Tablet
interfere	...	[DRAW 2] [SOLIDS] [Interfr:]	[Solids] [Interference]	Y 15

```
Command: interfere
Select the first set of solids: [pick]
Select the second set of solids: [pick]
Create interference solids? <N>: Y
Highlight pairs of interfering solids? <N>: Y
eXit/<Next pair>: X
```

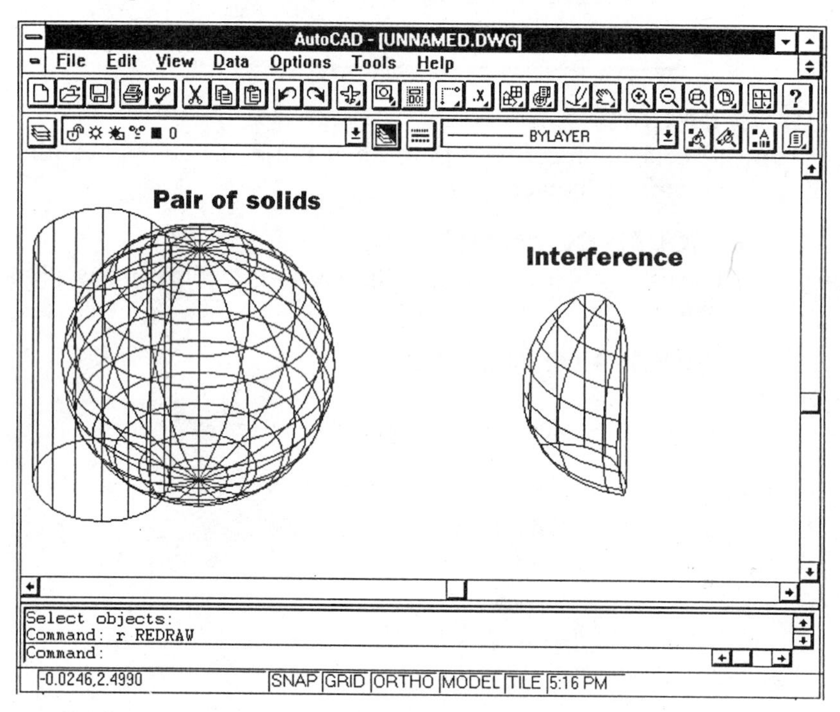

COMMAND OPTIONS

Select first set of solids All solids in a single selection set are checked for interference against each other.

Select second set of solids

All solids in the one selection set are checked for interference with solids in the second selection set.

Create interference solids

Creates a solid representing the volume of interference.

Highlight pairs of interfering solids

Useful when the drawing contains many interferences.

RELATED AUTOCAD COMMANDS

- **Intersect** Creates a new volume from the intersection of two volumes.
- **Section** Creates a 2D region from a 3D solid.
- **Slice** Slices a 3D solid with a plane.

Intersect

Rel. 11

Creates a 3D solid of 2D region from the intersection of two or more solids or regions; *(formerly the **SolInt** command; an external commad in Acis.Dll)*.

Command	Alias	Side Menu	Toolbar	Tablet
intersect	...	[DRAW 2]	[Modify]	Y 12
		[SOLIDS]	[Explode]	
		[Intersct:]	[Intersect]	

```
Command: intersect
Select objects: [pick]
```

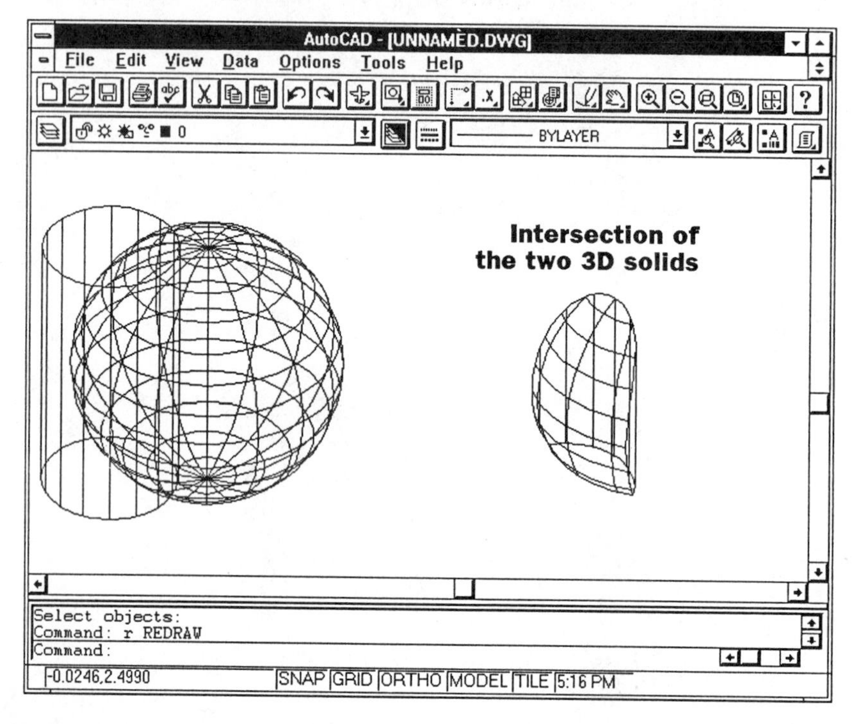

COMMAND OPTIONS

None

RELATED AUTOCAD COMMANDS

- **Interfere** Creates a new volume from the interference of two or more volumes.
- **Subtract** Subtracts one 3D solid from another.
- **Union** Joins D solids into a single body.

'Isoplane

Ver. 2.0

Switches the crosshairs between the three isometric drawing planes.

Command	Ctrl+	Side Menu	Function Key	Tablet
'isoplane	E	...	[F5]	V 19

```
Command: isoplane
Left/Top/Right/<Toggle>: [Enter]
Current Isometric plane is: Left
```

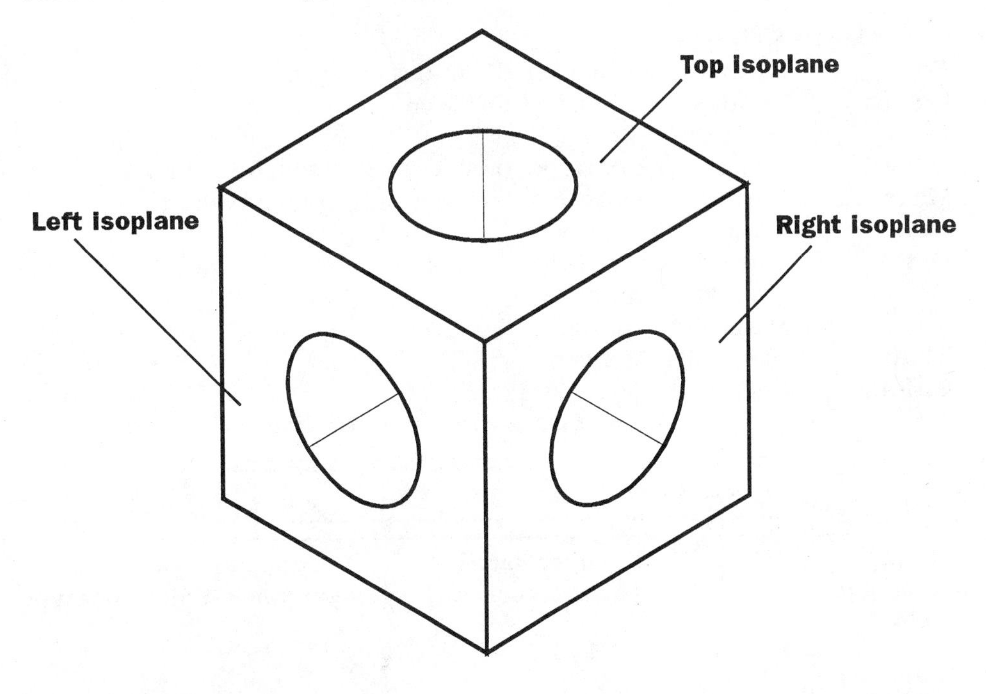

COMMAND OPTIONS

Left Switches to the left isometric plane.
Top Switches to the top isometric plane.
Right Switches to the right isometric plane.
<Toggle> Switches to the next isometric plane in the order of: left, top, right.

RELATED AUTOCAD COMMANDS

- **Ddrmodes** Dialogue box for setting isometric mode and planes.
- **Snap** Turns isometric drawing mode on.

RELATED SYSTEM VARIABLE

- **SnapIsoPair** Contains the current isometric plane.

'Layer

Controls the creation and visibility of layers.

Command	Alias	Side Menu	Menu Bar	Tablet
'layer	la	...	...	L 3
				M3 - O4

```
Command: layer
?/Make/Set/New/ON/OFF/Color/Ltype/Freeze/Thaw/LOck/Unlock:
```

COMMAND OPTIONS

Color Indicates the color for all entities drawn on the layer.
Freeze Disables the display of the layer.
LOck Locks the layer.
Ltype Indicates the linetype for all entities drawn on the layer.
Make Creates a new layer and make it the working layer.
New Creates a new layer.
OFF Turns the layer off.
ON Turns the layer on.
Set Makes the layer the working layer.
Thaw Un-freezes the layer.
Unlock Unlocks the layer.
? Lists the names of layers created in the drawing.

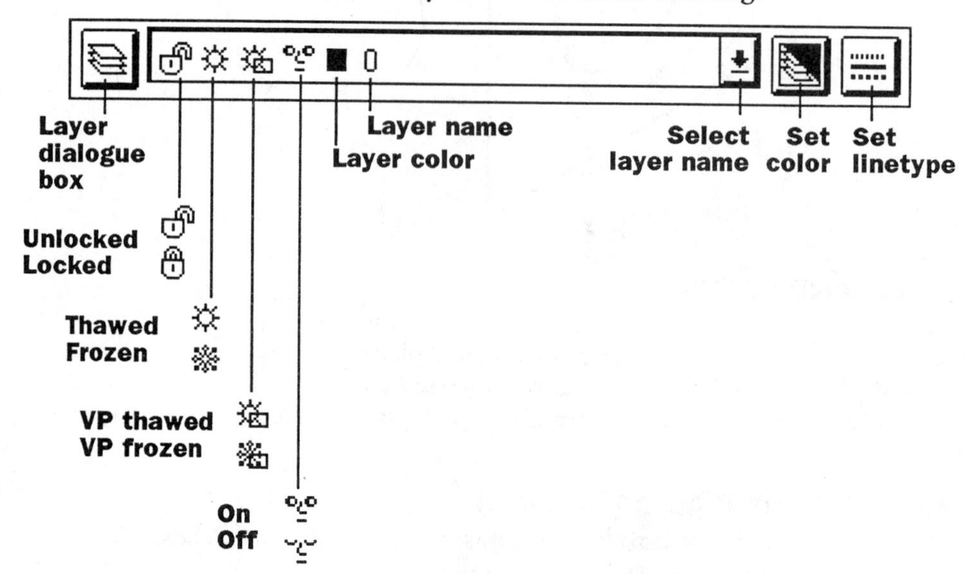

RELATED AUTOCAD COMMANDS

- **Change** Moves objects to a different layer.
- **ChProp** Moves objects to a different layer.
- **DdLModes** Uses a dialogue box to control layers.
- **Purge** Removes unused layers from the drawing.
- **Rename** Renames layer names.
- **VpLayer** Controls the visibility of layers in paper space viewports.
- **Xplode** Explodes blocks, hatches, meshes, and polylines and places them on a new layer.

RELATED SYSTEM VARIABLE

- **CLayer** Contains the name of the current layer.

TIP

- Layer **DimPts** is a non-plotting layer.

Leader

Rel. 13

Draws a leader line with a single line of text (*formerly the **Dim:LEAder** command*).

Command	Alias	Side Menu	Toolbar	Tablet
leader	lead	[DRAW DIM] [Leader:]	[Dimensioning] [Leader]	X 1

```
Command: leader
From point: [pick]
To point:  [pick]
To point (Format/Annotation/Undo) <Annotation>: F
Spline/STraight/Arrow/None/<eXit>:
Tolerance/Copy/Block/None/<MText>:
```

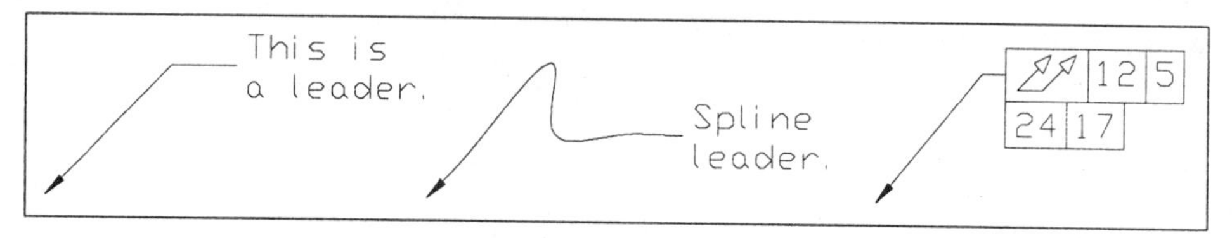

COMMAND OPTIONS

Format Specifies the style of leader:
- **Spline** Draws leader line as a NURBs curve.
- **STraight** Draws straight leader line (default).
- **Arrow** Draws leader with arrowhead (default).
- **None** Draws leader with no arrowhead.

Annotation Specifies the type of annotation:
- **Tolerance** Places one or more tolerance symbols.
- **Copy** Copies text from another part of the drawing.
- **Block** Places a block in the manner of the **Insert** command.
- **None** No annotation.
- **<MText>** Places an mtext note at the end of the leader line.

Undo Undoes the leader line to the previous vertex.

RELATED DIM VARIABLES

- **DimAsz** Arrowhead and hookline size.
- **DimBlk** Type of arrowhead.
- **DimClrd** Color of leader line and arrowhead.
- **DimGap** Gap between hookline and annotation; gap between box and text.
- **DimScale** Overall scale of leader.

TIPS

- The text in a leader is an MText (*multiline text*) entity.
- Use the '\P' metacharacter to create line breaks in leader text.

Lengthen

Lengthens and shortens open objects by four methods.

Command	Alias	Side Menu	Toolbar	Tablet
lengthen	...	[MODIFY] [Lengthn:]	[Modify] [Resize] [Lengthen]	Y 18

Command: **lengthen**
DElta/Percent/Total/DYnamic/<Select object>: **[pick]**
Current length: <>, included angle: <>

DElta option:
Angle/<Enter delta length>: **A**
Enter delta angle:
<Select angle to change>/Undo:

Percent option:
Enter percent length:
<Select angle to change>/Undo:

Total option:
Angle/<Enter total length>: **A**
Enter delta angle:
<Select angle to change>/Undo:

DYnamic option:
<Select angle to change>/Undo:

COMMAND OPTIONS

<Select object> Displays length and included angle; does not change object.
DElta Changes length by incremental length.
Percent Changes length by a percentage of the original length.
Total Changes by absolute value.
DYnamic Dynamically changes length by dragging.
  Angle Sets the angle of the selected arc.
  Undo Undoes the most-recent lengthening operation.

RELATED AUTOCAD COMMANDS

- **Extend** Lengthens an open object to a cutting edge.
- **Trim** Trims a open and closed objects back to a cutting edge.

TIPS

- Unlike the **Extend** and **Trim** commands, the **Lengthen** command does not require an object to work as a cutting edge.

- **Lengthen** command only works with open objects, such as lines, arcs, and polylines; it does not work with closed objects, such as circles, polygons, and regions.

- **DElta** option changes the length or angle using the following measurements:
  - Distance from endpoint of the selected object to the pick point.
  - For arc angles, changed by the incremental length measured from the endpoint of the arc..
  - Positive values lengthen; negative values shorten.

- **Percent** option works relative to 100%:
  - Less than 100% shortens the object; for example, 50% shortens object by half.
  - 100%: length does not change.
  - More than 100% lengthens the object; for example, 200% doubles the length.

Light

Rel. 12

Places four types of lights for the **Render** command (*an external command in Render.Arx*).

Command	Alias	Side Menu	Toolbar	Tablet
light	...	[TOOLS] [RENDER] [Lights:]	[Render] [Light]	M 2

`Command:` **`light`**
Displays dialogue box.

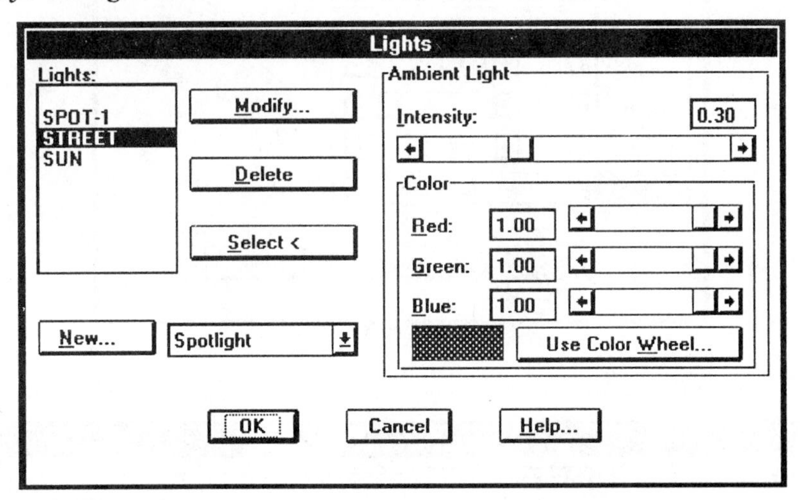

COMMAND OPTIONS

Modify Modifies an existing light in the drawing.
Delete Deletes the selected light.
Select Selects a light from the drawing.
Intensity Adjusts intensity of ambient light.
Color Adjusts color of ambient light; displays dialogue box:

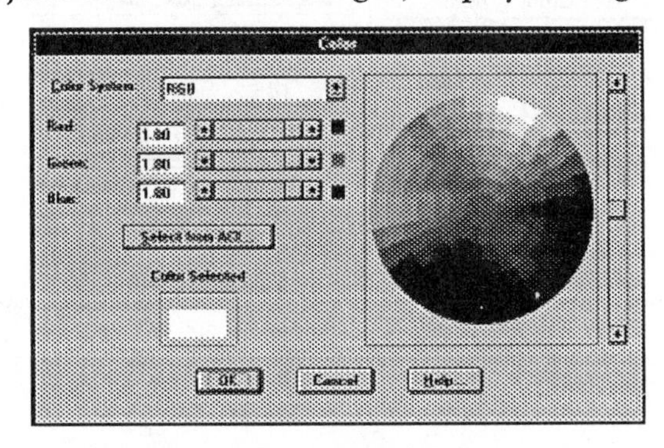

New Creates a new light; displays a dialogue box, depending on light type:

New point light:

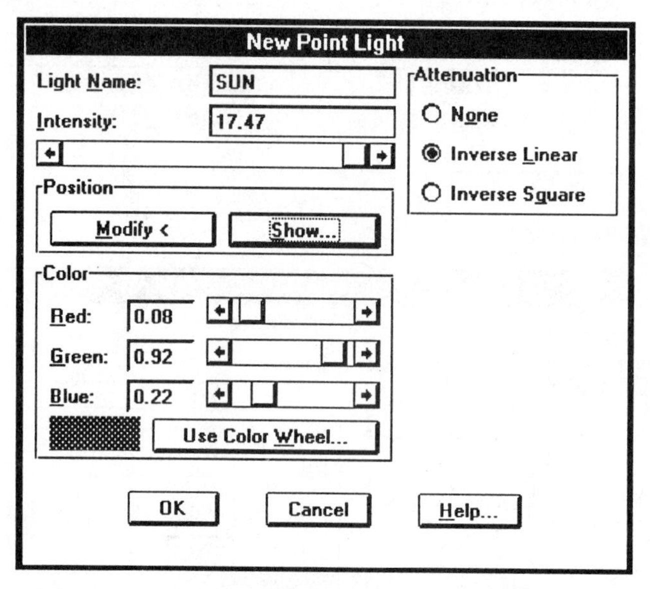

New distant light:

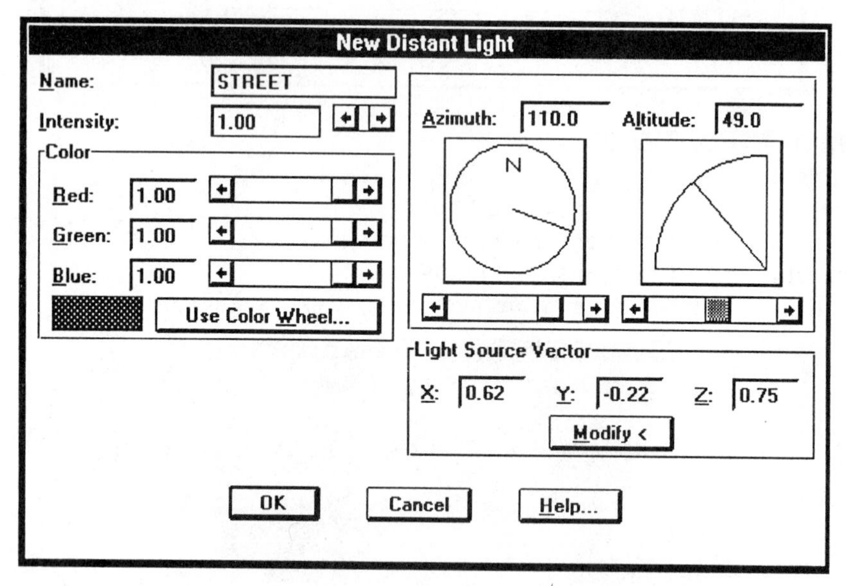

New spot light:

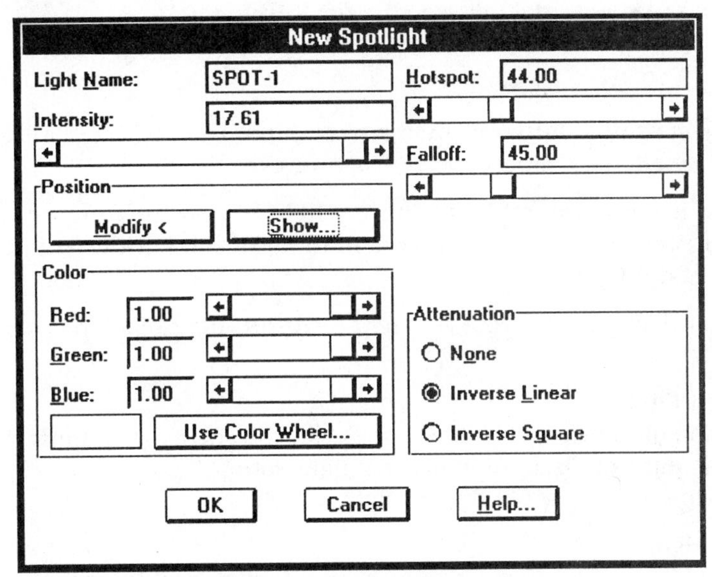

Postion Location of selected light:

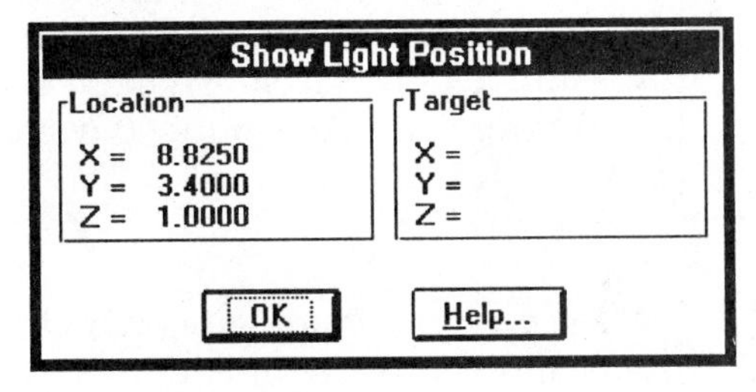

RELATED AUTOCAD COMMANDS

- **Render** Renders the drawing.
- **Scene** Specifies lights and view to use in rendering.

RELATED SYSTEM VARIABLE

- **Target** Coordinates of light's target point.

RELATED FILES

In \Acad13\Common\Support subdirectory:

- **Direct.Dwg** Direct light block.
- **Overhead.Dwq** Overhead drawing block.
- **Sh_Spot.Dwg** Spotlight drawing block.

In \Acad13\Dos subdirectory:

- **Render.Cfg** Render device configuration file.

- **Render.Mli** Rendering material library.
- **Render.Xmx** External message file for **Render**.
- **ReadMe.Ren** Last-minute updates concerning **Render**.

TIPS

- When you use the **Render** command with no lights defined, AutoCAD assumes a single light source located at your eye.

- While it is not necessary to define any lights to use the **Render** command, a light must be included in a **Scene** definition for the **Render** command to make use of the light.

- The light beam travels from the *light location* (light block placement) to the *light target*.

- Ambient light:
  - Ambient light ensures every object in the scene has illumination.
  - Ambient light is an omnipresent light source.
  - Set to 0 to turn off lights for night scenes.

- Distant lights:

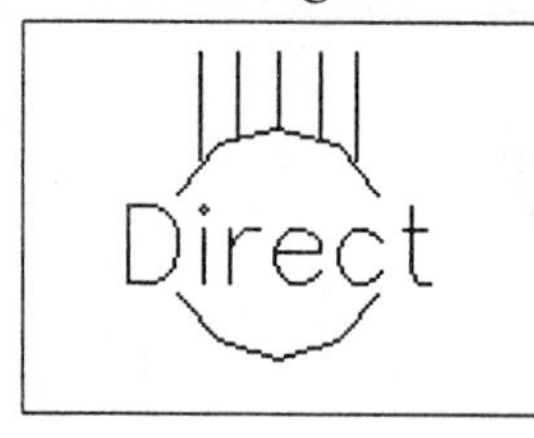

- Place one Distant light to simulate the Sun.
- Parallel light beams with constant intensity.
- Azimuth ranges from -180 to 180 degrees.
- Altitude ranges from -90 to 90 degrees (use the text box for values down to -90 degrees).

- Point lights:

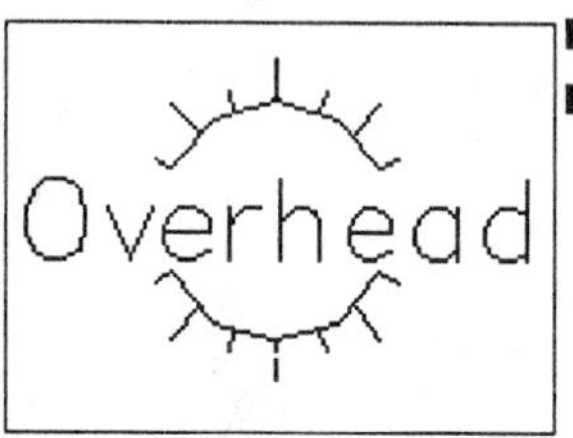

- Place several Point lights as light bulbs (lamps).
- Beams light in all directions, with inverse linear, inverse square, or constant intensity.

- Spot lights:

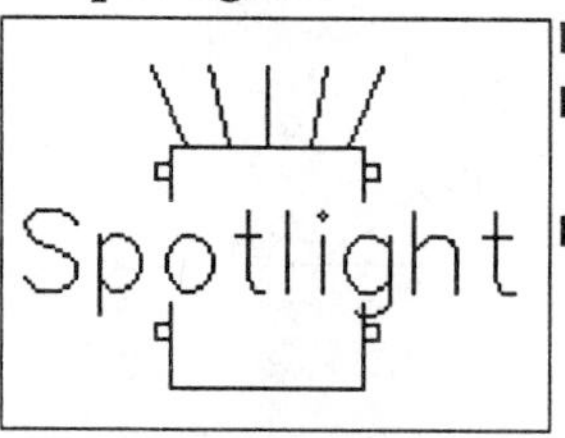

- Beams light in a cone.
- *Hotspot* is the brightest cone of light; beam angle ranges from 0 to 160 degrees (default: 45 degrees).
- *Falloff* is the angle of the full light cone; field angle ranges from 0 to 160 degrees (default: 45 degrees).

- Intensity of 0 turns light off.

DEFINITIONS

Constant light:

- Attenuation is 0.
- Default intensity is 1.0.

Inverse linear light:

- Light strength decreases to ½-strength two units of distance away, and ¼-strength four units away.
- Default intensity is ½ extents distance.

Inverse square light:

- Light strength decreases to ¼-strength two units away, and 1/8-strength four units away.
- Default intensity is ½ the square of the extents distance.

Extents distance:

- Distance from minimum lower-left coordinate to the maximum upper-right coordinate.

RGB color:

- The three primary colors – red, green, blue – shaded from black to white.

HLS color:

- Changes each color by hue (color), lightness (more white or more black), and saturation (less grey).

'Limits

Ver. 1.0

Defines the 2D limits in the WCS for the grid markings and the Zoom All command.

Command	Alt+	Side Menu	Menu Bar	Tablet
'limits	D,A	[DATA] [Limits:]	[Data] [Drawing Limits]	...

```
Command: limits
Reset Model space limits:
ON/OFF/<Lower left corner> <0.0000,0.0000>: [pick]
Upper right corner <12.0000,9.0000>: [pick]
```

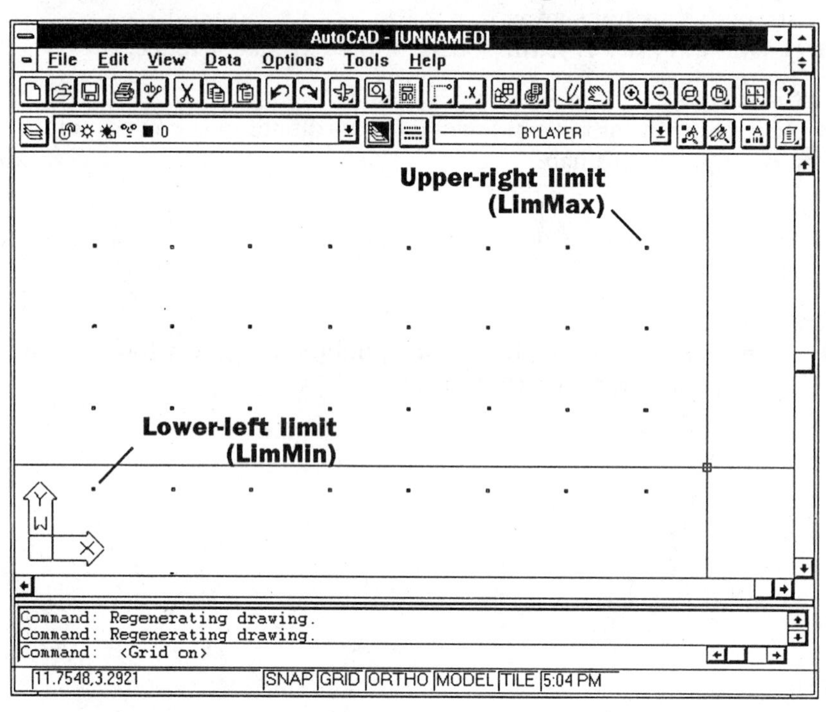

COMMAND OPTIONS

OFF Turns off limits checking.
ON Turns on limits checking.
[Enter] Retains current limits values.

RELATED AUTOCAD COMMANDS

- **Grid** Grid dots are bounded by limits.
- **Status** Lists the current drawing limits.
- **Zoom** Zoom All displays the drawing's extents or limits.

RELATED SYSTEM VARIABLES

- **LimCheck** Toggle for limit's drawing check:
  - **0** Off (*default*).
  - **1** On.
- **LimMin** Lower-right 2D coordinates of current limits.
- **LimMax** Upper-left 2D coordinates of current limits.

TIPS

- Use the **Limits** command to define the extents of grid markings.

- The limits determine the extents displayed by **Zoom All** command.

- If limits checking is turned on, AutoCAD will complain with an outside-limits error. Use this to limit drawing to the drawing extents.

- There are no limits in the z-direction.

- Model space and paper space have separate limits.

Line

Ver. 1.0

Draws straight 2D and 3D lines.

Command	Alias	Side Menu	Toolbar	Tablet
line	l	[DRAW 1]	[Draw]	J 10
		[Line:]	[Line]	
	3dline			

```
Command: line
From point: [pick]
To point: [pick]
To point: [Enter]
```

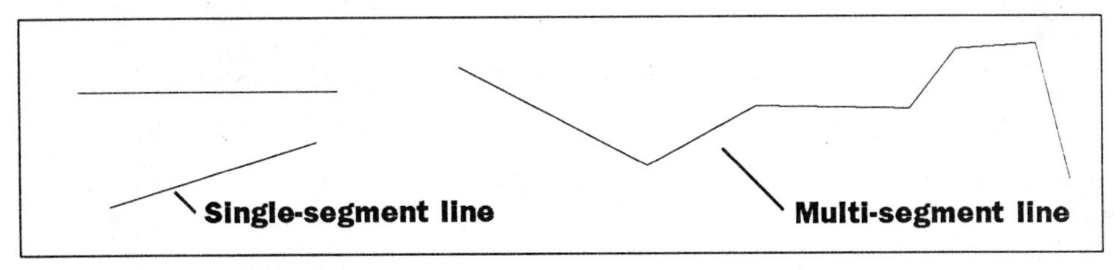

COMMAND OPTIONS

C	Closees the line from the current point to the starting point.
U	Undoes the last line drawn.
[Enter]	At the 'From point:' prompt, continues the line from the last endpoint; at the 'To point:' prompt, terminates the Line command.

RELATED AUTOCAD COMMANDS

- **MLine** Draws up to 16 parallel polylines.
- **PLine** Draws polylines and polyline arcs.
- **Trace** Draws lines with width.
- **Ray** Creates a semi-infinite construction line.
- **Xline** Creates an infinite construction line.

RELATED SYSTEM VARIABLES

- **Elevation** Distance above (*or below*) the x,y-plane a line is drawn.
- **Lastpoint** Last-entered coordinate triple.
- **Thickness** Determines thickness of the line.

TIPS

- To draw a 2D line, enter x,y-coordinate pairs; the z-coordinate takes on the value of the **Elevation** system variable.

- To draw a 3D line, enter x,y,z-coordinate triples.

- When system variable **Thickness** is not zero, the line has thickness, which makes it a plane.

'Linetype

Ver. 2.0

Loads linetype definitions into the drawing, creates new linetypes, and sets the working linetype.

Command	Alias	Side Menu	Menu Bar	Tablet
'linetype	lt	...	...	Y 15

```
Command: linetype
?/Create/Load/Set:
```

COMMAND OPTIONS

Create Creates a new user-defined linetype.
Load Loads a linetype from a LIN linetype definition file.
Set Sets the working linetype.
? Lists the linetypes loaded into the drawing.

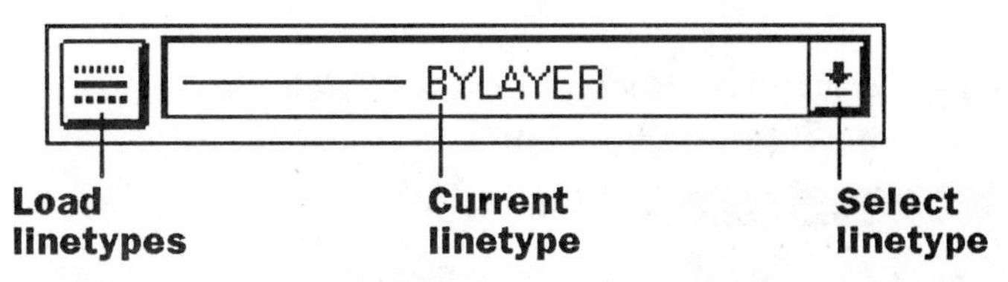

RELATED AUTOCAD COMMANDS

- **Change** Changes objects to a new linetype; changes linetype scale.
- **ChProp** Changes entities to a new linetype.
- **DdEModes** Sets the working linetype via a dialogue box.
- **DdLModes** Sets the linetype for all entities on a layer.
- **DdLtype** Sets the linetype via a dialogue box.
- **LtScale** Sets the scale of the linetype.
- **Rename** Changes the name of the linetype.

RELATED SYSTEM VARIABLES

- **CeLtype** The current linetype setting.
- **LtScale** The current linetype scale.
- **PsLtScale** Linetype scale relative to paper scale.
- **PlineGen** Controls how linetypes are generated for polylines.

TIPS

- The only linetype initially defined in an AutoCAD drawing is the CONTINUOUS linetype.

- Linetypes must be loaded from LIN definition files before being used in a drawing. When loading one or more linetypes, it's faster to load all linetypes, then use the **Purge** command to remove linetype definitions the drawing has not used.

- As of Release 13, objects can have independent linetype scales.

RELATED FILES

■ Linetypes stored in \Acad13\Common\Suppport\LTypeShp.Lin (*new to Release 13*):

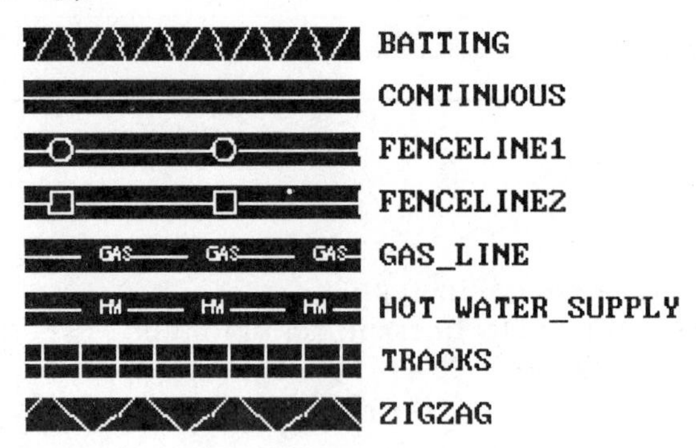

The following ISO-standard linetypes (*new to Release 13*) are stored in \Acad13\Common\Support\Acad.Lin:

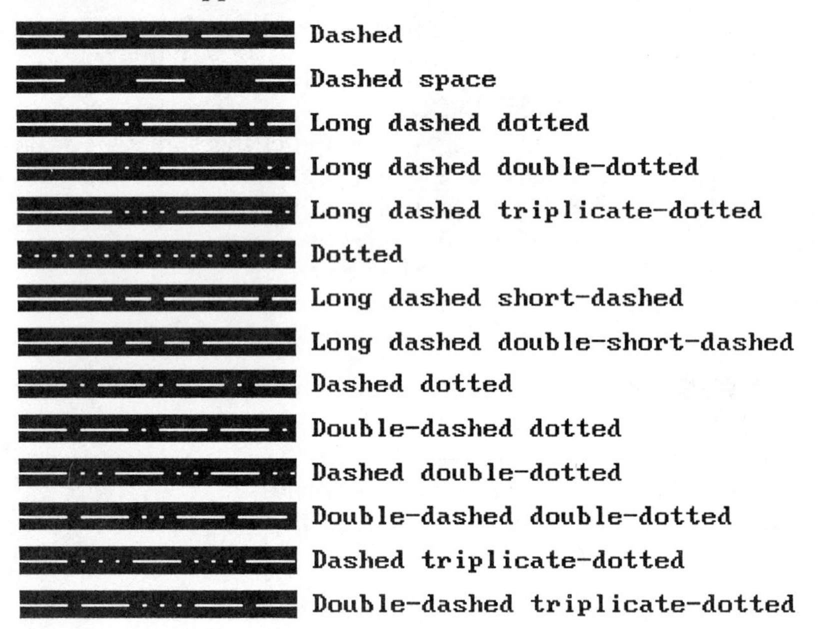

■ The following standard linetypes are supplied with AutoCAD; their definitions are stored in file \Acad13\Common\Support\Acad.Lin:

QUICK START: Create a Custom Linetype

To create a custom linetype on-the-fly:

1. Type the **Linetype Create** command:

```
Command: linetype
?/Create/Load/Set: C
```

2. Name the linetype in three steps:
   - First, the linetype name:

     `Name of linetype to create: [enter up to 31 characters]`
   - Second, the LIN filename.

     Append linetype description to Acad.Lin or create new LIN file.
   - Third, describe the linetype:

     `Descriptive text: [enter up to 47 characters]`

3. Define the linetype pattern by using five codes:
   - Positive number for dashes; 0.5 is a dash 0.5 units long.
   - Negative number for gaps; -0.25 is a gap 0.25 units long.
   - Zero is for dots; 0 is a single dot.
   - "A" forces the linetype to align between two endpoints (linetypes start and stop with a dash).
   - Commas (,) separate values.

Example:

```
*DASHDOT,__ . __ . __ . __ . __ . __ . __ . __ . __ .
A,.5,-.25,0,-.25 [Enter]
```

4. Press **[Enter]** to end linetype definition.

5. Use the **Linetype Load** command to load pattern into drawing.

   `Linetype to load: [type name]`

6. Use the **Linetype Set** command to set the linetype.

   `New object linetype (or ?) <>: [type name]`

Alternatively, use the **Change** command to change objects to the linetype.

List

Ver. 1.0

Lists information about selected entities in the drawing.

Command	Alias	Side Menu	Toolbar	Tablet
list	...	[ASSIST]	[Object Properties]	Q 2
		[INQUIRY]	[Inquiry]	
		[List:]	[List]	

```
Command: list
Select objects: [pick]
Select objects: [Enter]
```

Example output:

```
        LINE      Layer: 36
                  Space: Model space
         Color: BYLAYER    Linetype: CONTINUOUS
         Handle = 24A6
    from point, X=  10.0000  Y=   6.0000  Z=   0.0000
      to point, X=   9.0000  Y=   4.0000  Z=   0.0000
Length =   2.2361,  Angle in X-Y Plane =    243
Delta X =  -1.0000, Delta Y =   -2.0000, Delta Z =   0.0000
```

COMMAND OPTIONS

[Esc] Cancels the list display.
[Ctrl]+S Pauses the display; press any key to continue.
[F1] Returns to graphics screen.

RELATED AUTOCAD COMMANDS

- **Area** Calculates the area and perimeter of some entities.
- **DbList** Lists information about all entities in the drawing.
- **Dist** Calculates the 3D distance and angle between two points.

TIPS

- Use the **List** command as a faster alternative to using the **Dist** and **Area** commands for finding lengths and areas of objects.

- Conditional information listed by the **List** command:
  - Object's color and linetype, if not set BYLAYER.
  - Thickness, when not 0.
  - The elevation is not listed and must be interpolated from the z-coordinate.
  - Extrusion direction, if different from z-axis of current UCS.

- Object handles are described by hexadecimal numbers.

Load

Ver. 1.0

Loads SHX-format shape files into the drawing via a dialogue box.

Command	Alt+	Side Menu	Menu Bar	Tablet
load	D, S	[DRAW 2] [Shape:] [Load:]	[Data] [Shape File]	...

```
Command: load
Name of shape file to load (or ?):
```

COMMAND OPTION

? Lists the currently loaded shape files.

RELATED AUTOCAD COMMAND

- **Shape** Inserts shapes into the current drawing.

TIP

- Shapes are more efficient than blocks but are harder to create.

RELATED FILES

- *.SHP Source code for shape files.
- *.SHX Compile shape files.

In \Acad13\Common\Sample subdirectory:

- **Es.Shx** and **Es.Shp** Electronic component shapes.
- **Pc.Shx** and **Pc.Shp** Printed circuit board shapes.
- **St.Shx** and **St.Shp** Surface texture shapes for mechanical parts drawings.

In \Acad13\Common\Support subdirectory:

- **Gdt.Shx** and **Gdt.Shp**
  Geometric tolerancing shapes used by **Tolerance** command.
- **LtypeShp.Shx** and **LtypeShp.Shp**
  Linetype shapes used by the Linetype command.

QUICK START: Using shapes in your drawing.

1. (*When required*) Use the **Compile** command to compile the source code SHP file into an SHX file.

2. Use the **Load** command to load the SHX shape file into the drawing.

3. Use the **Shape** command to place shapes. The **?** option lists the names of shapes defined by the SHX file.

LogFileOff

Rel. 12

Closes the Acad.Log file.

Command	Alias	Side Menu	Menu Bar	Tablet
`logfileoff`	...	...	...	...

```
Command: logfileoff
```

COMMAND OPTIONS

None

RELATED AUTOCAD COMMANDS

- **LogFileOn** Turns on recording 'Command:' prompt text to file Acad.Log.
- [Ctrl]+Q Echos 'Command:' prompt text to the printer.

RELATED FILE

- **Acad.Log** Log file.

TIP

- AutoCAD places a dashed line at the end of each log file session.

LogFileOn

Rel. 12

Opens Acad.Log file and records Command: prompt text to the file.

Command	Alias	Side Menu	Menu Bar	Tablet
logfileon	...	...	...	...

Command: **logfileon**

COMMAND OPTIONS

None

RELATED AUTOCAD COMMANDS

- **LogFileOff** Turns off recording 'Command:' prompt text to file Acad.Log.
- **[Ctrl]+Q** Echoes 'Command:' prompt text to the printer.

RELATED FILE

- **Acad.Log** Log file.

TIPS

- If log file recording is left on, it resumes when AutoCAD is next loaded.
- AutoCAD places a dashed line at the end of each log file session.

'LtScale

Ver. 2.0

Sets the scale factor of linetypes (*short for Line Type SCALE*).

Command	Alt+	Side Menu	Menu Bar	Tablet
'ltscale	O,N,G	...	[Options] [Linetypes] [Global Linetype Scale]	Y19

```
Command: ltscale
New scale factor <1.0000>:
Regenerating drawing.
```

Dashed Dashed2 DashedX2

COMMAND OPTIONS

None

RELATED AUTOCAD COMMANDS

- **Change** Changes linetype scale of objects.
- **Linetype** Loads, creates, and sets the working linetype.

RELATED SYSTEM VARIABLES

- **LtScale** Contains the current linetype scale factor.
- **PlineGen** Controls how linetypes are generated for polylines.
- **PsLtScale** Linetype scale relative to paper space.

TIPS

- If the linetype scale is too large, the linetype appears solid.

- If the linetype scale is too small, the linetype appears as a solid line that redraws very slowly.

- In addition to setting the scale with the **LtScale** command, the Acad.Lin contains each linetype in three scales: normal, half-size, and double-size.

MakePreview

Rel. 13

Creates a BMP-format thumbnail image of the drawing.

Command	Alias	Side Menu	Menu Bar	Tablet
makepreview	...	...	...	...

```
Command: makepreview
```

Sample image:

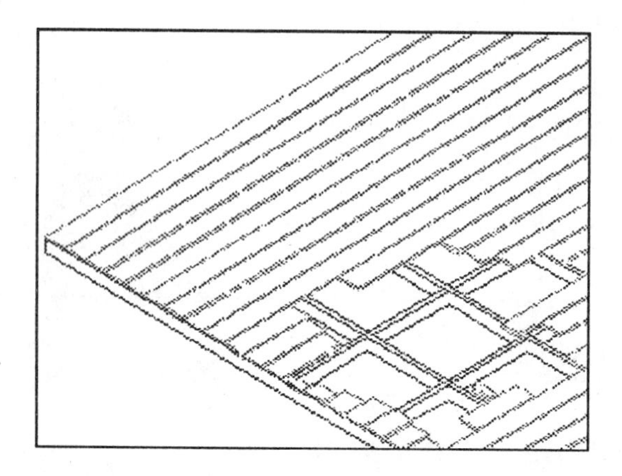

COMMAND OPTIONS

None

RELATED AUTOCAD COMMANDS

- **SaveAsR12** Saves drawing in Release 12 format.
- **SaveImg** Saves rendering in TIFF, Targa, or GIF format.
- **Plot** Outputs drawing in one of many raster formats.

RELATED SYSTEM VARIABLE

- **RasterPreview** Controls output of the **MakePreview** command.

RELATED FILES

- ***.DWG** Bitmap can be created for any AutoCAD drawing file loaded into Release 13.
- ***.BMP** Bitmap file created by **MakePreview** command.

TIPS

■ **MakePreview** creates a compressed BMP file (*short for bitmap, a raster image*) in 255 x 188 pixels by 256 colors.

■ The BMP file created by Release 13 appears to be an oddball format, unable to be read by most common graphics programs; for example, Windows Paintbrush complains, "The format of this file is not supported."

■ The two programs that read **MakePreview**'s BMP are PaintShop Pro and Word 6; once loaded into either, use the Windows Clipboard to move the image into other applications.

■ AutoCAD automatically stores a preview image of the drawing in Release 13 drawings, displayed by the **Open File** dialogue box.

■ Use the **MakePreview** command to provide preview images of drawings saves in formats other than Release 13 DWG; the BMP files can be used by other programs to display a snapshot image of the drawing.

MassProp

Rel. 11

Reports the mass properties of a 3D solid model, body, or 2D region (*short for MASS PROPerties; an external command in Acis.Dll; formerly the **SolMassP** command.*).

Command	Alias	Side Menu	Toolbar	Tablet
massprop	...	[ASSIST]	[Standard]	P 1
		[INQUIRY]	[Inquiry]	
		[MassPro:]	[MassProp]	

```
Command: massprop
Select objects: [pick]
Select objects: [Enter]
```

Example output of a solid sphere:

```
————————    SOLIDS    ————————
Mass:                   112.6241
Volume:                 12.6241
Bounding box:           X: 4.7910  -  10.7826
                        Y: -1.0540  -  4.9376
                        Z: -2.9958  -  2.9958
Centroid:               X: 7.7868
                        Y: 1.9418
                        Z: 0.0000
Moments of inertia:     X: 828.9818
                        Y: 7233.2389
                        Z: 7657.9057
Products of inertia:    XY: 1702.9437
                        YZ: 0.0000
                        ZX: 0.0000
Radii of gyration:      X: 2.7130
                        Y: 8.0140
                        Z: 8.2459
Principal moments and X-Y-Z directions about centroid:
                      I: 404.3150 along [1.0000 0.0000 0.0000]
                      J: 404.3150 along [0.0000 1.0000 0.0000]
                      K: 404.3150 along [0.0000 0.0000 1.0000]
Write to a file <N>? y
```

COMMAND OPTIONS

Y Writes mass property report to an MPR file.

N Does not write report to file.

RELATED AUTOCAD COMMAND

■ **Area** Calculates area of non-ACIS objects.

RELATED FILE

■ *.MPR **MassProp** writes its results to an MPR mass properties report file.

TIPS

■ As of Release 13, AutoCAD's solid modeling no longer allows you to apply a material density to any solid model. All solids and bodies have a density of 1.

■ AutoCAD only analyses regions coplanar with the first region selected.

DEFINITIONS

Area

■ Total surface area of selected 3D solids, bodies, or 2D regions.

Bounding Box

■ The lower-right and upper-left coordinates of a rectangle enclosing the 2D region.
■ The x,y,z-coordinate pair of a 3D box enclosing the 3D solid or body.

Centroid

■ The x,y,z-coordinates of the center of the 2D region.
■ The center of mass for 3D solids and bodies.

Mass

■ Not calculated for regions.
■ Ten times the volume, since density = 10.

Moment of Inertia

■ For 2D regions = Area * Radius^2
■ For 3D solids and bodies = Mass * Radius^2

Perimeter

■ Total length of inside and outside loops of 2D regions.
■ Not calculated for 3D solids and bodies.

Product of Inertia

■ For 2D regions = Mass * Distance (*of centroid to y,z-axis*) * Distance (*of centroid to x,z-axis*)
■ For 3D solids and bodies = Mass * Distance (*of centroid to y,z-axis*) * Distance (*of centroid to x,z-axis*)

Radius of Gyration

■ For 2D regions and 3D solids = (MomentOfInertia / Mass)$^{1/2}$

Volume

■ Not calculated for regions.
■ The amount of 3D space occupied by a 3D solid or body.

MatLib

Rel. 13

Imports and exports material-look definitions for use by the RMat command (*short for MATerial LIBrary; an external command in Render.Arx*).

Command	Alias	Side Menu	Toolbar	Tablet
matlib	...	[TOOLS]	[Render]	M 1
		[RENDER]	[MatLib]	
		[MatLib:]		

Command: **matlib**

Displays dialogue box.

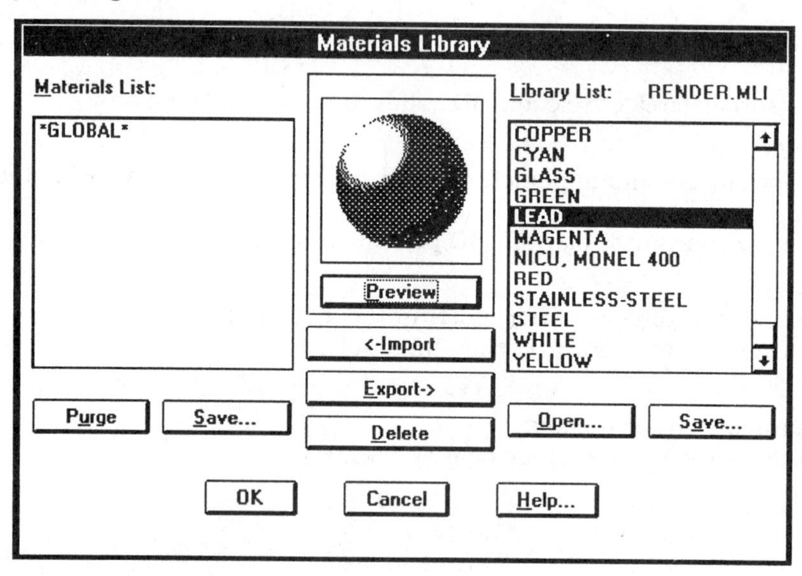

COMMAND OPTIONS

Import Brings a material definition into the drawing. If there is a conflict; displays dialogue box:

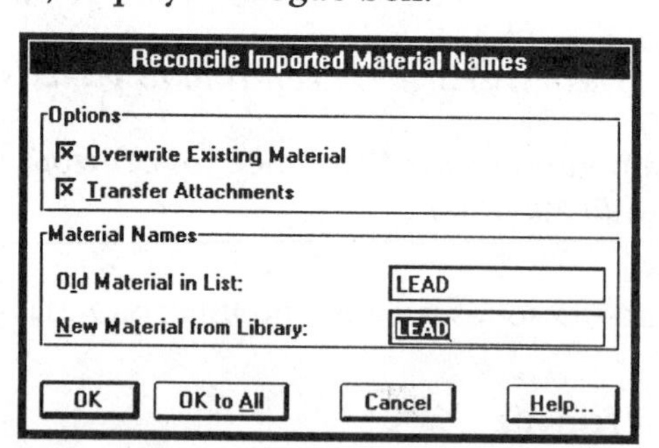

Preview	Previews the selected material mapped to a sphere object.
Export	Adds material definition to MLI library file.
Purge	Deletes unattached material definitions from the **Materials** list.
Save	Saves to an MLI file.
Delete	Deletes selected material definitions from the **Materials** or **Library** lists.
Open	Loads material definitions from an MLI file.

RELATED AUTOCAD COMMAND

■ **RMat** Attaches a material definition to objects, colors, and layers.

RELATED FILES

■ **Render.Mli** Material LIbrary; contains the material definitions.

■ **AutoVis.Mli** AutoVision material library file.

TIPS

■ **MatLib** only loads and purges material definitions; use **RMat** to attach the definitions to objects.

■ "Materials" define the look of a rendered object: coloring, reflection (*shinyness*), roughness, and ambience reflection.

■ When you import materials from an AutoVision MLI file, only the following parameters are read: Color, Ambient, Reflection, and Roughness.

■ Materials do not define the density of 3D solids and bodies.

■ By default, a drawing contains a single material definition, called *GLOBAL*, with the default parameters for color, roughness, ambient, and reflection.

■ The AutoVis.Mli file contains 144 more definitions than Render.Mli; find the file in subdirectory \Autovis\Avis_sup of the R13 distribution CD-ROM.

Measure

Ver. 2.5

Divides lines, arcs, circles, and polylines into equi-distant segments, placing a point or a block at each segment.

Command	Alias	Side Menu	Toolbar	Tablet
measure	...	[DRAW 2] [Measure:]	[Draw] [Point] [Measure]	X 22

```
Command: measure
Select object to measure: [pick]
&lt;Segment length&gt;/Block: B
Block name to insert:
Align block with object? <Y>
Segment length:
```

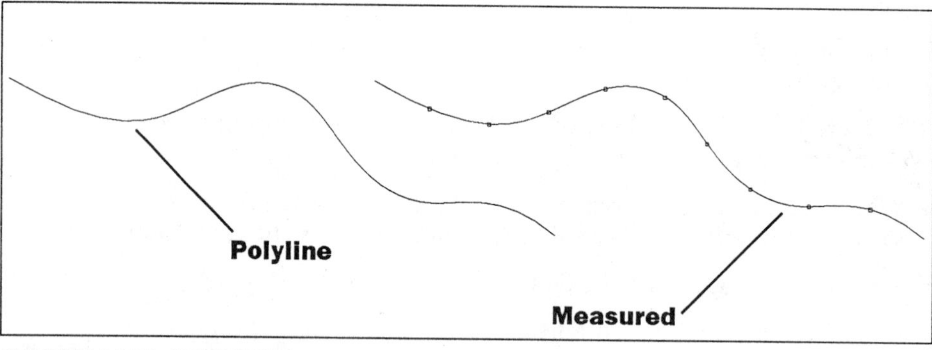

COMMAND OPTIONS

Block Indicates the name of the block to use as a marker.

<Segment length>
Indicates the distance between markers.

RELATED AUTOCAD COMMANDS

- **Block** Creates blocks that can be used with **Measure**.
- **Divide** Divides an entity into a number of segments.

RELATED SYSTEM VARIABLES

- **PdMode** Controls the shape of a point.
- **PdSize** Controls the size of a point.

TIPS

- You must define the block before it can be used with the **Measure** command.

- The **Measure** distance does not place a point or block at the beginning of the measured entity.

Menu

Ver. 1.0

Loads a MNX menu file into the drawing editor; compiles the MNU and MNL source files.

Command	Alias	Side Menu	Menu Bar	Tablet
menu	...	[TOOLS]	[Tools]	X 25
		[Menu:]	[Menus]	

```
Command: menu
Menu file name or . for none <acad>:
```

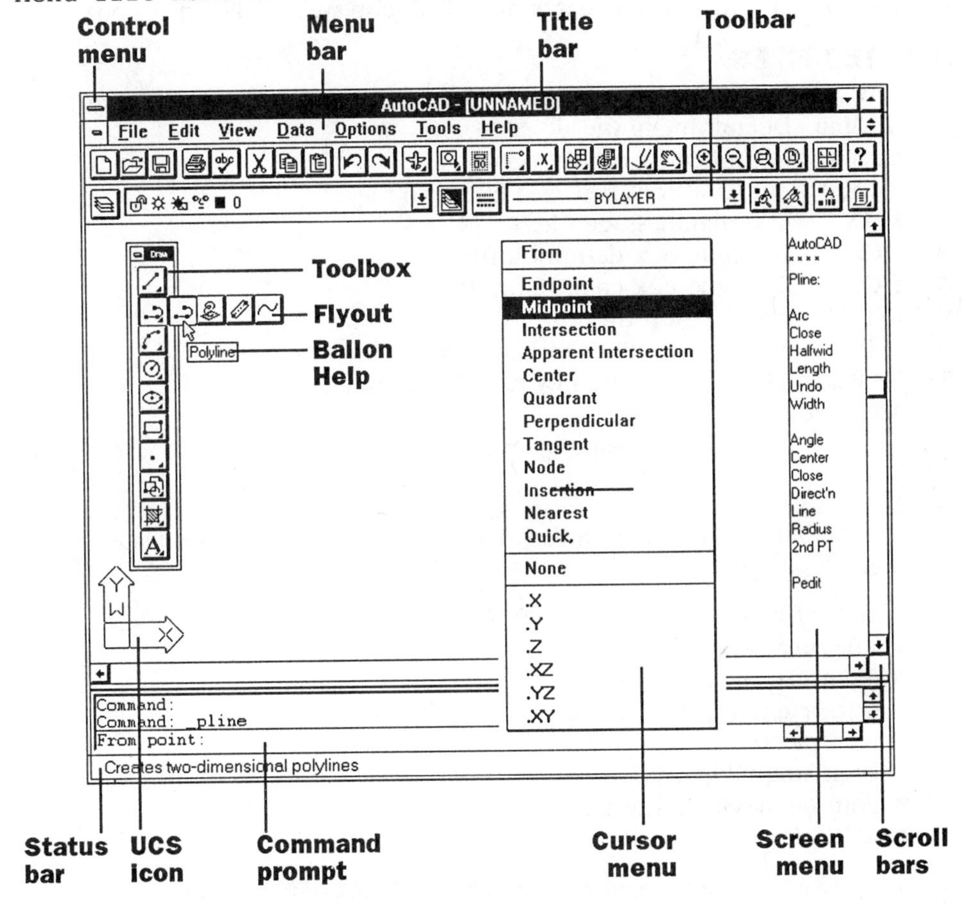

COMMAND OPTIONS

.	(*Dot*) Removes current menu from AutoCAD.
[Enter]	Reloads current menu.

RELATED AUTOCAD COMMAND

- **Tablet** Configures digitizing tablet for use with overlay menus.

RELATED AUTODESK PROGRAM

- **MC.Exe** Stand-alone menu compiler.

RELATED SYSTEM VARIABLES

- **MenuName** The currently loaded menu file.
- **MenuEcho** Suppresses menu echoing.
- **ScreenBoxes** Specifies the number of menu lines displayed on the side menu.

RELATED FILES

Files are located in the \Common\Support or \Dos\Support subdirectories:

- **Acad.Mnu** Default menu file.
- **Acad.Mnx** Compiled menu file.
- **Acad.Mnl** AutoLISP routines used by menu files.
- ***.XMX** External message file.
- ***.DCL** Dialogue box definition files.
- ***.DCC** Dialogue box color definitions.
- ***.INI** Initialization files.
- ***.LSP** AutoLISP routines for external commands.
- ***.EXP** ADS routines used by external commands.
- ***.Arx** *and* ***.Dll**
  ARx routines used by external commands.

TIPS

- AutoCAD automatically compiles MNU files into MNX files for faster loading.

- The MNU file defines the function of:
  - The screen menu (*see figure*).
  - Menu Bar menu (*see figure*).
  - Cursor menu (*see figure*).
  - Icon menus.
  - Digitizing tablet menus.
  - Pointing device buttons.
  - AUX: device.

- To create a large drawing area, use the **Config** command to remove the screen menu area.

- The DOS version of Release 13 does not support the Toolbar and Toolbox icons found in the Windows version.

- The DOS version of Release 13 does not support the **MenuLoad** and **MenuUnload** commands, for loading (*and unloading*) part of the menu file.

MenuLoad

Rel. 13

Loads a part of a menu file.

Command	Alt+	Side Menu	Menu Bar	Tablet
menuload	T,M	...	[Tools] [Customize Menus]	...

Command: **menuload**

Displays tabbed dialogue box.

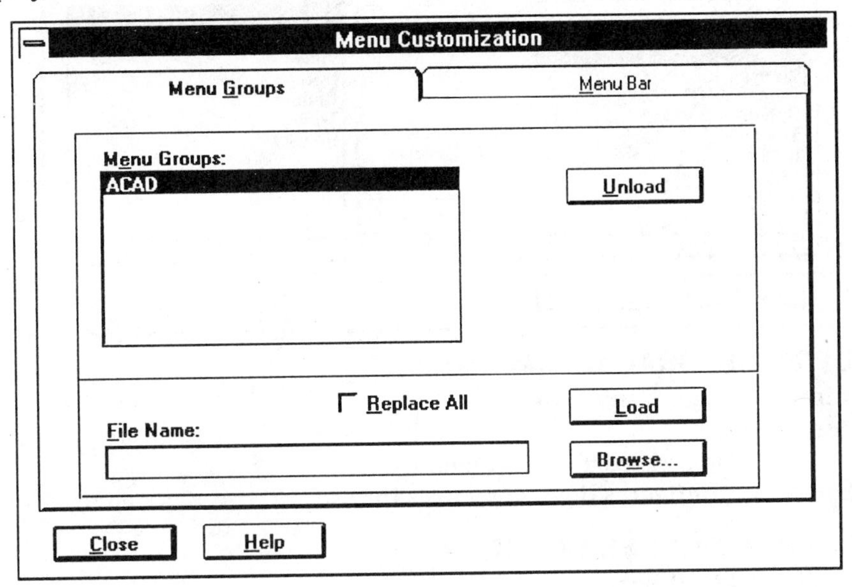

COMMAND OPTIONS

Load Loads selected menu group into AutoCAD.

Unload Unloads selected menu group.

Menu Bar Displays dialogue box:

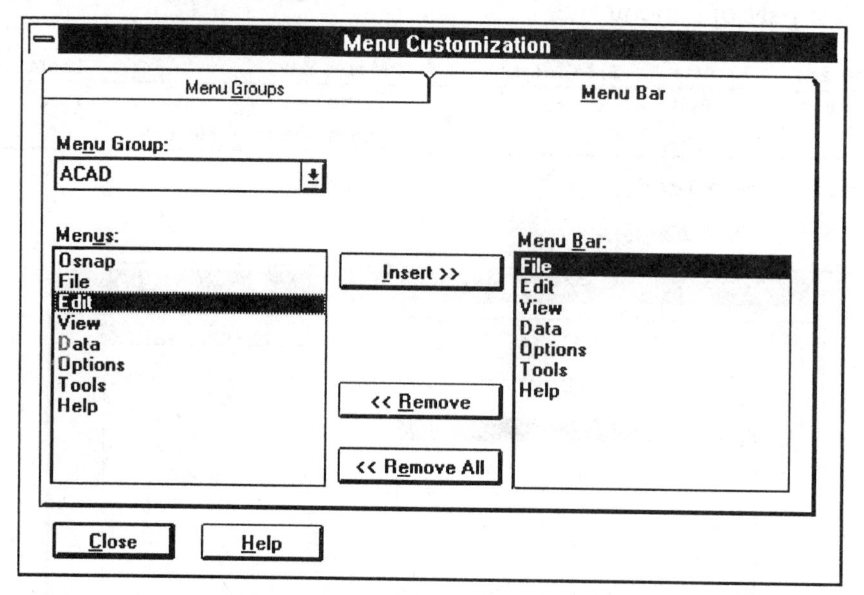

RELATED AUTOCAD COMMANDS

- **Menu** Loads a full menu.
- **MenuUnload**
  Unloads part of the menu file.
- **Tablet** Configures digitizing tablet for use with overlay menus.

RELATED AUTODESK PROGRAM

- **MC.Exe** Stand-alone menu compiler.

RELATED SYSTEM VARIABLES

- **MenuName** The currently loaded menu file.
- **MenuEcho** Suppresses menu echoing.
- **ScreenBoxes**
  Specifies the number of menu lines displayed on the side menu.

MenuUnLoad

Rel. 13

Unloads a part of the menu file.

Command	Alt+	Side Menu	Menu Bar	Tablet
menuunload	T,M	...	[Tools] [Customize Menus]	...

```
Command: menuunload
Enter the name of the MENUGROUP to unload:
```

COMMAND OPTIONS

None

RELATED AUTOCAD COMMANDS

- **Menu** Loads a full menu.
- **MenuLoad** Loads part of a menu file.
- **Tablet** Configures digitizing tablet for use with overlay menus.

RELATED AUTODESK PROGRAM

- **MC.Exe** Stand-alone menu compiler.

RELATED SYSTEM VARIABLES

- **MenuName** The currently loaded menu file.
- **MenuEcho** Suppresses menu echoing.
- **ScreenBoxes** Specifies the number of menu lines displayed on the side menu.

MInsert

Ver. 2.5

Inserts an array of blocks as a single block (*short for Multiple INSERT*).

Command	Alias	Side Menu	Menu Bar	Tablet
minsert	...	[DRAW 2] [Minsert:]	[Miscellaneous] [MInsert]	...

```
Command: minsert
Block name (or ?) <>:
Insertion point: [pick]
X scale factor <1> / Corner / XYZ: [pick]
Y scale factor (default=X): [pick]
Rotation angle <0>: [pick]
Number of rows (—) <1>:
Number of columns (|||) <1>:
Unit cell or distance between rows (—):
Distance between columns (|||):
```

COMMAND OPTIONS

Block name Indicates the name of the block to be inserted.

? Lists the names of blocks stored in the drawing.

X scale factor <1>
Indicates the x-scale factor.

Corner Indicates the x- and y-scale factors by pointing on the screen.

XYZ Displays the x-, y- and z-scale submenu.

P Supplies predefined block name, scale, and rotation values.

INPUT OPTIONS

- In response to the 'Block Name:' prompt, you can enter:
  - ~ Displays a dialogue box of blocks stored on disk.
  - = Redefines existing block with a new block, as in:
    `Block name: oldname=newname`

- In response to the 'Insertion point:' prompt, you can enter:
  - **Scale** Specifies x-, y-, z-scale factors.
  - **PScale** Presets the x-, y-, and z-scale factors.
  - **Xscale** Specifies x-scale factor.
  - **PsScale** Presets x-scale factor.
  - **Yscale** Specifies y-scale factor.
  - **PyScale** Presets y-scale factor.
  - **Zscale** Specifies z-scale factor.
  - **PzScale** Presets the z-scale factor.
  - **Rotate** Specifies the rotation angle.
  - **PRotate** Presets the rotation angle.

RELATED AUTOCAD COMMANDS

- **3dArray** Creates 3D rectangular and polar arrays.
- **Array** Creates 2D rectangular and polar arrays.
- **Block** Creates a block.

TIP

- The array placed by the **MInsert** command is a single block.

Mirror

Ver. 2.0

Creates a mirror copy of a group of entities in 2D space.

Command	Alias	Side Menu	Menu Bar	Tablet
mirror	...	[Constrct] [Mirror:]	[Modify] [Copy] [Mirror]	X 12

```
Command: mirror
Select objects: [pick]
Select objects: [Enter]
First point of mirror line: [pick]
Second point: [pick]
Delete old objects? <N>
```

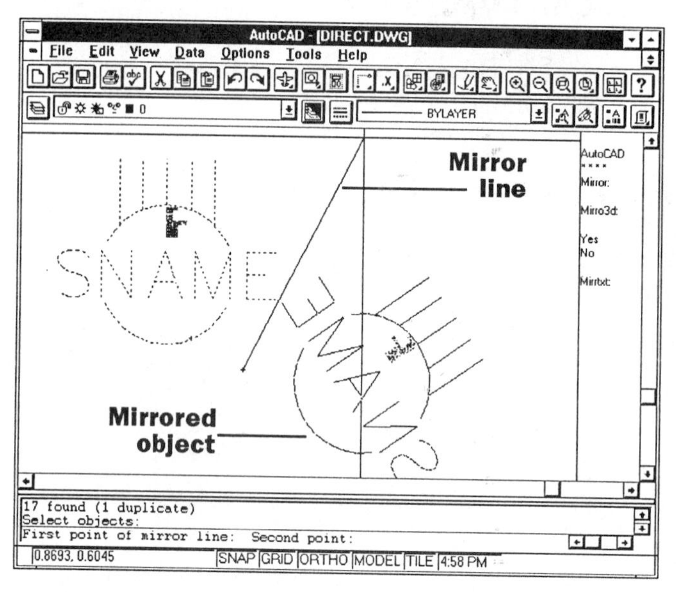

COMMAND OPTIONS

N Does not delete selected objects.

Y Deletes selected objects.

RELATED AUTOCAD COMMANDS

- **Copy** Creates a non-mirrored copy of a group of entities.
- **Mirror3d** Mirrors objects in 3D-space.

RELATED SYSTEM VARIABLE

- **MirrText** Determines whether text is mirrored by the **Mirror** command:
  - **0** Text is not mirrored about horizontal axis.
  - **1** Text is mirrored.

Mirror3d

Rel. 11

Mirrors objects about a plane in 3D space (*an external command in Geom3d.Exp*).

Command	Alias	Side Menu	Toolbar	Tablet
mirror3d	...	[CONSTRCT] [Mirr3D:]	[Modify] [Copy] [Mirror3D]	Y 21

```
Command: mirror3d
Select objects: [pick]
Select objects: [Enter]
Plane by Object/Last/Zaxis/View/XY/YZ/ZX/<3 points>:
Delete old objects? <N>
```

COMMAND OPTIONS

Object Select objects to specify mirroring plane:
Pick a circle, arc or 2D-polyline segment:

Last Selects last-picked mirroring plane.
View Current view plane is the mirror plane:
Align on view plane <0,0,0>:

XY/YZ/ZX X,y-, y,z- or z,x-plane is the mirror plane:
Point on XY plane <0,0,0>:

Zaxis Defines mirroring plane by a point on the plane and on the normal to the plane (*z-axis*):
Point on plane:
Point on Z-axis (normal) of the plane:

<3 points> Defines three points on mirroring plane:
1st point on plane:
2nd point on plane:
3rd point on plane:

RELATED AUTOCAD COMMANDS

- **Mirror** Mirrors objects in 2D space.
- **Rotate3d** Rotates objects in 3D space.

RELATED SYSTEM VARIABLE

- **MirrText** Determines whether text is mirrored by the **Mirror** command.

MlEdit

Edits multilines (*short for MultiLine EDITor; replaces the **DLine** command*).

Command	Alias	Side Menu	Menu Bar	Tablet
mledit	...	[MODIFY] [MlEdit:]	[Modify] [Special Edit] [MlEdit]	...

Command: **mledit**

Displays dialogue box.

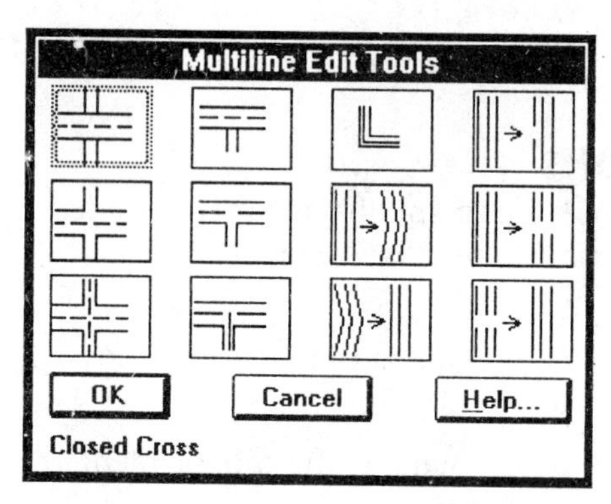

COMMAND OPTIONS

U Undoes the most-recent multiline edit.

Closed Cross
Closes the intersection of two multilines.

Open Cross
Opens the intersection of two multilines.

Merged Cross
Merges a pair of multilines: opens exterior lines; closes interior lines.

Closed Tee
Closes a T-intersection.

Open Tee
Opens a T-intersection.

Merged Tee
Merges a T-intersection: opens exterior lines; closes interior lines.

Corner Joint
Creates a corner joint of a pair of intersecting multilines.

Add Vertex
Adds a vertex (*joint*) to a multiline segment.

Delete Vertex
Removes a vertex from a multiline segment.

Cut Single
Places a gap in a single line of a multiline.

Cut All
Places a gap in all lines of a multiline.

Weld All
Removes a gap in a multiline.

RELATED AUTOCAD COMMANDS

- **MLine** Draws up to 16 parallel lines.
- **MlProp** Defines the properties of a multiline.

RELATED SYSTEM VARIABLES

- **CMlJust** Current multiline justification:
  - 0 Top (default)
  - 1 Middle
  - 2 Bottom
- **CMlScale** Current multiline scale factor (*default = 1.0*).
- **CMlStyle** Current multiline style name (*default = " "*).

RELATED FILE

- ***.MLN** Multiline style definition file in \Acad13\Common\Support.

TIPS

- Use the **Cut All** option to open up a gap before placing door and window symbols in a multiline wall.

- Use the **Weld All** option to close up a gap after removing the door or window symbol in a multiline.

- Use the **Stretch** command to move a door or window symbols in a multiline wall.

MLine

Rel. 13

Draws up to 16 parallel lines (*short for Multiple LINE; replaces the **DLine** command*).

Command	Alias	Side Menu	Toolbar	Tablet
mline	...	[DRAW 1] [Mline:]	[Draw] [Polyline] [MLine]	J 9

```
Command: mline
Justification=Top, Scale=1.0000, Style=STANDARD
Justification/Scale/STyle/<From point>: [pick]
Undo/<To point>: [pick]
Close/Undo/<To point>:
```

COMMAND OPTIONS

Undo Backs up by one segment.

Close Closes the multiline to its start point.

Justification options:

Top Draws top line of multiline at cursor; remainder of multiline is "below" cursor.

Zero Draws center (*zero offset point*) of multiline at cursor.

Bottom Draws bottom line of multiline at cursor; remainder of multiline is "above" cursor.

Scale option examples:

1.0 Default scale factor.

2.0 Draw multiline twice as wide.

-1.0 Flip multiline.

0 Collapse multiline to a single line.

STyle options:

Multiline style name:
Specifies the name of multiline style.

? Lists names of multiline styles defined in drawing.

RELATED AUTOCAD COMMANDS

- **MlEdit** Edits multilines.
- **MlProp** Defines the properties of a multiline.

RELATED SYSTEM VARIABLES

- **CMlJust** Current multiline justification:
  0 Top (*default*).
  1 Middle.
  2 Bottom.
- **CMlScale** Current multiline scale factor (*default* = *1.0*).
- **CMlStyle** Current multiline style name (*default* = " ").

RELATED FILE

- ***.MLN** Multiline style definition file in \Acad13\Common\Support.

TIP

- Multiline styles are stored in MLN files in DXF-like format.

MlStyle

Rel. 13

Defines the characteristics of multilines (*short for MultiLine STYLE*).

Command	Alt+	Side Menu	Menu Bar	Tablet
mline	D,M	[DATA]	[Data]	...
		[MlStyle:]	[Multiline Style]	

Command: **mlstyle**
Displays dialogue box.

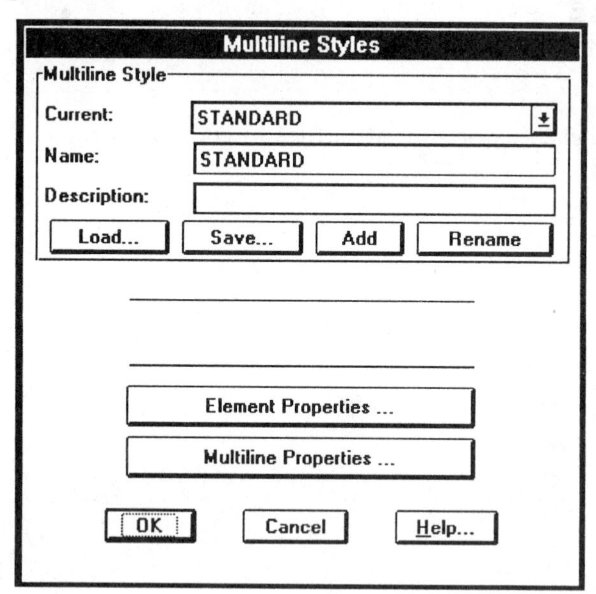

COMMAND OPTIONS

Current Lists currently loaded multiline style names (*default = STANDARD*).

Name Gives a new multiline style a name or renames an existing style.

Description Describes multiline style, with up to 255 characters.

Load Loads style from the multiline library file Acad.Mln; displays dialogue box:

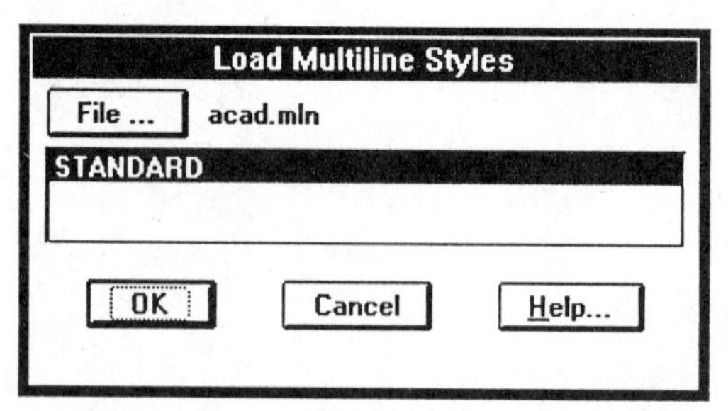

Save Saves a multiline style or rename a style.

Add Adds multiline style from the **Name** box to the **Current** list.

Remove Removes the multiline style from the **Current** list.

Element Properties

Specifies the properties for the multiline style; displays a dialogue box:

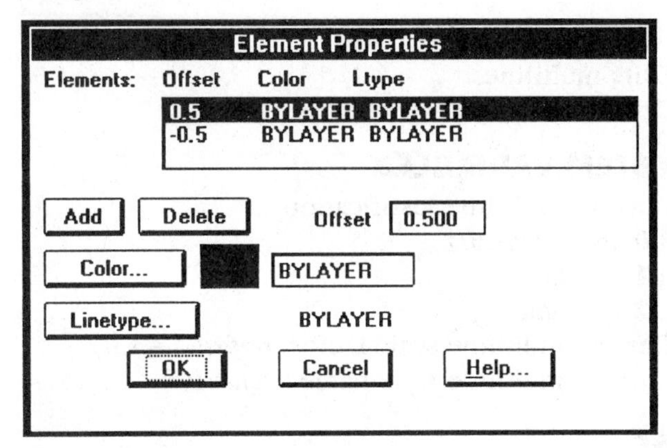

This dialogue box controls:

- Add and remove lines (*maximum = 16*).
- Offset distance (*negative and positive distance from origin*).
- Color of each line.
- Linetype of each line.

Multiline Properties Specifies additional properties for multilines; displays dialogue box:

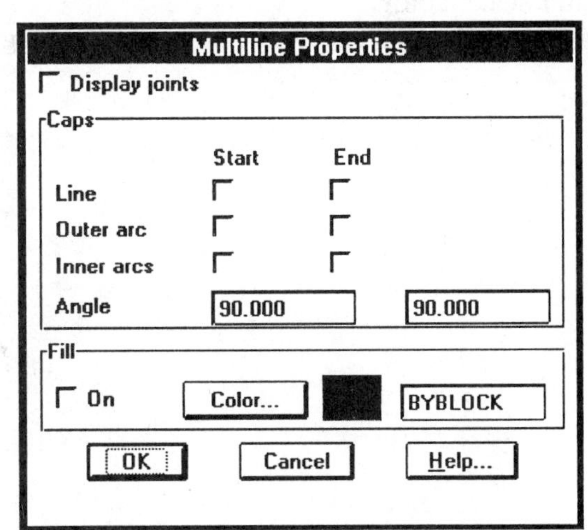

This dialogue box controls:

- Toggle display of joints (*cross-segment at vertices*).
- Endcaps at start and endpoints (*line, angled line, outer arc, and/or inner arc*).
- Toggle fill of multiline.
- Select solid color for fill.

RELATED AUTOCAD COMMANDS

- **MlEdit** Edits multilines.
- **MLine** Draws up to 16 parallel lines.

RELATED SYSTEM VARIABLES

- **CMlJust** Current multiline justification:
  - **0** Top (*default*).
  - **1** Middle.
  - **2** Bottom.
- **CMlScale** Current multiline scale factor (*default* = *1.0*).
- **CMlStyle** Current multiline style name (*default* = " ").

RELATED FILE

- **Acad.Mln** Multiline style definition file in \Acad13\Common\Support.

TIPS

- Use the **MlEdit** command to create – or close up – gaps to place door and window symbols in multiline walls.

- The multiline scale factor has the follow effect on the look of a multiline:
  - **1.0** Default scale factor.
  - **2.0** Draw multiline twice as wide, not twice as long.

Move

Ver. 1.0

Moves a group of entities to a new location.

Command	Alias	Side Menu	Toolbar	Tablet
move	m	[MODIFY]	[Modify]	W 15
		[Move:]	[Move]	

```
Command: move
Select objects: [pick]
Select objects: [Enter]
Base point or displacement: [pick]
Second point of displacement: [pick]
```

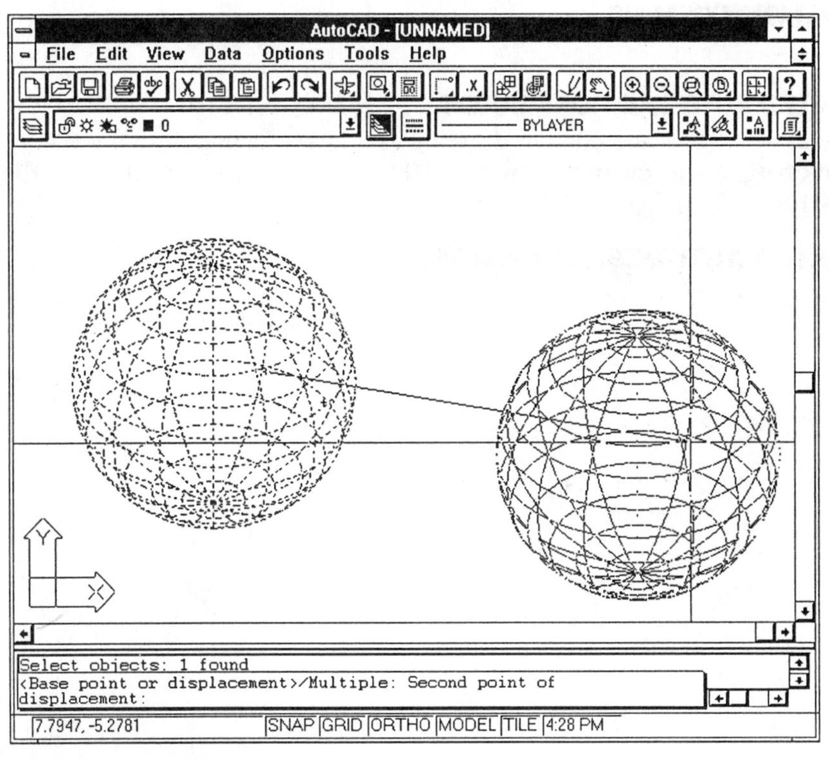

COMMAND OPTIONS

Base point Indicates the starting point for the move.
Displacement Indicates the distance to move .

RELATED AUTOCAD COMMANDS

- **Copy** Makes a copy of selected objects.
- **MlEdit** Moves the vertices of a multiline.
- **PEdit** Moves the vertices of a polyline.

MSlide

Ver. 2.0

Save the current view as an SLD-format slide file on disk (*short for Make SLIDE*).

Command	Alt+	Side Menu	Menu Bar	Tablet
mslide	T,D,S	[TOOLS]	[Tools]	...
		[SLIDES]	[Slide]	
		[MSlide:]	[Save]	

```
Command: mslide
Slide file:
```

COMMAND OPTIONS

None

RELATED AUTOCAD COMMANDS

- **Save** Saves the current drawing as a DWG-format drawing file.
- **SaveImg** Saves current view as a TIFF, Targa, or GIF-format raster file.
- **VSlide** Displays an SLD-format slide file in AutoCAD.

RELATED AUTODESK PROGRAM

- **Slidelib.Exe**

  Compiles a group of slides into an SLB-format slide library file.

MSpace

Rel. 11

Switches the drawing from paper space back to model space (*short for Model SPACE*).

Command	Alias	Side Menu	Status bar	Tablet
mspace	ms	[VIEW] [MSpace:]	[MODEL]	V 14
			[View] [Floating Model Space]	

Command: **mspace**

COMMAND OPTIONS

None

RELATED AUTOCAD COMMANDS

- **MView** Creates viewports in paper space.
- **MvSetup** Sets up the configuration of a new drawing.
- **PSpace** Switches from model space to paper space.
- **VpLayer** Sets independent visibility of layers.

RELATED SYSTEM VARIABLES

- **MaxActVp** Maximum number of viewports with visible entities (default = 16).
- **PsLtScale** Linetype scale relative to paper space.
- **Tilemode** The current setting of tilemode:
  - **0** Off (*default*).
  - **1** On.

TIPS

■ **Tilemode** must be set to zero to switch to paper space and use the **MSpace** command.

■ To switch from paper space back to model space, at least one viewport must be active; create the viewport with the **MView** command.

■ Objects in the current selection set are ignored if they were not collected in the current space.

■ AutoCAD clears the selection set when moving between paper and model space.

■ Double-click on **PAPER** on the status bar to switch to model space.

MText

Rel. 13

Places paragraph text in a boundary box (*short for Multline TEXT*).

Command	Alias	Side Menu	Toolbar	Tablet
mtext	t	...	[Draw] [MText]	...

Command: **mtext**
Attach/Rotation/Style/Height/Direction/<Insertion point>: **[pick]**
Attach/Rotation/Style/Height/Direction/Width/2Points/<Other corner>: **[pick]**
Height <>:
Loads text editor.

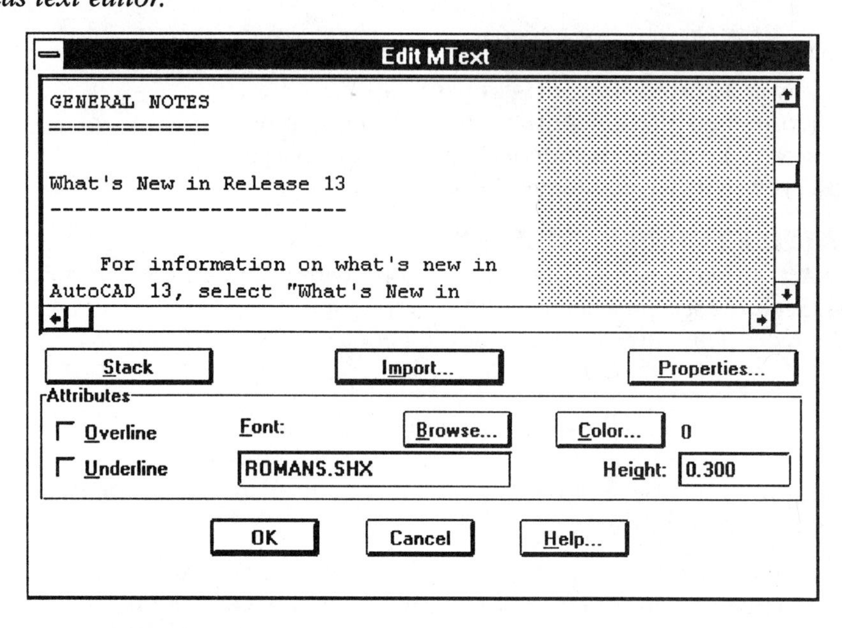

COMMAND OPTIONS

Attach boundary box options:

TL	Top left.
TC	Top center.
TR	Top right.
ML	Middle left.
MC	Middle center.
MR	Middle right.
BL	Bottom left.
BC	Bottom center.
BR	Bottom right.

Rotation Rotation angle of boundary box.
Style Text style for multiline text (*default = STANDARD*).
Height Height of UPPERCASE text (*default = 0.2 units*).
Direction Drawing direction of multiline text: horizontal or vertical.
Width Width of boundary box.
2Points Pick two points to define the boundary box.

Stack Stacks pair of characters to create a fraction:
Import Import text:

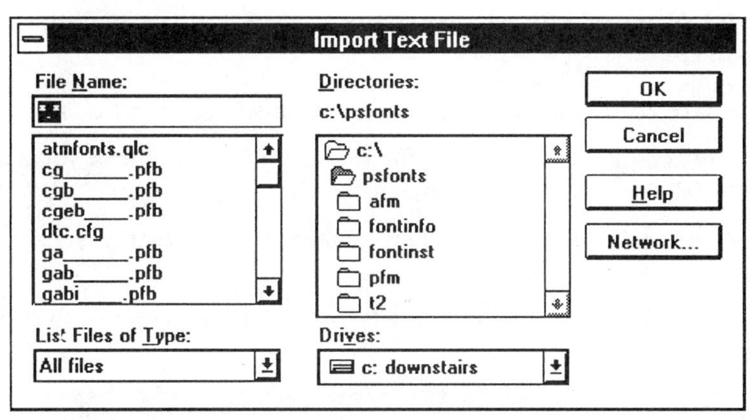

Spell Spell check text.
Properties Specify MText properties:

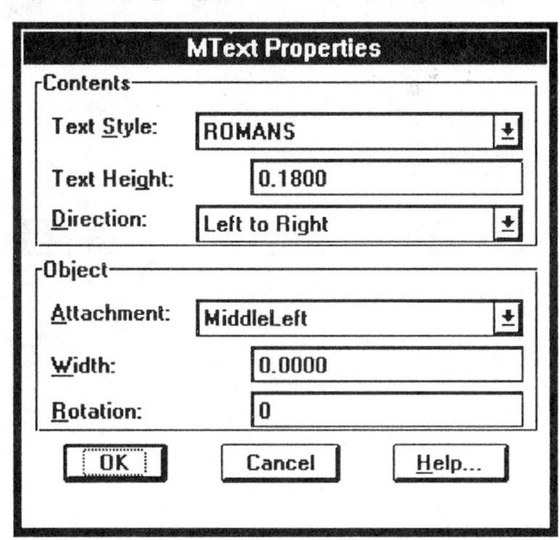

RELATED AUTOCAD COMMANDS

- **DdEdit** Edits multiline text via text editor.
- **DText** Places several lines of text.
- **MtProp** Changes properties of multiline text.
- **Style** Creates a named text style from a font file.
- **Text** Places a single line of text.

RELATED SYSTEM VARIABLE

- **MTextEd** Name of external text editor to place and edit multiline text.

TIPS

- Use the **MTextEd** system variable to define a different text editor.
- The **MText** command fails when the replacement text editor you specify with **MTextEd** does not accept filenames at the command line.
- Use the **Direction** option for languages that read vertically, such as Chinese and Japanese.
- The **MText** editor prompts you to select a font that it can use to display text in the editor:

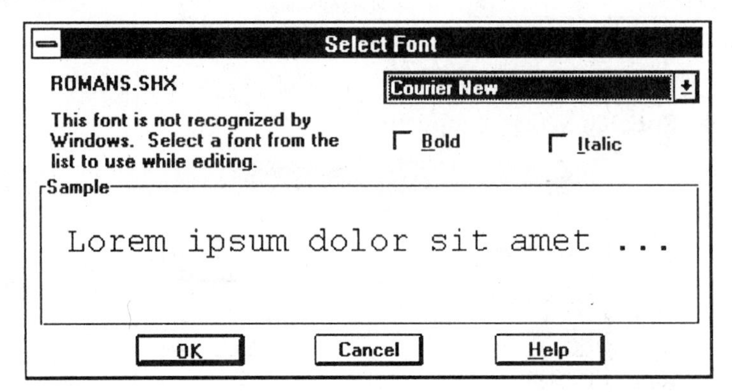

'MtProp

Rel. 13

Changes the properties of multiline text (*short for Multiline Text PROPerties*).

Command	Alias	Side Menu	Menu Bar	Tablet
mtprop	...	...	...	...

```
Command: mtprop
Select an MText object: [pick]
```

Displays dialogue box.

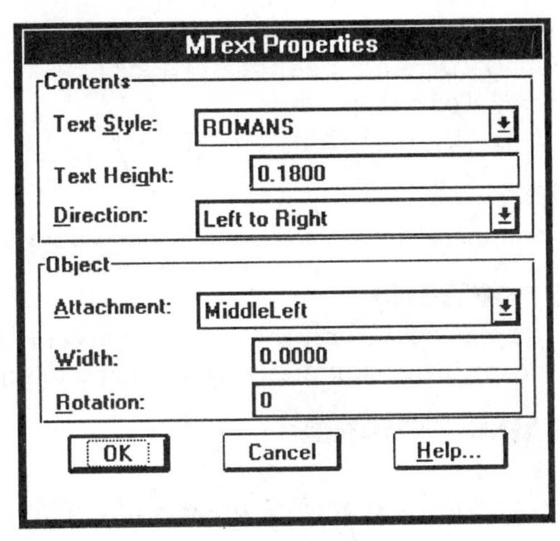

COMMAND OPTIONS

Text Style Selects a named text style (*default = STANDARD*).
Text Height Height of UPPERCASE text (*default = 0.2 units*).
Direction Directions of text (*default = left to right*).
Attachment Alignment of text boundary box at the insertion point.
Width Width of text boundary box.
Rotation Rotation of text boundary box.

RELATED AUTOCAD COMMANDS

- **DdEdit** Edits multiline text via external text editor.
- **DText** Places several lines of text.
- **MText** Places multiline text.
- **Style** Creates a named text style from a font file.
- **Text** Places a single line of text.

TIP

- Use the **Direction** option for languages that read vertically, such as Chinese and Japanese.

Multiple

Ver. 2.5

A command modifier that automatically repeat commands; not a command.

Command	Alias	Side Menu	Menu Bar	Tablet
`multiple`	...	...	...	...

Example usage:

```
Command: multiple circle
3P/2P/TTR/<Center point>: [pick]
Diameter/<Radius>: [pick]
circle 3P3P/2P/TTR/<Center point>: [pick]
Diameter/<Radius>: [pick]
circle 3P3P/2P/TTR/<Center point>: [Esc]
```

COMMAND OPTION

[Esc] — Stops command from automatically repeating itself.

COMMAND INPUT OPTIONS

[Space] — Press the spacebar to repeat the previous command.
[Click] — Click on any blank spot on the tablet menu to repeat a command.

RELATED AUTOCAD COMMANDS

- **Redo** — Undoes a undo.
- **U** — Undoes previous command; undoes one multiple command at a time.

RELATED COMMAND MODIFIERS

- ' — (*Apostrophe*) Allows use of certain transparent commands in another command.
- ~ — (*Tilde*) Forces display of dialogue box.
- - — (*Dash*) Forces display of prompts on command line.

TIPS

- **Multiple** is not a command but a command modifier; it does nothing on its own.

- **Multiple** only repeats the command name; it does not repeat command options.

- Some commands automatically repeat, including **Point** and **Donut**.

Creates and manipulates overlapping viewports (*short for Make VIEWports*).

Command	Alt+	Side Menu	Menu Bar	Tablet
mview	V, G	[VIEW] [MView:]	[View] [Floating Viewports]	P 3-5
				Q 3-5
				R4

```
Command: mview
ON/OFF/Hideplot/Fit/2/3/4/Restore/<First Point>: [pick]
Other corner: [pick]
Regenerating drawing.
```

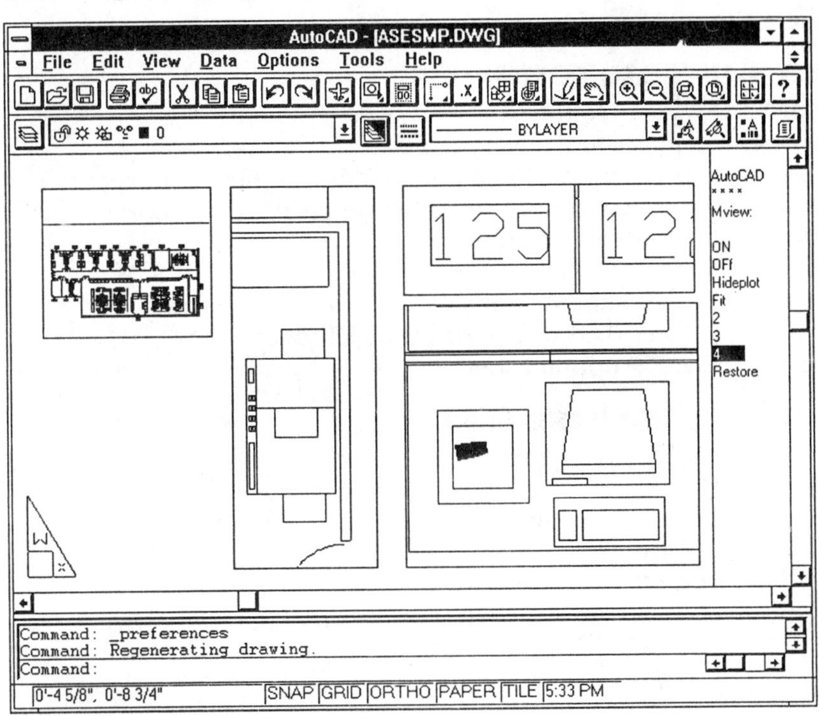

COMMAND OPTIONS

<First Point>	Indicates the first point of a single viewport.
Fit	Creates a single viewport that fits the screen.
Hideplot	Creates a hidden-line view during plotting and printing.
OFF	Turns off a viewport.
ON	Turns on a viewport.
Restore	Restore a saved viewport configuration.

2 Displays the submenu of viewport orientations:

- **Horizontal** Stacks the two viewports.
- **<Vertical>** Side-by-side viewports.

3 Displays the submenu of viewport orientations:

- **Horizontal** Stacks the three viewports.
- **Vertical** Side-by-side viewports.
- **Above** Two viewports above the third.
- **Below** Two viewports below the third.
- **Left** Two viewports to the left of the third.
- **<Right>** Two viewports to the right of the third.

4 Displays the submenu of viewport orientations:

- **Fit** Creates four same-size viewports to fit the screen.
- **<First Point>** Indicates the area for the four viewports.

RELATED AUTOCAD COMMANDS

- **[Ctrl]+V** Switches to the next viewport.
- **MSpace** Switches to model space.
- **MvSetup** Sets the up configuration of a drawing in paper space.
- **PSpace** Switches to paper space before creating viewports.
- **Redrawall** Redraws all viewports.
- **RegenAll** Regenerates all viewports.
- **VpLayer** Controls the visibility of layers in each viewport.
- **VPorts** Creates tiled viewports in model space.
- **Zoom** The **XP** option zooms a viewport relative to paper space.

RELATED SYSTEM VARIABLES

- **CvPort** Current viewport.
- **MaxActVp** Controls the maximum number of visible viewports:
  - **1** Minimum.
  - **16** Default.
  - **32767** Maximum.
- **Tilemode** Controls the availability of overlapping viewports:
  - **0** Off.
  - **1** On.

TIPS

- Although system variable **MaxActVp** limits the number of simultaneously visible viewports, the **Plot** command plots all viewports.

- **Tilemode** must be set to zero to switch to paper space and use the **MSpace** command.

- **Snap**, **Grid**, **Hide**, **Shade**, etc., can be set separately in each viewport.

MvSetup

Rel. 11

Inserts predefined title blocks, creates a set of viewports, sets a global scale factor (*short for Model View SETUP; an external command in MvSetup.Lsp*).

Command	Alias	Alt+	Menu Bar	Tablet
mvsetup	mvs	V,G,S	[View] [Floating Viewports] [MV Setup]	...

```
Command: mvsetup
```

Command prompt when Tilemode = 1:

```
Enable paper space? (No/<Yes>): N
Units type (Scientific/Decimal/Engineering/Architectural/
Metric):
Enter the scale factor:
Enter the paper width:
Enter the paper height:
```

Command prompt when Tilemode = 0:

```
Align/Create/Scale viewports/Options/Title block/Undo: A
Angles/Horizontal/Vertical alignment/Rotate view/Undo: A
Base point: [pick]
Other point: [pick]
Distance from basepoint:
Angle from basepoint:
```

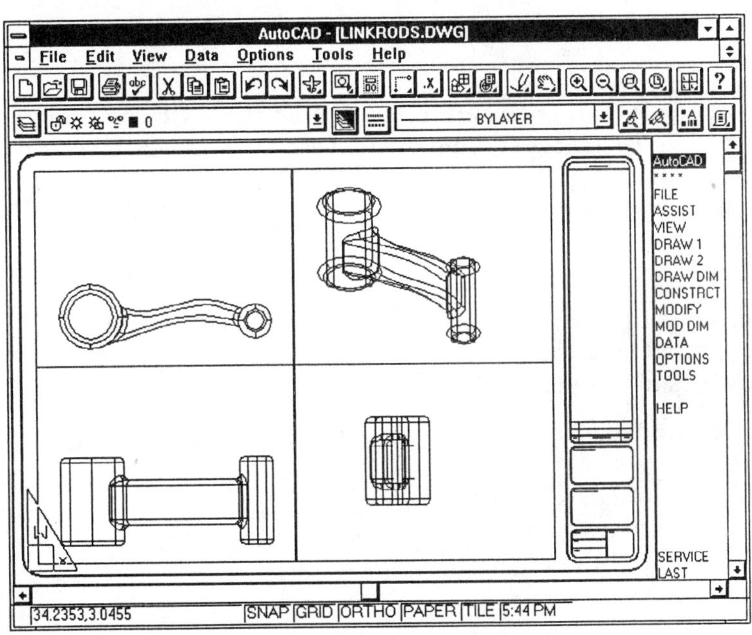

COMMAND OPTIONS

Align	Aligns new viewport with base point of existing viewports.
Create	Creates viewports in four layouts:
0	No layout.
1	Single viewport.
2	Standard engineering layout.
3	Array viewports along x- and y-axes.
Scale viewports	Scales border with respect to drawing entities.
Options	Specifies options:
Set layer	Specifies layer for title block.
Limits	Specifies whether to reset limits after title block insertion.
Units	Specifies inch or millimeter paper units.
Xref	Specifies whether title is inserted as a block or as an xref.
Title block	Specifies title block style.
Undo	Undoes **MvSetup** operations in reverse order.

RELATED SYSTEM VARIABLE

■ **Tilemode** The current setting of tilemode:
0 Off (*default*).
1 On.

RELATED FILES

■ **MvSetup.Dfs** The **MvSetup** default settings file.
■ **AcadIso.Dwg** Prototype drawing with ISO defaults.

TIPS

■ When option 2 (Std. Engineering) is selected at the **Create** option, the following views are created (counterclockwise from upper left):
- Top view.
- Isometric view.
- Front view.
- Right view.

■ To create the title block, **MvSetup** searches the path specified by the AcadPrefix variable. If the appropriate drawing cannot be found, **MvSetup** creates the default border.

■ **MvSetup** produces the following predefined title blocks:
- None.
- ISO A0 through A4 (*mm, metric*).
- ANSI A through E, ANSI V (*in, imperial*).
- Architectural and engineering D-size.
- Generic D-size.

■ The metric A0 size is similar to the imperial E-size, while the metric A4 is similar to A-size.

■ You can add your own title block with the **Add** option. Before doing so, create the title block as an AutoCAD drawing.

QUICK START: Using MvSetup

MvSetup has many options but does not present them in a logical fashion. To set up a drawing with **MvSetup**, follow these basic steps:

1. Start the **MvSetup** command:

```
Command: mvsetup
```

2. Place the title block with the **Title** option.

3. Setup the viewports with the **Create** option. For standard drawings, select option #2, Std. Engineering.

4. Make the object the same size in all four viewports with the **Scale** option. When you are prompted to 'Select objects:', select the four viewports, not the objects in the drawing.

5. You can interrupt the **MvSetup** command at any time with the **[Esc]** key, then resume the command complete the setup.

New

Rel. 12

Names and starts a new drawing.

Command	Ctrl+	Side Menu	Menu Bar	Tablet
new	N	[FILE]	[File]	U 24
		[New:]	[New]	

```
Command: new
```

Displays dialogue box.

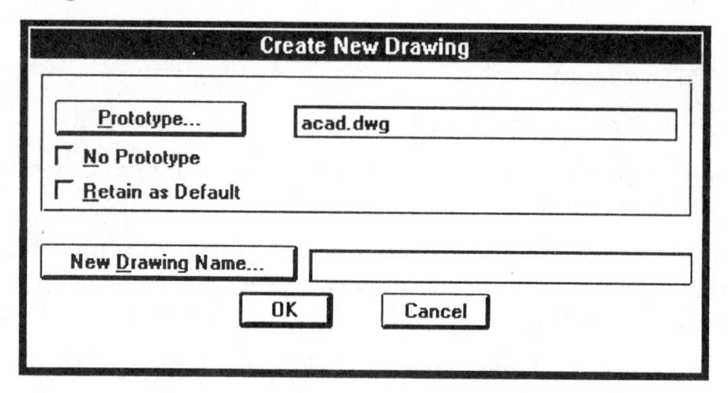

COMMAND OPTIONS

Prototype Specifies the name of the prototype drawing.

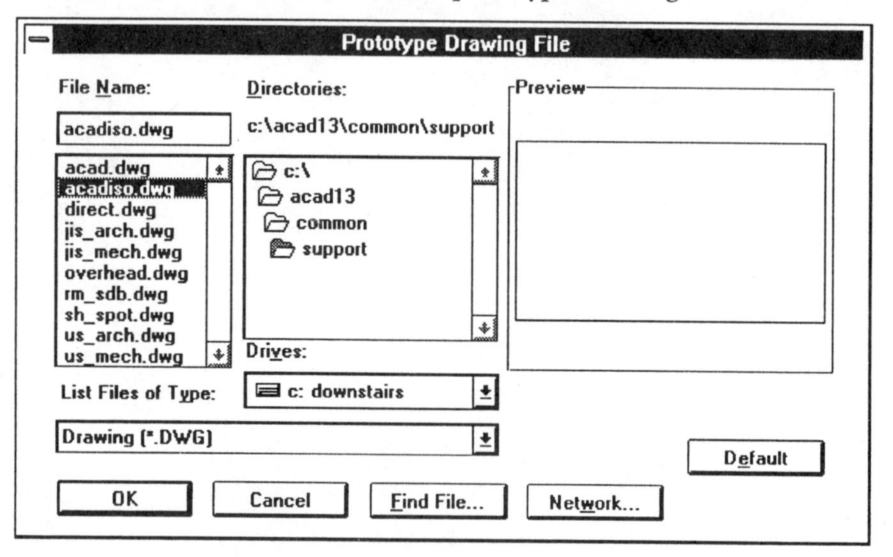

Find File Searches for file on selected drives and directories:

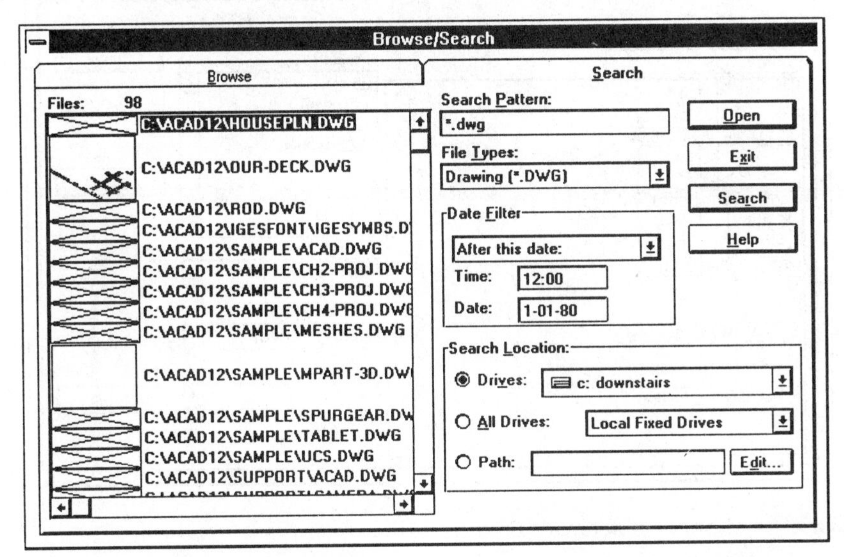

Edit Edits the search path:

Edit Path

Path:
c:\acad13\common\support

Directories:
c:\
acad13
common
support

Search Path:

Drives:
c: downstairs

Add
Delete
Delete All
Close
Network...

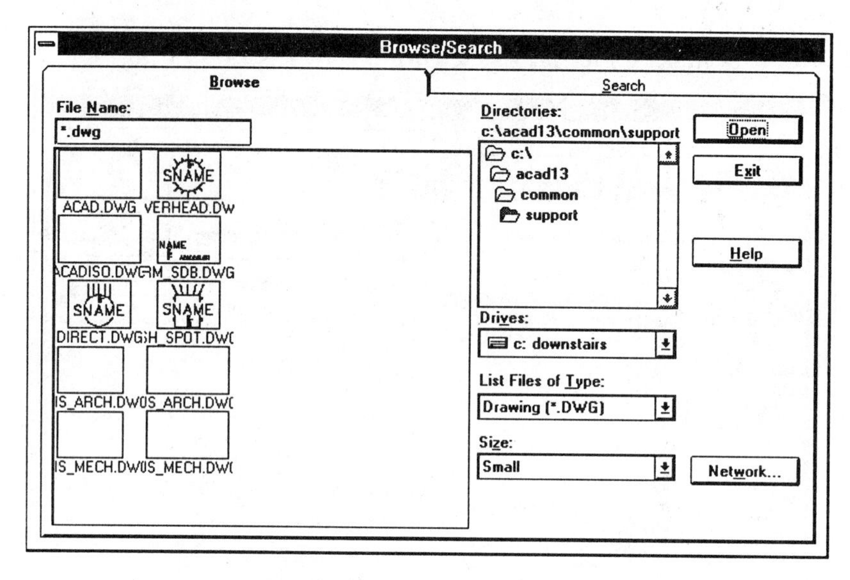

Network Connects network drives:

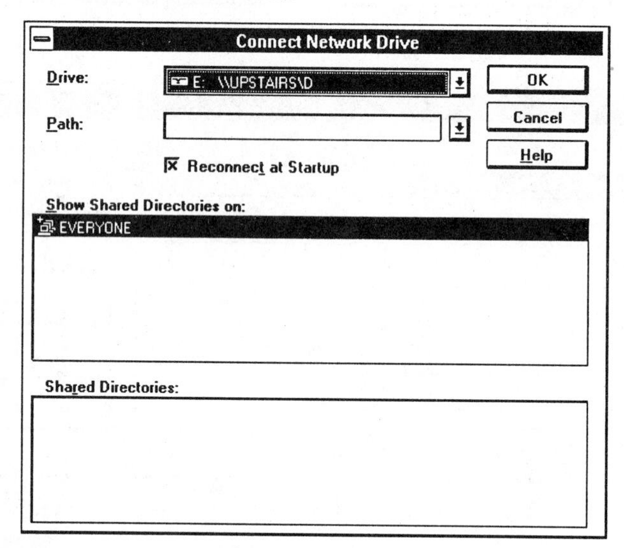

No prototype Does not use a prototype drawing.

Retain as default

Retains drawing as the default prototype drawing.

New drawing name

Specifies the filename of the new drawing:

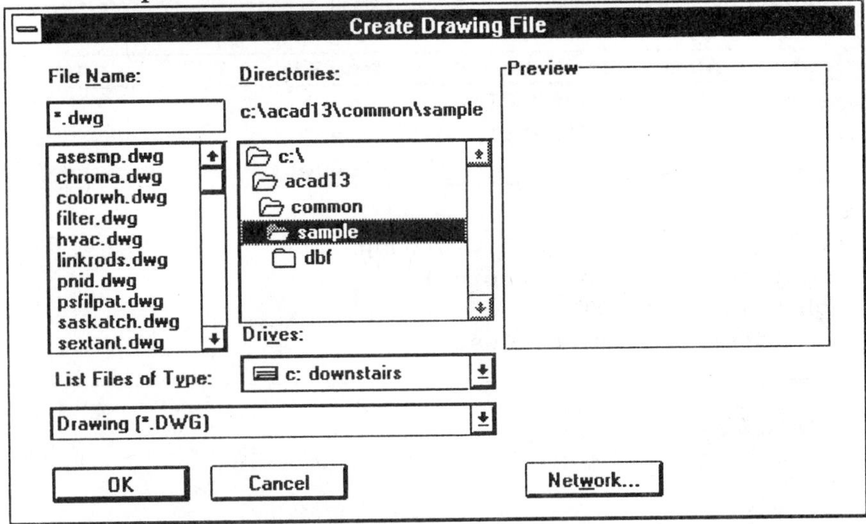

RELATED AUTOCAD COMMANDS

- **FileOpen** Opens drawing without displaying a dialogue box.
- **SaveAs** Saves drawing with a different name.

RELATED SYSTEM VARIABLES

- **DbMod** Indicates whether drawing has changed since being loaded.
- **DwgName** Name of current drawing.

RELATED FILES

Found in \Acad13\Common\Support subdirectory:

- **Acad.Dwg** The default prototype drawing.
- **AcadIso.Dwg** The ISO/DIN prototype drawing.
- **Jis_Arch.Dwg** The JIS architectural prototype drawing.
- **Jis_Mech.Dwg** The JIS mechanical prototype drawing.
- **Us_Arch.Dwg** The US architectural prototype drawing.
- **Us_Mech.Dwg** The US mechanical prototype drawing.

TIPS

- AutoCAD allows you to save your work before using the **New** command.

- Until you give the drawing a name, AutoCAD names it Untitled.Dwg.

- The **Prototype** button lets you select a different prototype drawing from a dialogue box.

- The default prototype drawing is Acad.Dwg; edit and save Acad.Dwg to change the defaults in new drawings.

Offset

Ver. 2.5

Draws parallel lines, arcs, circles and polylines; repeats automatically until cancelled.

Command	Alias	Side Menu	Menu Bar	Tablet
offset	...	[CONSTRCT] [Offset:]	[Construct] [Copy] [Offset]	W 14

```
Command: offset
Offset distance or Through <Through>: t
Select object to offset: [pick]
Through point: [pick]
Select object to offset: [Esc]
```

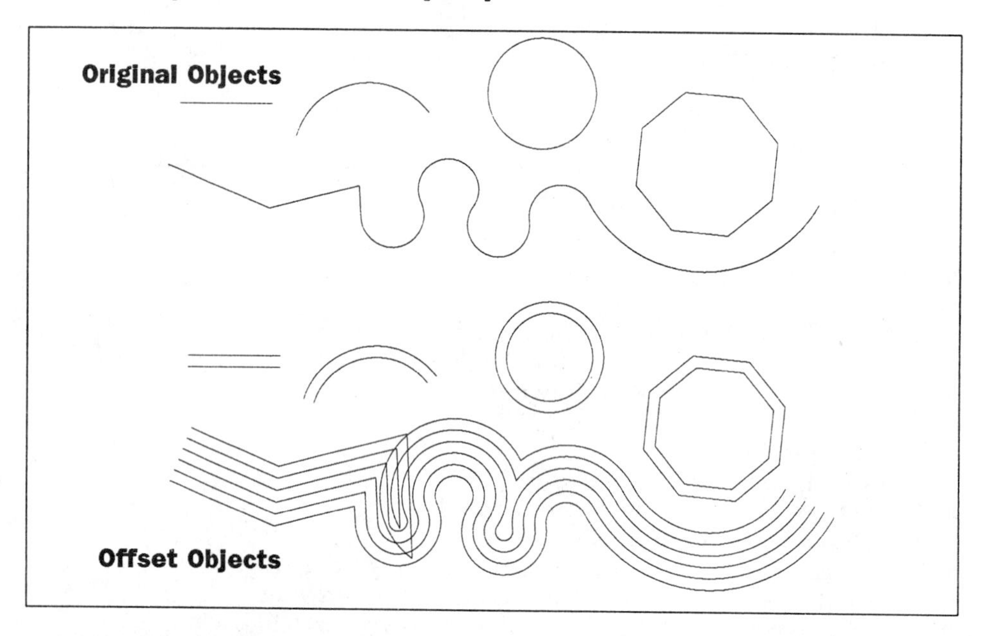

COMMAND OPTIONS

Through Indicates the offset distance.
[Esc] Exits the **Offset** command.

RELATED AUTOCAD COMMANDS

- **Copy** Creates copies of a group of entities.
- **Mirror** Creates a mirror copy of a group of entities.
- **MLine** Draws up to 16 parallel lines.

RELATED SYSTEM VARIABLE

- **OffsetDist** Current offset distance.

OleLinks

Rel. 13

Changes, updates, and cancels OLE (object linking and embedding) links between the drawing and other Windows applications.

Command	Alt+	Screen menu	Menu Bar	Tablet
olelinks	E,L	...	[Edit] [Links]	...

Command: **olelinks**
Displays dialogue box.

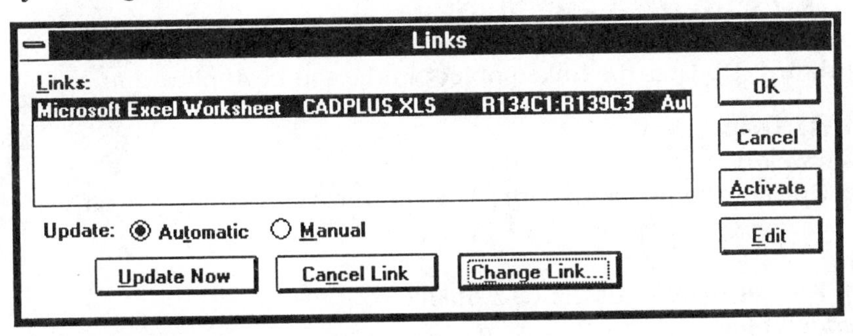

COMMAND OPTIONS

Links Displays a list of linked objects:
- Name of object.
- Source filename.
- Portion of file.
- Update mode: automatic or manual.

Update Automatic or manual updates.

Update Now Updates selected links.

Cancel Link Cancesl the OLE link; keeps the object in place.

Change Link Specifies a new object:

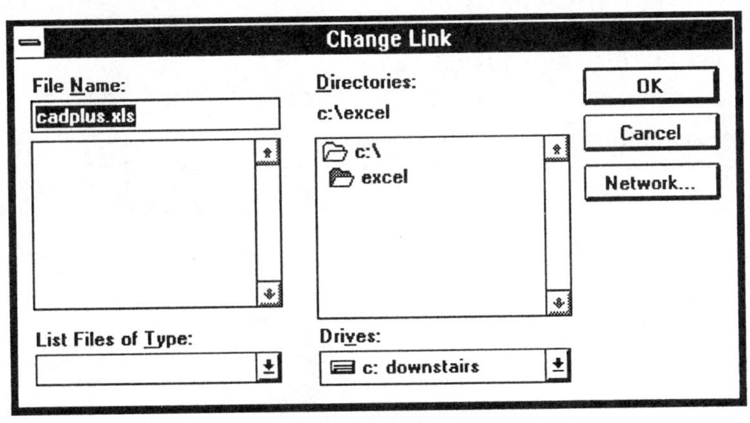

Done	Closes dialogue box.
Activate	Activates the object, if an activatable object such as a WAV sound or AVI video file.
Edit	Edits the object back in the source application.

RELATED AUTOCAD COMMANDS

- **InsertObj** Places an OLE object in the drawing.
- **PasteSpec** Places objects from Clipboard as linked objects in drawing.

RELATED WINDOWS COMMANDS

- **Copy** Copies object from source application to the Windows Clipboard.
- **Update** Updates the linked object in the source application.

DEFINITIONS

Automatic link:

- Updates the link automatically each time the source application changes the object.

Client:

- The destination for the OLE object.

Embedded object:

- Object placed in a document from the Windows Clipboard but loses knowledge of its source.

Linked object:

- Object placed in a document from the Windows Clipboard and retains knowledge of its source.

Manual link:

- You are prompted whether you want to update the object each time you open the document.

Object:

- A document page, spreadsheet range, drawing, bitmap, sound recording, video clip, etc.

OLE:

- Object linking and embedding.

Server:

- The source of the OLE object.

Oops

Ver. 1.0

Unerases the last-erased group of objects; returns the group of objects after the **Block** command.

Command	Alias	Side Menu	Menu Bar	Tablet
oops	...	[MODIFY] [Oops:]	[Miscellaneous] [Oops]	W 18

Command: **oops**

COMMAND OPTIONS

None

RELATED AUTOCAD COMMANDS

■ **Undo** Undoes the most-recent command.

RELATED AUTOLISP PROGRAMS

■ *None*

RELATED SYSTEM VARIABLES

■ *None*

TIPS

■ **Oops** only unerases the most recently erased entity; use the **Undo** command to unerase earlier entities.

■ Use **Oops** to bring back entities after turning them into a block with the **Block** and **WBlock** commands.

Open

Rel. 12

Loads a drawing into AutoCAD.

Command	Ctrl+	Side Menu	Menu Bar	Tablet
open	O	[FILE] [Open:]	[File] [Open]	U 25

Command: **open**

Displays dialogue box.

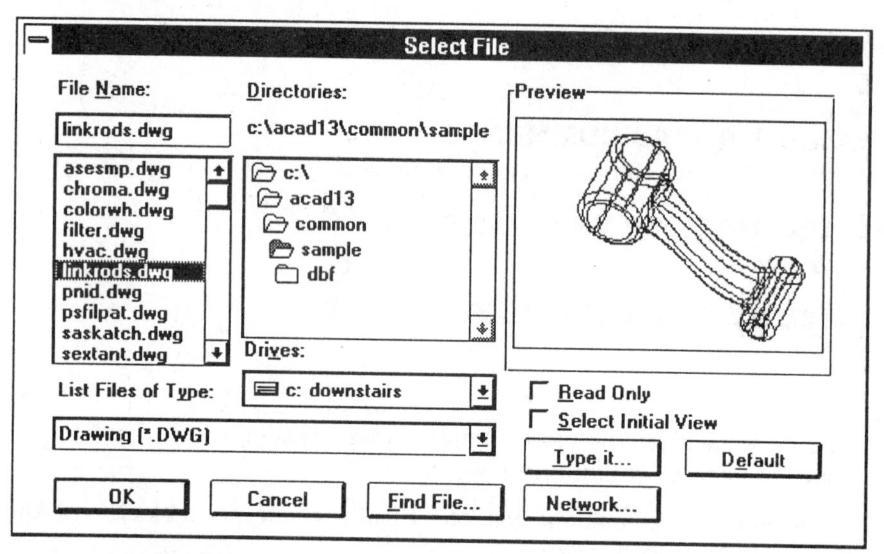

COMMAND OPTIONS

Default Loads current drawing again.

Pattern Specifies the filename pattern.

Read only mode Displays drawing but you cannot edit it.

Select initial view

Selects a named view from a dialogue box:

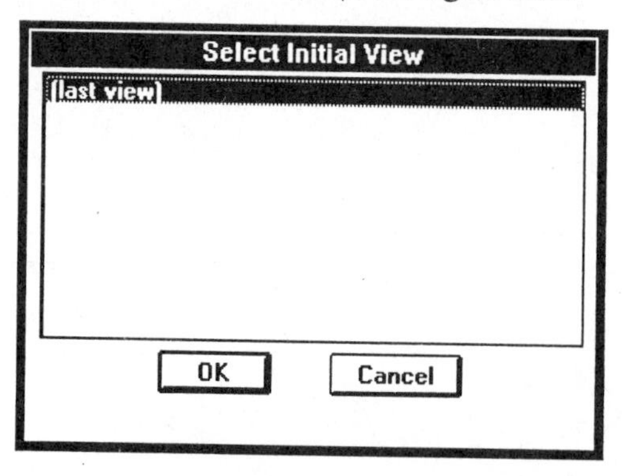

RELATED AUTOCAD COMMANDS

- **FileOpen** Opens a drawing without displaying dialogue box.
- **New** Starts a new drawing.
- **SaveAs** Save drawing with a new name.

TIPS

- The **Open** command loads DWG drawing files for Release 13 and earlier.

- When a pre-Release 13 drawing is loaded, it is converted with the message, "Converting old drawing."

- To retain a drawing in a pre-Release 13 DWG format, use the **SaveAs** command to rename the file or store it in another subdirectory. As an alternative, use the **SaveAsR12** command to save the drawing in Release 12 format.

- The **Open** command does not load any other file format into AutoCAD. Instead, use the **Import** command or these commands:
    - **DxfIn** ASCII and binary DXF file format.
    - **DxbIn** DXB file format.
    - **PsIn** EPS (*encapsulated PostScript*).
    - **TiffIn** TIFF (*tagged image file format*).
    - **PcxIn** PCX
    - **GifIn** GIF (*graphics interchange format*).
    - **Replay** TIFF, GIF, and TGA (*Targa*).
    - **AcisIn** ASCII-format SAT (*save as text*) ACIS; new to Release 13.
    - **VlConv** Visual Link files.
    - **3dsIn** 3D Studio files.
    - **VSlide** SLD (*slide*) files.

'Ortho

Ver. 1.0

Constrains drawing and editing commands to the vertical and horizontal directions only (*short for ORTHOgraphic*).

Command	Status bar	Function Key	Menu Bar	Tablet
'ortho	ORTHO	[F8]	...	V 15

```
Command: ortho
ON/OFF <Off>:
```

COMMAND OPTIONS

OFF Turns off ortho mode.

ON Turns on ortho mode.

RELATED AUTOCAD COMMAND

■ **DdRModes** Toggles ortho mode via a dialogue box.

RELATED SYSTEM VARIABLE

■ **OrthoMode** The current state of ortho mode.

TIPS

■ Use ortho mode when you want to constrain your drawing and editing to right angles.

■ Rotate the angle of ortho with the **Snap** command's **Rotate** option

■ In isoplane mode, ortho mode constraints the cursor to the current isoplane.

■ Ortho mode is ignored when entering coordinates by keyboard and in perspective mode.

■ Double click on **OTHO** on the status bar to toggle ortho mode.

'OSnap

Ver. 2.0

Sets and turns off object snap modes (*short for Object Snap*).

Command	Button	Side Menu	Toolbar	Tablet
'osnap	[#2]	[* * * *] [Osnap:]	[Standard] [OSnap]	T 12-22
		[SERVICE] [Osnap:]		U 12-13

```
Command: osnap
Object snap modes:
```

COMMAND OPTIONS

As abbreviation, enter only the first three letters:

APParent intersection: Snaps to the imaginary intersection of two objects.

CENter: Snaps to center point of arcs and circles.

ENDpoint: Snaps to end point of lines, polylines, traces, and arcs.

from: Extends from a point by a given distance.

INSertion: Snaps to insertion point of blocks, shapes, and text.

INTersection: Snaps to intersection of two entities.

MIDpoint: Snaps to middle point of lines and arcs.

NEArest: Snaps to entity nearest to crosshairs.

NODe: Snaps to a point entity.

NONe: Temporarily turns off all object snap modes.

OFF Turns off all object snap modes.

PERpendicular: Snaps perpendicularly to entities.

QUAdrant: Snaps the quadrant points of circles and arcs.

 **QUIck:** Snaps to the first entity found in the database.

 **TANgent:** Snaps tangent to arcs and circles.

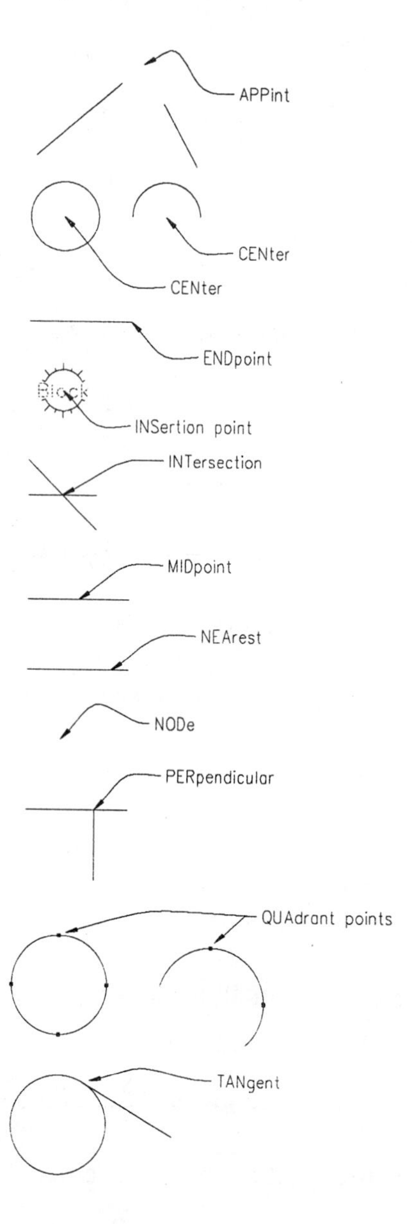

RELATED AUTOCAD COMMANDS

- **Aperture** Controls the size of the object snap cursor up to 50 pixels.
- **DdOSnap** Dialogue box for selecting object snap modes.

RELATED SYSTEM VARIABLES

- **Aperture** Controls the size of the object snap cursor to any size.
- **OsMode** Current object snap mode settings:
  - **0** No object snap modes set.
  - **1** ENDpoint.
  - **2** MIDpoint.
  - **4** CENter.
  - **8** NODe.
  - **16** QUAdrant.
  - **32** INTersection.
  - **64** INSertion point.
  - **128** PERpendicular.
  - **256** TANgent.
  - **512** NEArest.
  - **1024** QUIck.
  - **2048** APPint.

TIPS

- The **Aperture** command controls the snap area AutoCAD searches through.
- If AutoCAD finds no snap matching the current modes, the pick point is selected.
- The **APPint** and **from** object snap modes are new to Release 13.

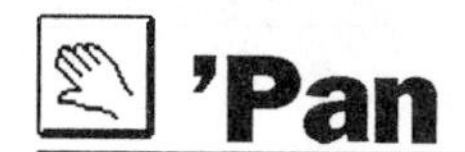

'Pan

Ver. 1.0

Moves the view in the current viewport to a different position.

Command	Alias	Side Menu	Menu Bar	Tablet
'pan	p	[VIEW] [Pan:]	[View] [Pan]	Q 10

```
Command: pan
Displacement: [pick]
Second point: [pick]
```

Before panning left:

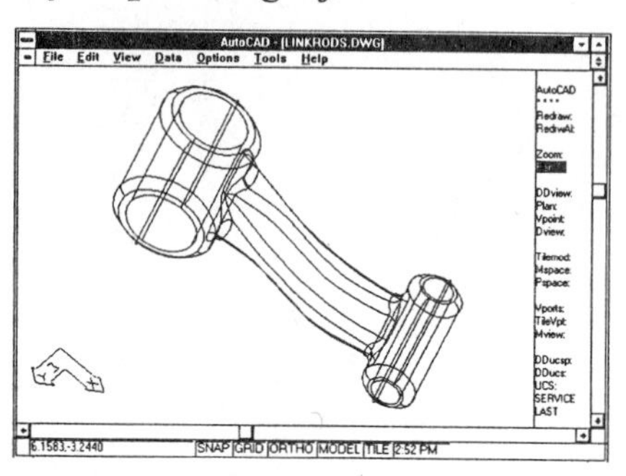

After panning to the left:

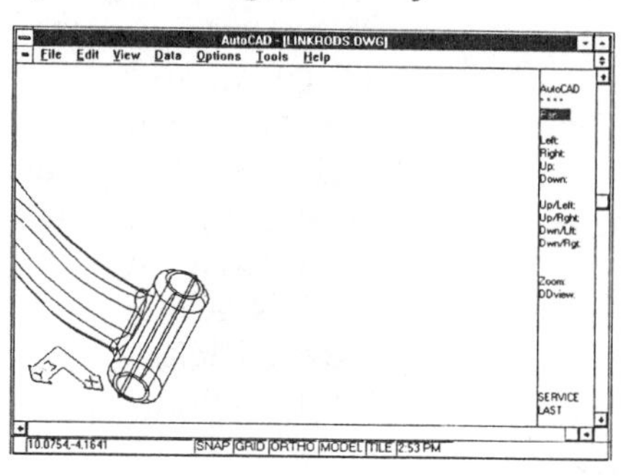

COMMAND OPTIONS

None

RELATED AUTOCAD COMMANDS

- **DsViewer** Aerial View pans in an independent window.
- **DView** Pans during perspective mode.
- **View** Saves and restorse named views.
- **ViewRes** Toggles whether pans are redrawn or regenerated.
- **Zoom** The **Dynamic** option includes a pan option.

RELATED SYSTEM VARIABLES

- **RegenAuto** Determines how regenerations are handled.
- **ViewCtr** The x,y-coordinate of the view's center.
- **ViewDir** View direction relative to UCS.
- **ViewSize** Height of view in units.

TIPS

- You pan each viewport independently.

- You can use the **Pan** command transparently to start drawing an object in one area of the drawing, pan over, then continue drawing in another area of the drawing.

- Change the **Static** button (of the AV window) to **Dynamic** to perform 'real-time' panning: the drawing pans as quickly as you move the mouse.
- You cannot use transparent pan during:
  - Paper space.
  - Perspective mode.
  - **VPoint** command.
  - **DView** command.
  - Another **Pan**, **View**, or **Zoom** command.
- The **DView** command has its own **Pan** option.
- The vertical and horizontal scroll bars are used for panning at any time.
- The meaning of Pan icons:

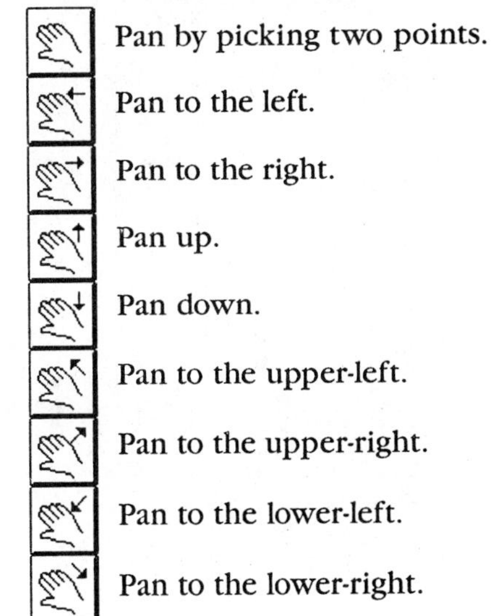

PasteClip

Rel. 13

Places object from Windows Clipboard in the drawing (*short for PASTE from CLIPboard*).

Command	Ctrl+	Sidebar menu	Menu Bar	Tablet
`pasteclip`	V	...	[Edit] [Paste]	...

```
Command: pasteclip
```

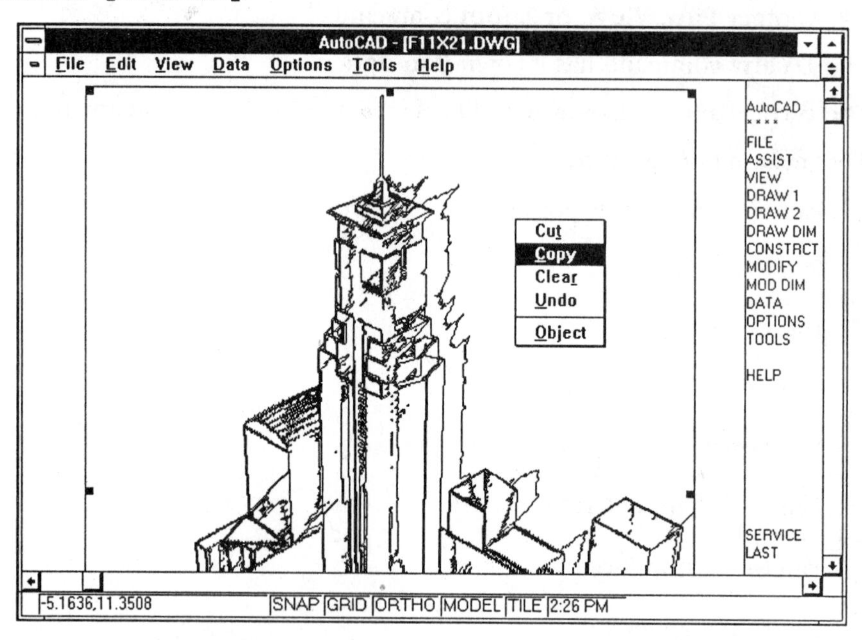

COMMAND OPTIONS

None

RELATED AUTOCAD COMMANDS

- **CopyClip** Copies drawing to the Windows Clipboard.
- **InsertObj** Inserts an OLE object in the drawing.
- **PasteSpec** Places Clipboard object as pasted or linked object.

TIPS

- The **PasteClip** command places all objects in the upper-left corner of the current viewport.
- Graphical objects are placed in the drawing as an OLE object.
- Text is placed in the drawing as an MText object.

- Once the object is placed, click on the object to edit it:
  - Drag to move the object.
  - Drag one of the eight rectangles to resize the object.
  - Right-click to display cursor menu:

Copy	Copies object to the Clipboard.
Cut	Cuts object to the Clipboard.
Clear	Erases the object.
Undo	Undoes the last editing command.
Object	Edits object back in source application.

- Use the **PasteSpec** command to convert the object to an AutoCAD block.

PasteSpec

Rel. 13

Places the Clipboard object in the drawing as a linked, pasted, or converted object (*short for PASTE SPECial*).

Command	Alt+	Side Menu	Menu Bar	Tablet
pastespec	E,S	...	[Edit] [Paste Special]	...

Command: **pastespec**

Displays dialogue box.

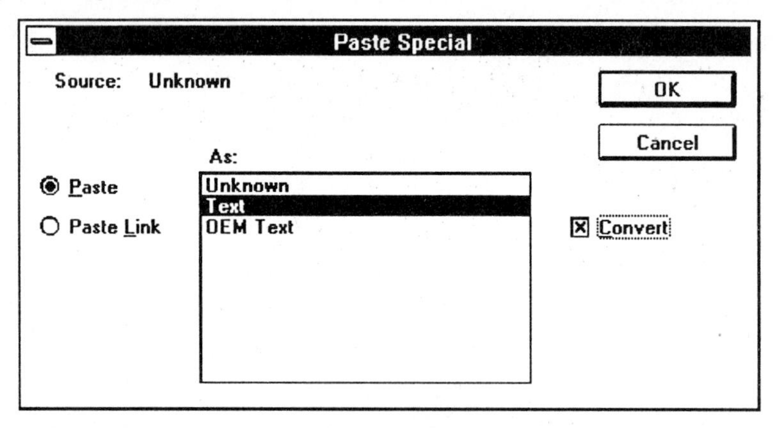

COMMAND OPTIONS

Paste Pastes object as embedded object.
Paste Link Pastes object as a linked object.
Convert Convert object into an AutoCAD block.

RELATED COMMANDS

- **CopyClip** Copies drawing to the Windows Clipboard.
- **InsertObj** Inserts an OLE object in the drawing.
- **OleLinks** Edits OLE link data.
- **PasteClip** Places Clipboard object as pasted object.

'Pcxin

Rel. 12

Imports PCX raster files into the drawing as a block *(an external file in Raster.Exp)*.

Command	Alt+	Side Menu	Menu Bar	Tablet
'pcxin	F,I	[FILE]	[File]	...
		[IMPORT]	[Import]	
		[PCXin:]	[PCX]	

```
Command: pcxin
PCX filename:
Insertion point <0,0,0>:
Scale factor:
```

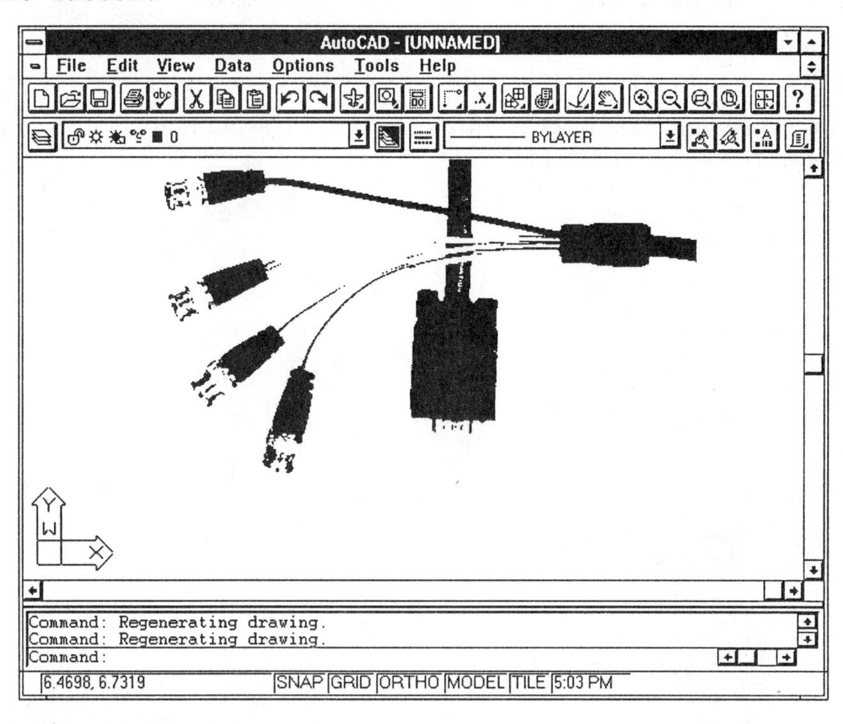

COMMAND OPTIONS

None

RELATED AUTOCAD COMMANDS

- **PsIn** Imports EPS files.
- **GifIn** Imports GIF raster files.
- **Import** Dialogue box shell for the **PcxIn** comand.
- **Replay** Displays GIF, TIFF, and Targa files.
- **TiffIn** Imports TIFF raster files.

RELATED SYSTEM VARIABLES

- **RiAspect** Adjust image's aspect ration.
- **RiBackG** Change the image's background color.
- **RiEdge** Outline edges.
- **RiGamut** Specify number of colors.
- **RiGrey** Import as a grey scale image.
- **RiThresh** Control brightness threshold.

TIPS

■ The PCX format was devised by Z-Soft for their PC Paintbrush graphics program.

■ **PcxIn** is limited to displaying a maximum of 256 colors.

■ Exploding an imported PCX block doubles the drawing file size, since each run of same-color pixels is defined as a solid.

■ Turn off system variable **GripBlock** (*set to 0*) to avoid highlighting all the solid objects making up the block.

PEdit

Ver. 2.1

Edits a 2D polyline, 3D polyline, or 3D mesh – depending on which object is picked (*short for Polyline EDIT*).

Command	Alias	Side Menu	Toolbar	Tablet
pedit	...	[MODIFY] [Pedit:]	[Modify] [Special Edit] [Pedit]	W 19

For 2D polylines:

```
Command: pedit
Select polyline: [pick a 2D polyline]
Close/Join/Width/Edit vertex/Fit/Spline/Decurve/Ltype gen
   /Undo/eXit <X>: e
Next/Previous/Break/Insert/Move/Regen/Straighten/Tangent
   /Width/eXit <N>: b
Next/Previous/Go/eXit <N>:
```

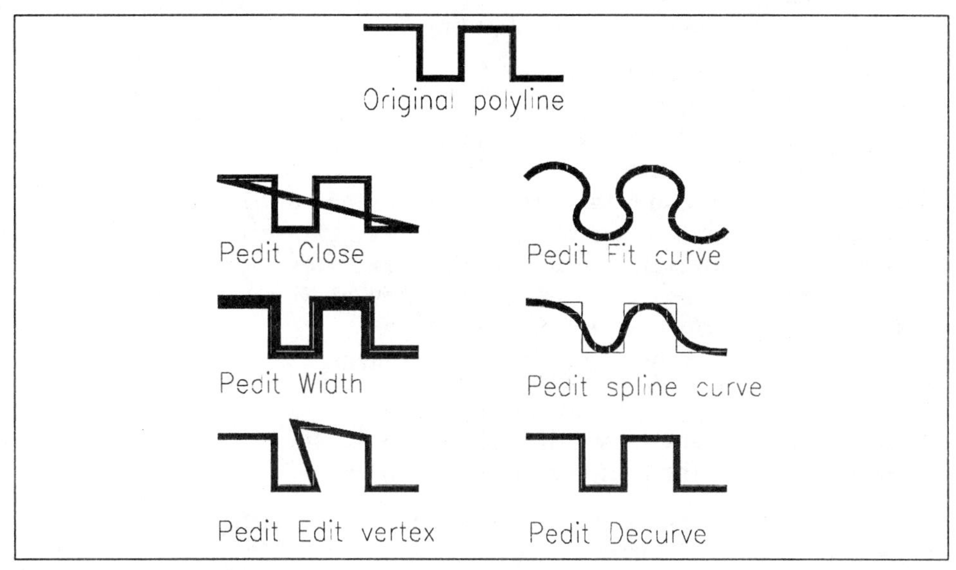

COMMAND OPTIONS

Close Closes an open polyline by joining the two endpoints with a single segment.

Decurve Reverses the effects of a Fit-curve or Spline-curve operation.

Edit vertex Edits individual vertices and segments (*see figure below*):

- **Break** Removes a segment or break the polyline at a vertex.
  - **<Next>** Moves the x-marker to the next vertex.
  - **Previous** Moves the x-marker to the previous vertex.

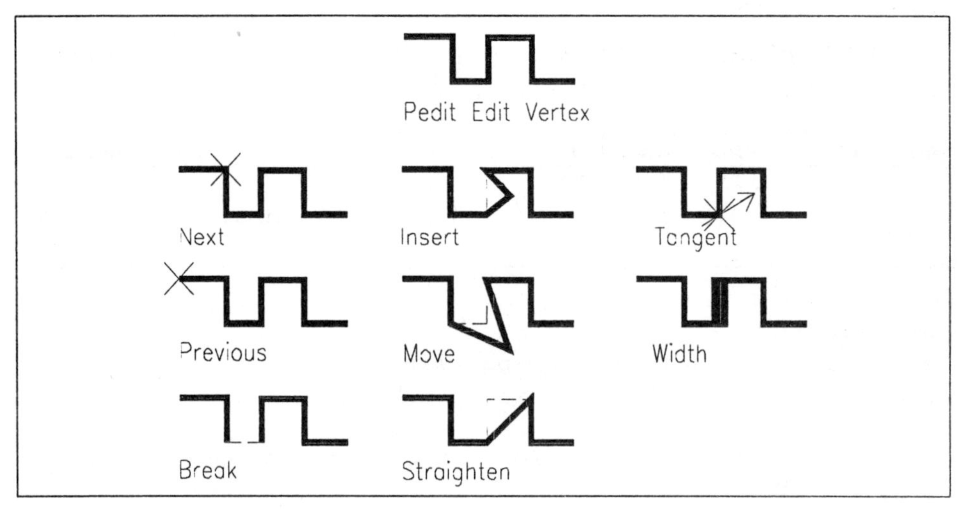

Go	Performs the break.
eXit	Exits the **Break** sub-submenu.

Insert	Inserts another vertex.
Move	Relocates a vertex.
Next	Moves the x-marker to the next vertex.
Previous	Moves the x-marker to the previous vertex.
Regen	Regenerates the screen to show effect of **PEdit** commands.
Straighten	Draws a straight segment between two vertices:
<Next>	Moves the x-marker to the next vertex.
Previous	Moves the x-marker to the previous vertex.
Go	Performs the straightening.
eXit	Exits the **Straighten** sub-submenu.
Tangent	Shows tangent to current vertex.
Width	Changes the width of a segment.
<eXit>	Exits the **Edit-vertex** submenu.

Fit	Fits a curve to the tangent points of each vertex.
Ltype gen	Specifies linetype generation style.
Join	Adds other polylines to the current polyline.
Open	Opens a closed polyline by removing the last segment.
Spline	Fits a splined curve along the polyline.
Undo	Undoes the most-recent **PEdit** operation.
Width	Changes the width of the entire polyline.
<eXit>	Exits the **PEdit** command.

For 3D polylines:

```
Command: pedit
Select polyline: [pick a 3D polyline]
Close/Edit vertex/Spline curve/Decurve/Undo/eXit <X>: E
Next/Previous/Break/Insert/Move/Regen/Straighten/eXit <N>:
```

COMMAND OPTIONS

Close Closes an open polyline.

Decurve Reverses the effects of a **Fit-curve** or **Spline-curve** operation.

Edit vertex Edits individual vertices and segments:

- **Break** Removes a segment or break the polyline at a vertex.
  - **<Next>** Moves the x-marker to the next vertex.
  - **Previous** Moves the x-marker to the previous vertex.
  - **Go** Performs the break.
  - **eXit** Exits the **Break** sub-submenu.
- **Insert** Inserts another vertex.
- **Move** Relocates a vertex.
- **<Next>** Moves the x-marker to the next vertex.
- **Previous** Draws a straight segment between two vertices:
  - **<Next>** Moves the x-marker to the next vertex.
  - **Previous** Moves the x-marker to the previous vertex.
  - **Go** Performs the straightening.
  - **eXit** Exits the **Straighten** sub-submenu.
- **eXit** Exits the **Edit-vertex** submenu.

Open Removes the last segment of a closed polyline.

Spline curve Fits a splined curve along the polyline.

Undo Undoes the most-recent **PEdit** operation.

<eXit> Exits the **PEdit** command.

For 3D meshes:

```
Command: pedit
Select polyline: [pick a 3D mesh]
Edit vertex/Smooth surface/Desmooth/Mclose/Nclose/Undo
   /eXit<X>: E
Vertex (0,0). Next/Previous/Left/Right/Up/Down/Move/REgen
   /eXit <N>:
```

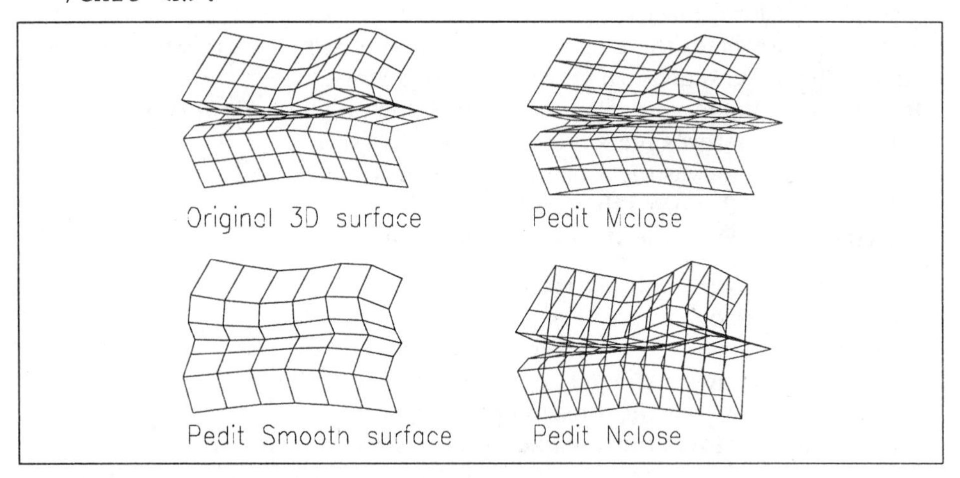

COMMAND OPTIONS

Desmooth Reverses the effect of the Smooth surface options.

Edit vertex Edits individual vertices with the following submenu:

- **Down** Moves x-marker down the mesh by one vertex.
- **Left** Moves x-marker along the mesh by one vertex left.
- **Move** Relocates the vertex to a new position.
- **<Next>** Moves x-marker along the mesh to the next vertex.
- **Previous** Moves x-marker along the mesh to the previous vertex.
- **REgen** Regenerates the drawing to show the effects of **PEdit**.
- **Right** Moves x-marker along the mesh by one vertex right.
- **Up** Moves x-marker up the mesh by one vertex.
- **eXit** Exits the **Edit-vertex** submenu.

Mclose Closes the mesh in the m-direction.

Mopen Opens the mesh in the m-direction.

Nclose Closes the mesh in the n-direction.

Nopen Opens the mesh in the n-direction.

Smooth surface Smoothes the mesh with a B-spline.

Undo Undoes the most recent **PEdit** operation.

<eXit> Exits the **PEdit** command.

RELATED AUTOCAD COMMANDS

- **Break** Breaks a 2D polyline at any position.
- **Chamfer** Chamfers all vertices of a 2D polyline.
- **EdgeSurf** Draws 3D mesh.
- **Fillet** Fillets all vertices of a 2D polyline.
- **PLine** Draws a 2D polyline.
- **RevSurf** Draws a 3D surface of revolution mesh.
- **RuleSurf** Draws a 3D ruled surface mesh.
- **TabSurf** Draws a 3D tabulated surface mesh.
- **3D** Draws 3D surface objects
- **3dPoly** Draws a 3D polyline.

RELATED SYSTEM VARIABLES

- **Splframe** Determines visibility of a polyline spline frame.
- **SplineSegs** Number of lines used to draw a splined polyline.
- **SplineType** Determines B-spline smoothing for 2D and 3D polylines.
- **SurfType** Determines the smoothing using the **Smooth-surface** option.

TIP

■ During vertex editing, button #2 (*left button on a two-button mouse*) moves the x-marker to the next vertex.

PFace

Rel. 11

Draws multi-sided 3D meshes; meant for use by AutoLISP, ADS, and ARx programs (*short for Poly FACE*).

Command	Alias	Side Menu	Menu Bar	Tablet
pface	...	[DRAW 2] [SURFACES] [Pface:]	...	...

```
Command: pface
Vertex 1: [pick]
Vertex 2: [pick]
Face 1, vertex 1: 1
Face 1, vertex 2: 2
Face 2, vertex 1: 1
Face 2, vertex 2: 2
```

COMMAND OPTIONS

None

RELATED AUTOCAD COMMAND

- **3dFace** Draws three- and four-sided 3D meshes.

RELATED SYSTEM VARIABLE

- **PFaceVmax** Maximum number of vertices per polyface.

TIP

- Maximum number of vertices in the m- and n-direction:
  - 256 vertices, when entered from the keyboard.
  - 32,767 vertices, when entered from a DXF file or created by programming.

Plan

Rel. 10

Displays the plan view of the WCS or the UCS.

Command	Alt+	Side Menu	Menu Bar	Tablet
plan	V,I,P	[VIEW]	[View]	M5
		[Plan:]	[3D Viewpoint Presets]	
			[Plan View]	

```
Command: plan
<Current UCS>/Ucs/World: W
Regenerating drawing.
```

Example 3D view:

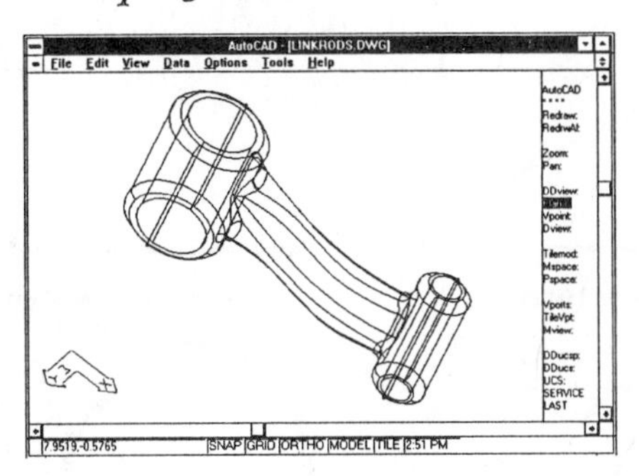

*After using **Plan World** command:*

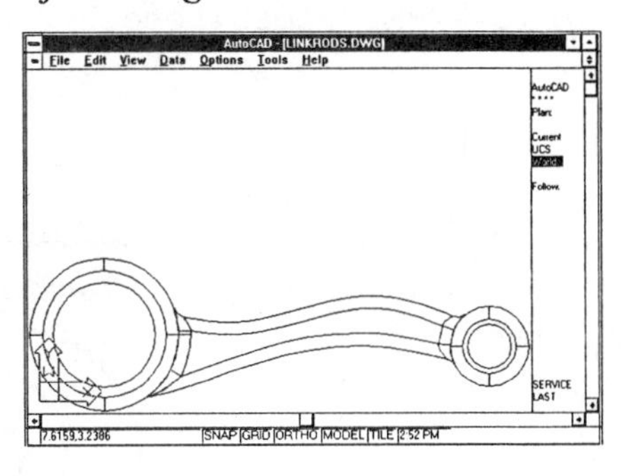

COMMAND OPTIONS

<Current UCS> Shows the plan view of the current UCS.
Ucs Shows the plan view of a named UCS.
World Shows the plan view of the WCS.

RELATED AUTOCAD COMMANDS

- **UCS** Creates new UCS views.
- **VPoint** Changes the viewpoint of 3D drawings.

RELATED SYSTEM VARIABLE

- **UcsFollow** Automatic plan view display for UCS or WCS.

TIPS

- Typing **VPoint 0,0,0** is an alternative command to the **Plan** command.
- The **Plan** command turns off perspective mode and clipping planes.
- **Plan** does not work in paper space.
- The **Plan** command is an excellent method for turning off perspective mode.

PLine

Ver. 1.4

Draws a complex 2D line made of straight and curved sections of constant and variable width; treated as a single object (*short for Poly LINE*).

Command	Alias	Side Menu	Toolbar	Tablet
pline	pl	[DRAW 1]	[Draw]	K 10
		[Pline:]	[Polyline]	

```
Command: pline
From point: [pick]
Current line-width is 0.0
Arc/Close/Halfwidth/Length/Undo/Width/<Endpoint of line>:
```

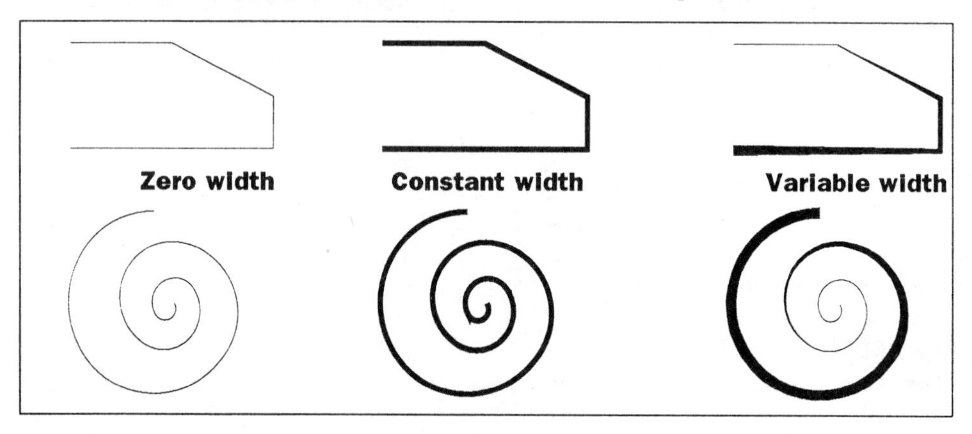

COMMAND OPTIONS

Arc Displays the submenu for drawing arcs:

- **Angle** Indicates the included angle of the arc.
- **CEnter** Indicates the arc's center point.
- **CLose** Uses an arc to close a polyline.
- **Direction** Indicates the arc's starting direction.
- **Halfwidth** Indicates the halfwidth of the arc.
- **Line** Switches back to the menu for drawing lines.
- **Radius** Indicates the arc's radius.
- **Second pt** Draws a three-point arc.
- **Undo** Erases the last drawn arc segment.
- **Width** Indicates the width of the arc.
- **<Endpoint of arc>** Indicates the arc's endpoint.

Close Closes the polyline with a line segment.

Halfwidth Indicates the halfwidth of the polyline.

Length Draws a polyline tangent to the last segment.
Undo Erases the last-drawn segment.
Width Indicates the width of the polyline.
<Endpoint of line>:
Indicates the polyline's endpoint.

RELATED AUTOCAD COMMANDS

- **Boundary** Draws a polyline boundary.
- **Donut** Draws solid-filled circles as polyline arcs.
- **Ellipse** Draws ellipses as polyline arcs when **PEllipse** = 1.
- **Explode** Reduces a polyline to lines and arcs with zero width.
- **Fillet** Fillets polyline vertices with a radius.
- **PEdit** Edits the polyline's vertices, widths, and smoothness.
- **Polygon** Draws polygons as polylines of up to 1,024 sides.
- **Rectang** Draws a rectangle out of a polyline.
- **Sketch** Draws polyline sketches, when **SkPoly** = 1.
- **Xplode** Explodes a group of polylines into line and arcs of zero width.
- **3dPoly** Draws 3D polylines.

RELATED SYSTEM VARIABLES

- **PlineGen** Style of linetype generation:
  0 Vertex to vertex (*default*).
  1 End to end.
- **PlineWid** Current width of polyline.

TIPS

- Use the **Boundary** command to automatically outline a region; then use the **List** command to find its area.

- If you cannot see a linetype on a polyline, change system variable **PlineGen** to 1; this regenerates the linetype from one end of the polyline to the other.

- If the angle between a joined polyline and polyarc is less than 28 degrees, the transition is chamfered; at greater than 28 degrees, the transition is not chamfered.

- Use the object snap mode **INTersection** to snap to the vertices of a polyline.

Plot

Ver. 1.0

Creates a copy of the drawing on a vector, raster, or PostScript plotter or printer via the serial or parallel ports; or plots to file on disk.

Command	Ctrl+	Side Menu	Menu Bar	Tablet
plot	P	[FILE] [Print:]	[File] [Print]	W 24

Command: **plot**

Displays dialogue box.

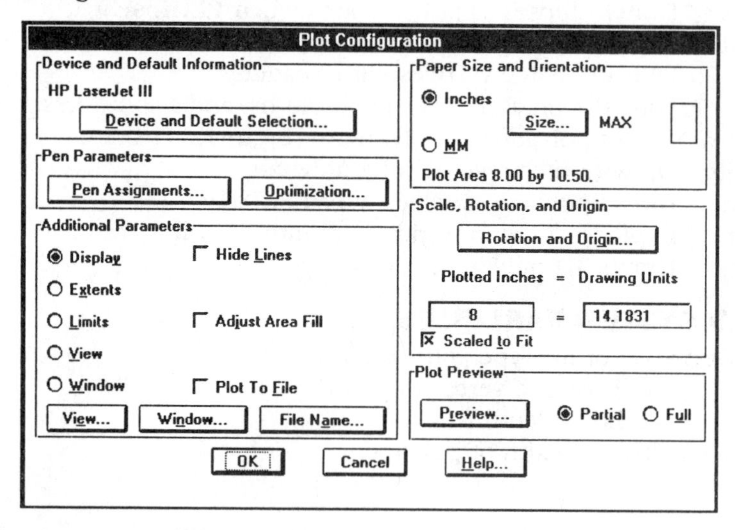

COMMAND OPTIONS

Device and default selection

Selects and configures output devices; displays a dialogue box:

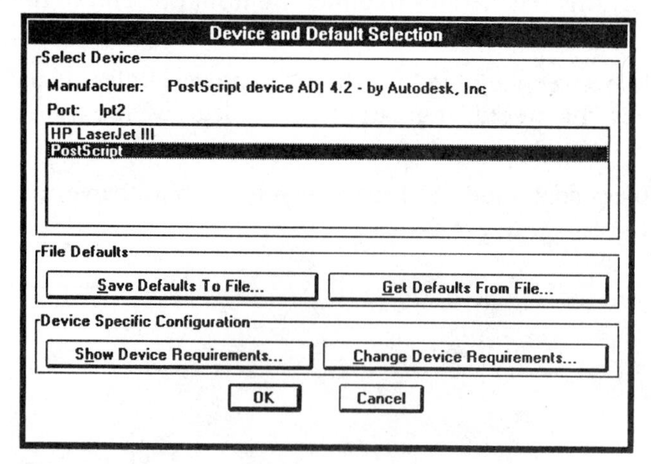

Feature legend

Data varies, depending on device capabilities.

Pen assignments

Assign pen numbers; displays dialogue box.

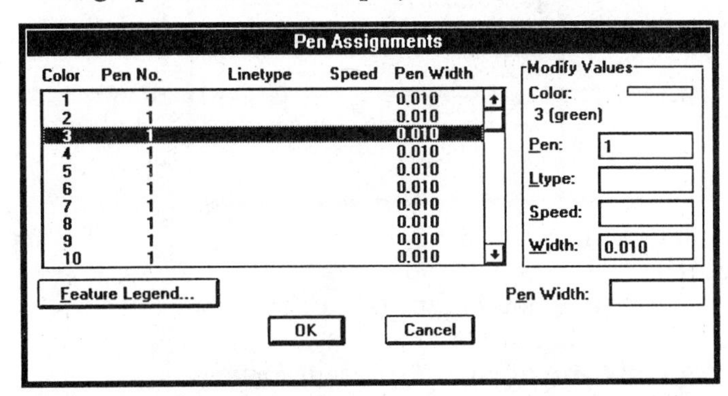

Optimization

Select pen motion optimization; displays dialogue box.

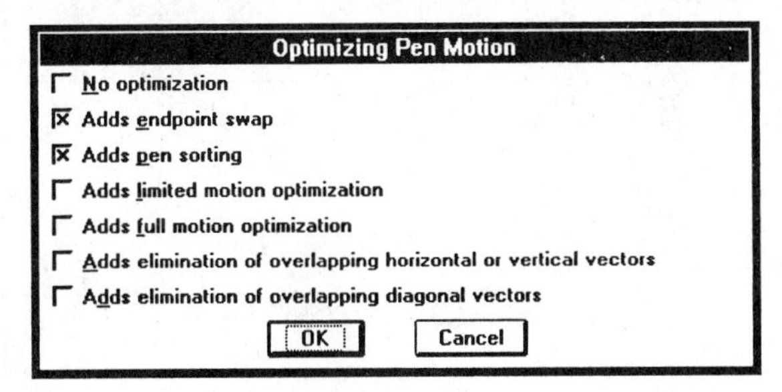

Display Plot current display.

Extents Plot drawing extents.

Limits Plot drawing limits.

View Plot named view; displays dialogue box if drawing contains saved views.

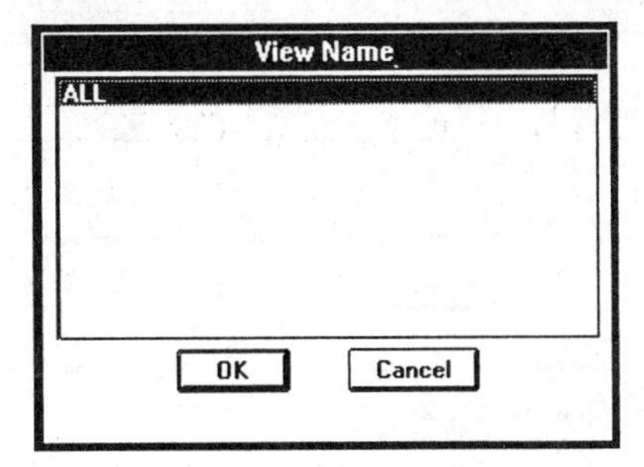

Window Plot windowed area; displays dialogue box.

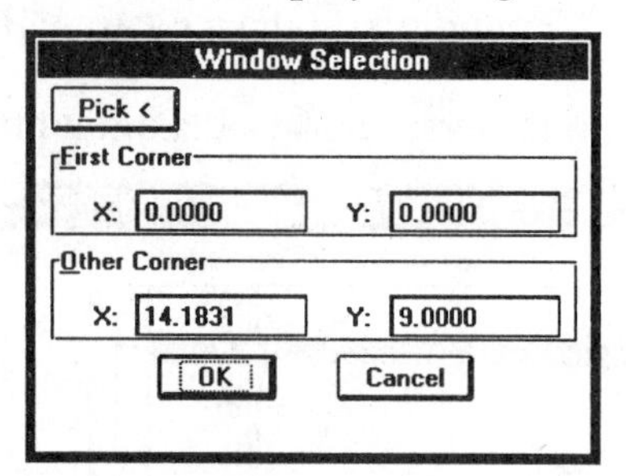

Hide lines Remove hidden lines.

Adjust area fill
Adjust pen motion for filled areas.

Plot to File Plot drawing to file.

Size Specify size of plot; displays dialogue box.

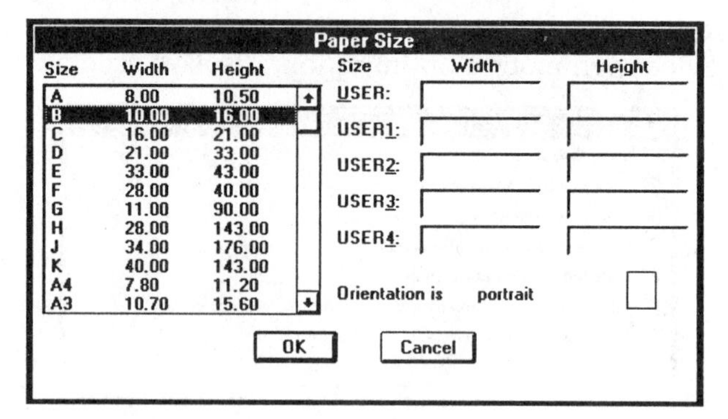

Inches Plot in inches.

Mm Plot in millimeters.

Rotation and origin
Specify plot origin and rotation; displays dialogue box.

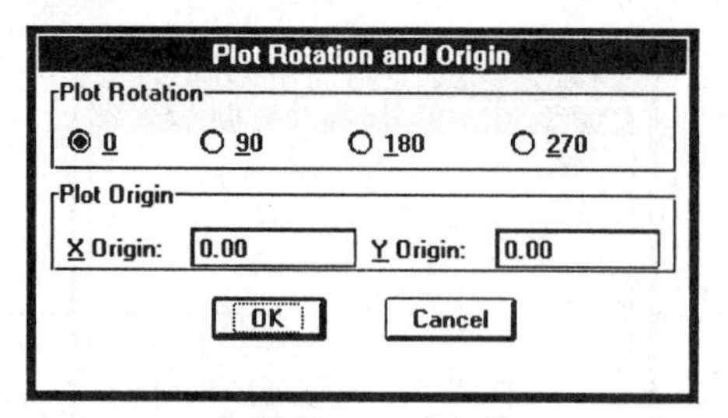

Scaled to fit Scale plot to fit paper size.

Preview Preview the plot.

Partial Quick plot preview; displays dialogue box.

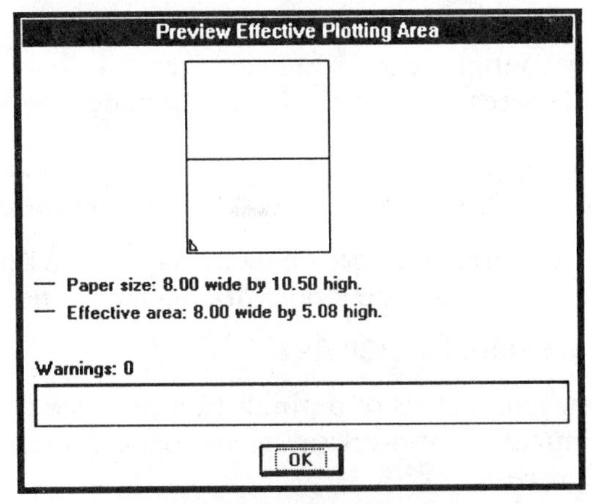

Full Full plot preview; displays viewport.

RELATED AUTOCAD COMMANDS

- **Config** Selects one or more plotter devices.
- **HpMPlot** Plots combines raster and vector to HPGL/2 devices.
- **PsOut** Saves a drawing in EPS format.

RELATED SYSTEM VARIABLES

- **CmdDia** Determines the **Plot** command's interface:
  - **0** Command-line interface; compatible with script files.
  - **1** Dialogue box interface.
- **PlotId** Currently selected plotter number.
- **Plotter** Currently selected plotter name.

RELATED DOS VARIABLE

- AcadPlCmd Plot spooler support.

RELATED FILES

- *.PCP Plotter configuration parameter files.
- *.PLT Plot files created with the **Plot** command.

TIPS

- As of Release 12, the **Plot** command replaces the **PrPlot** command.

- As of Release 13, the 'freeplot' feature (starting AutoCAD with the **-p** parameter to plot without using up a network license) is no longer available.

- Plot parameters are stored in PCP files.

- Don't assume that more levels of optimization produce faster plots. In particularly, the elimination of overlapping vectors can dramatically slow down the plotting process.

Point

Ver. 1.0

Draws a 3D point.

Command	Alias	Side Menu	Toolbar	Tablet
`point`	...	[DRAW 2] [Point:]	[Draw] [Point]	O 10

```
Command: point
Point: [pick]
```

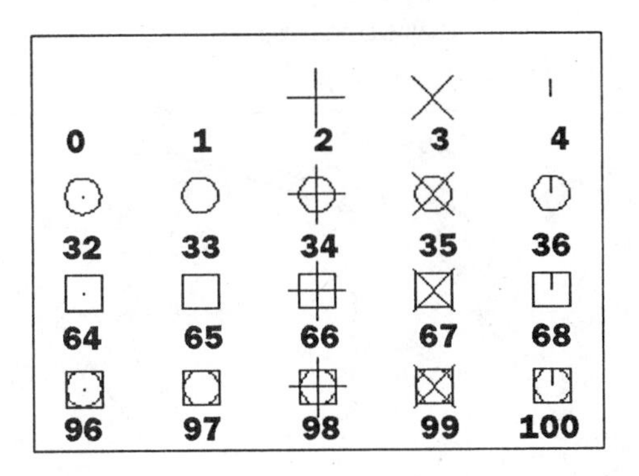

COMMAND OPTIONS

None

RELATED AUTOCAD COMMANDS

■ **DdPType** Dialogue box for selecting **PsMode** and **PdSize**.

RELATED SYSTEM VARIABLES

■ **PdMode** Determines the look of a point (*see figure*).

■ **PdSize** Determines the size of a point:

- **0** Point is 5% of height of ScreenSize system variable.
- **1** No display.
- **-10** Ten percent of viewport size.
- **10** Ten pixels in size.

TIPS

■ The size and shape of the point is determined by **PdSize** and **PdMode**; changing these values changes the look and size of all points in the drawing with the next regeneration.

■ Entering only x,y-coordinate places the point at a z-coordinate of the current elevation; setting **Thickness** to a value draws the point as a line in 3D space.

■ Prefix the coordinate with * (*asterisk*) to place a point in the WCS, rather than the current UCS.

■ Use the object snap mode **NODe** to snap to a point.

Polygon

Ver. 2.5

Draws a 2D polygon of between three to 1,024 sides.

Command	Alias	Side Menu	Toolbar	Tablet
polygon	...	[DRAW 1] [Polygon:]	[Draw] [Polygon]	N 9

```
Command: polygon
Number of sides <4>:
Edge/<Center of polygon>: [pick]
Inscribed in circle/Circumscribed about circle (I/C): I
Radius of circle: [pick]
```

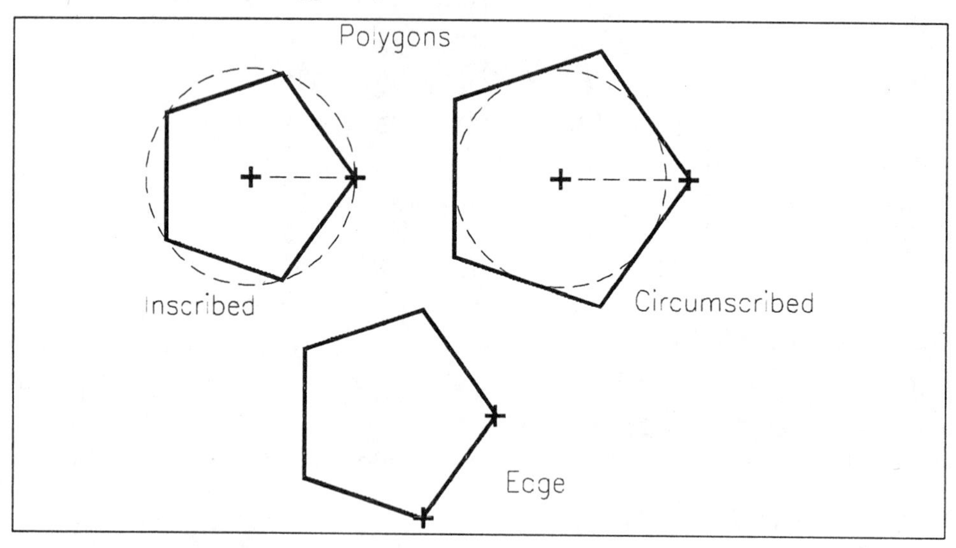

COMMAND OPTIONS

<Center of polygon>

Indicates the center point of the polygon; then:

- **C** — Fits the polygon outside of a circle.
- **I** — Fits the polygon inside a circle.

Edge — Draws the polygon based on the length of one edge.

RELATED AUTOCAD COMMANDS

- **Donut** Draws solid-filled circles with a polyline.
- **Ellipse** Draws ellipsis with a polyline, when **PEllipse** = 1.
- **PEdit** Edits polylines, include polygons.
- **PLine** Draws polylines and polyline arcs.
- **Rectang** Draws a rectangle from a polyline.

RELATED SYSTEM VARIABLE

■ **PolySides** Most-recently specified number of sides; default is 4.

TIPS

■ Polygons are drawn from polylines; use the **PEdit** command to change the polygon, such as the width of the polyline.

■ The pick point determines the location of polygon's first vertex; polygons are drawn counter-clockwise.

■ Use the system variable **PolySides** to preset the default number of polygon sides.

■ Use the **Snap** command to precisely place the polygon.

■ Use object snap mode **INTersection** to snap to the polygon's vertices.

Preferences

Lets you set a couple of user preferences.

Command	Alt+	Side Menu	Menu Bar	Tablet
preferences	O,P	...	[Options] [Preferences]	...

Command: **preferences**

Displays tabbed dialogue boxes.

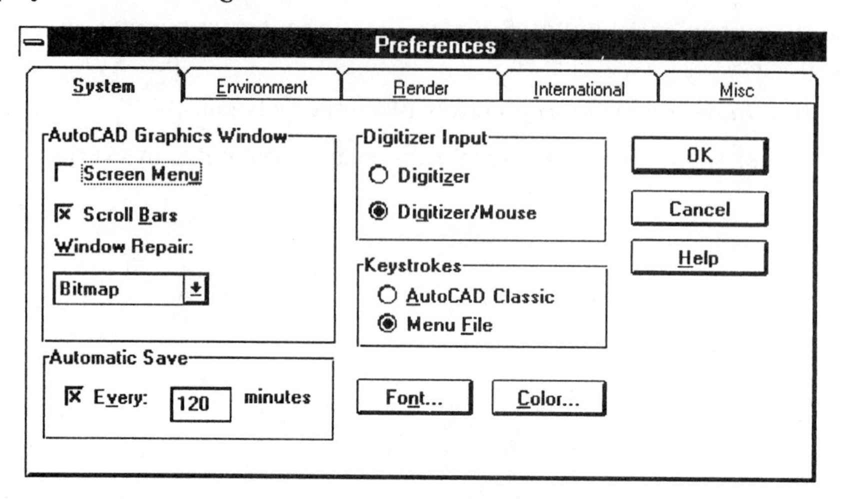

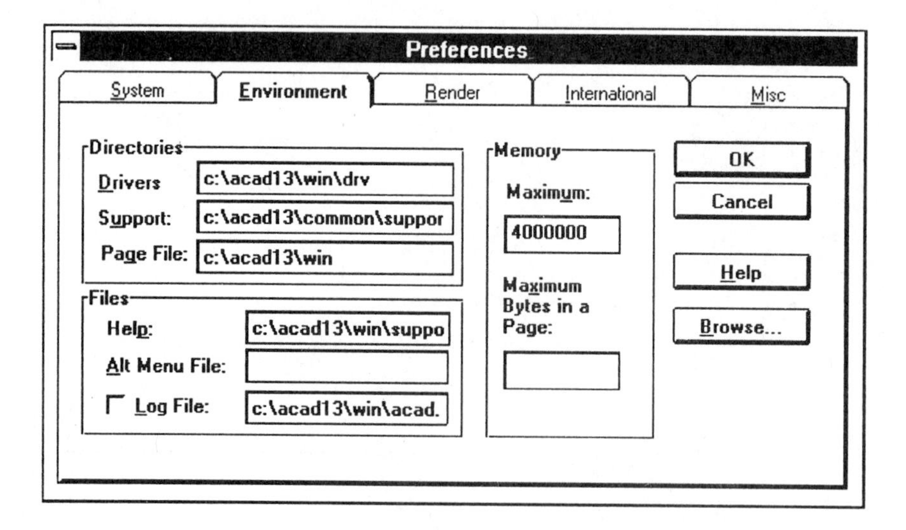

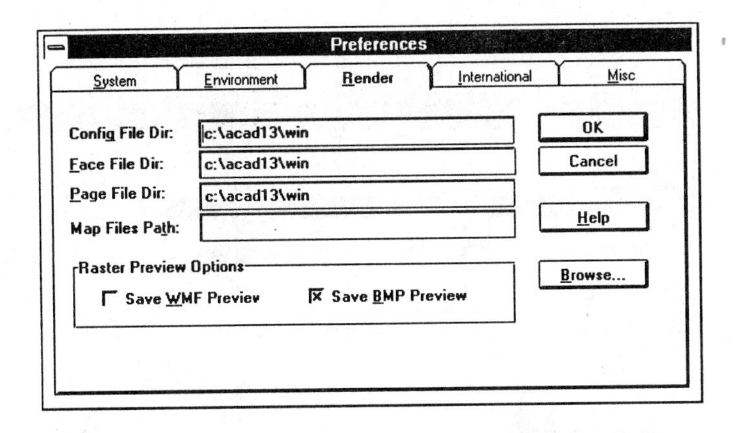
Preferences
System
Environment
Render
International
Misc
Config File Dir:
c:\acad13\win
Face File Dir:
c:\acad13\win
Page File Dir:
c:\acad13\win
Map Files Path:
Raster Preview Options
Save WMF Preview
Save BMP Preview
OK
Cancel
Help
Browse...

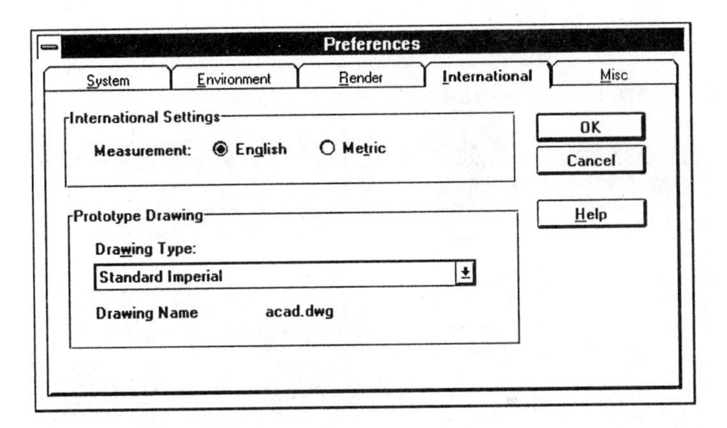
Preferences
System
Environment
Render
International
Misc
International Settings
Measurement:
English
Metric
Prototype Drawing
Drawing Type:
Standard Imperial
Drawing Name
acad.dwg
OK
Cancel
Help

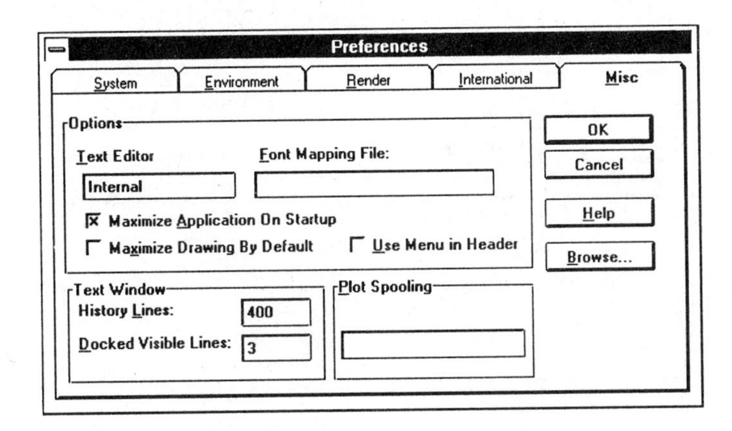
Preferences
System
Environment
Render
International
Misc
Options
Text Editor
Internal
Font Mapping File:
Maximize Application On Startup
Maximize Drawing By Default
Use Menu in Header
Text Window
History Lines:
400
Docked Visible Lines:
3
Plot Spooling
OK
Cancel
Help
Browse...

COMMAND OPTIONS

Font Selects fonts for AutoCAD menu and prompt areas:

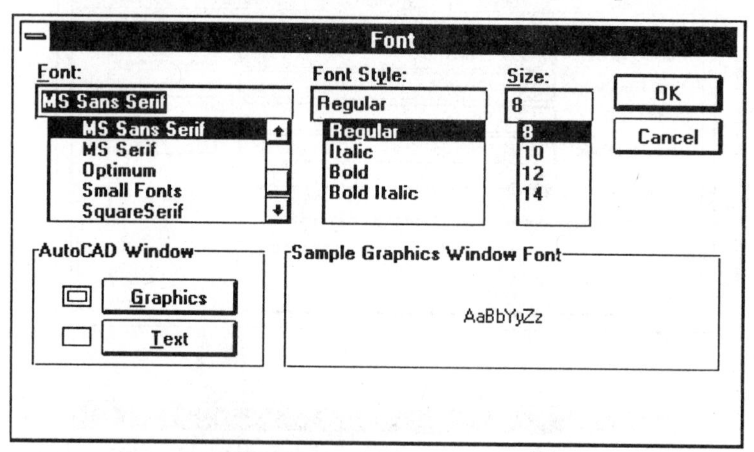

Color Specifies colors of the AutoCAD window.

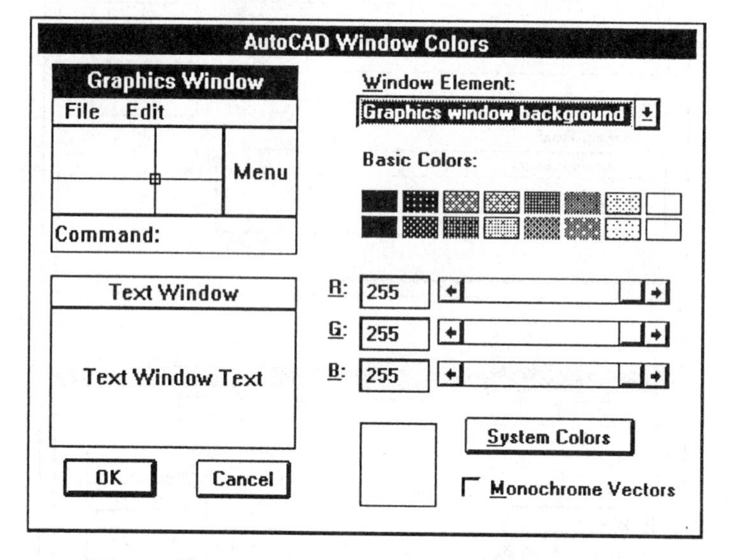

Browse Searches for filename.

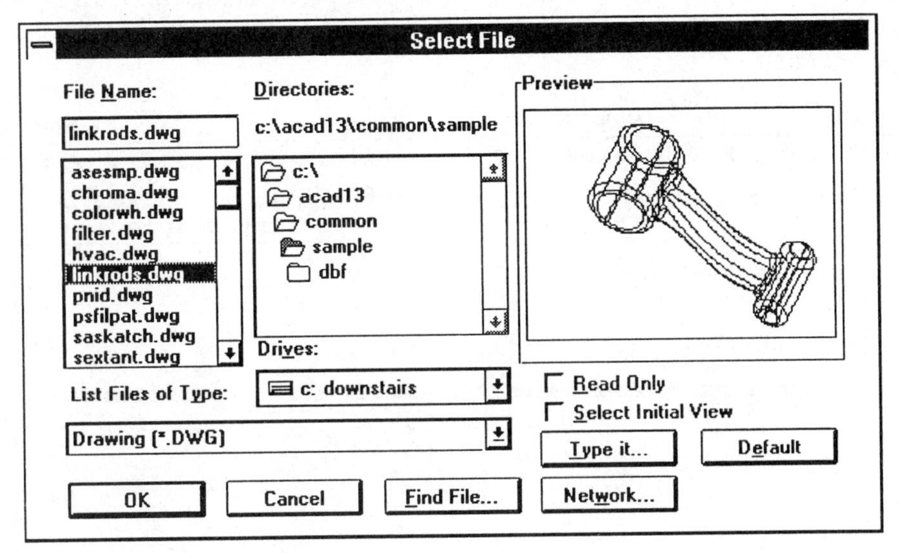

RELATED FILES

- ***.DWG** Prototype drawing files in \Acad13\Common\Support subdirectory.
- **Acad.Ini** Stores settings from **Preferences** command in \Acad13\Dos subdirectory.

PsDrag

Rel. 12

Controls the appearance of the PostScript image during the PsIn command (*short for PostScript DRAG; an external command in AcadPs.Exp*).

Command	Alt+	Side Menu	Menu Bar	Tablet
psdrag	F,T,D	[FILE]	[File]	...
		[IMPORT]	[Options]	
		[PsDrag:]	[PostScript Display]	

```
Command: psdrag
PSIN drag mode <0>:
```

***PsDrag** set to 0:*

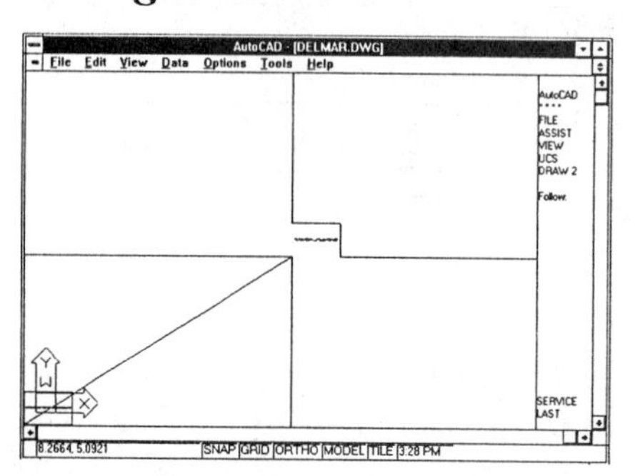

***PsDrag** set to 1:*

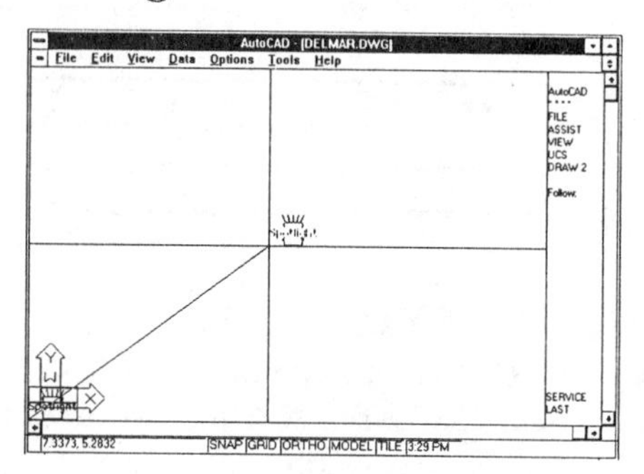

COMMAND OPTIONS

0 Turns off drag.

1 Turns on drag.

RELATED AUTOCAD COMMAND

■ **PsIn** Imports a PostScript file.

RELATED SYSTEM VARIABLE

■ **PsQuality** Display options for PostScript files:

- **75** Display filled at 75dpi (*default*)
- **0** Display bounding box and filename; no image.
- **-75** Display image outline at 75dpi; no fill.

TIP

■ The **PsDrag** command is set to zero if the system variable **PsQuality** is set to zero.

PsFill

Fills a 2D polyline outline with a raster PostScript pattern (*short for PostScript FILL; an external command in AcadPs.Exp*).

Command	Alias	Side Menu	Toolbar	Tablet
psfill	...	...	[Draw] [Hatch] [PsFill]	...

```
Command: psfill
Select polyline: [pick]
PostScript pattern (. = none) <.>/?
```

COMMAND OPTIONS

.	Selects no fill pattern.
?	Lists available fill patterns.
*	Does not outline pattern with polyline.

RELATED AUTOCAD COMMAND

■ **BHatch** Fills an area with a vector harch pattern.

RELATED SYSTEM VARIABLE

■ **PsQuality** Display options for PostScript files:

75 Display filled at 75dpi (*default*)

0 Display bounding box and filename; no image.

-75 Display image outline at 75dpi; no fill.

TIP

■ The following PostScript fill patterns are defined in file Acad.Psf (*pattern name is followed by parameters and default values*):

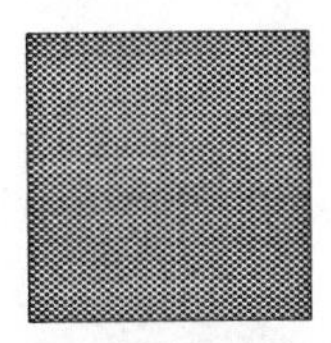

Grayscale Grayscale = 50

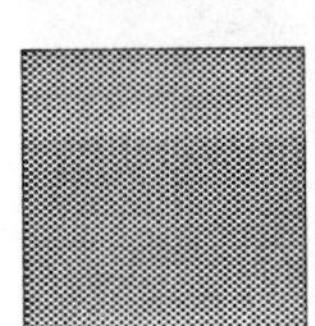

RGBcolor Red = 50
Green = 50
Blue = 50

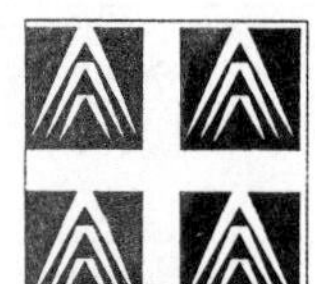

AIlogo

Frequency = 1.0
Separation = 25
Linewidth = 0
ForegroundGray = 100
BackgroundGray = 0

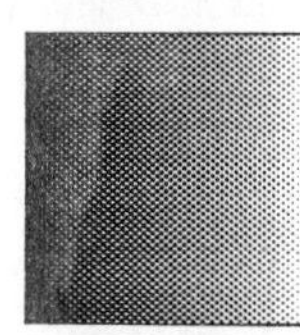

Lineargray

Levels = 256
Cycles = 1
Angle = 0.0
ForegroundGray = 100
BackgroundGray = 0

Radialgray

Levels = 256
ForegroundGray = 100
BackgroundGray = 0

Square

Scale = 1.0
Separation = 25
LineWidth = 1
ForegroundGray = 100
BackgroundGray = 0

Waffle

Scale = 1.0
Proportion = 30
LineWidth = 1
UpLeftGray = 100
BotRight Gray = 50
TopGray = 0

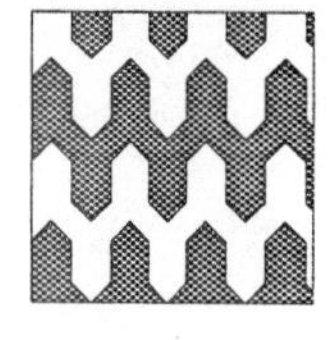

ZigZag

Scale = 1.0
LineWidth = 1
ForegroundGray = 100
BackgroundGray = 0

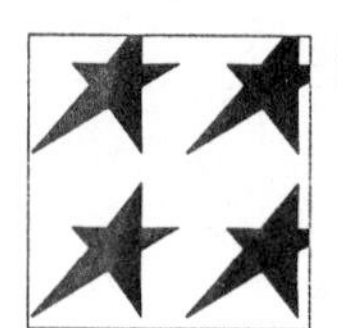

Stars

Scale = 1.0
LineWidth = 1
ForegroundGray = 100
BackgroundGray = 0

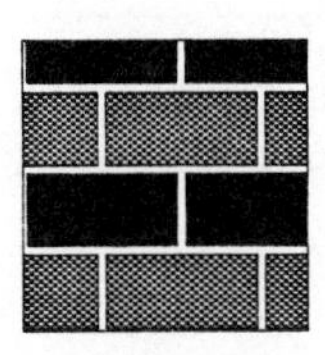

Brick

Scale = 1.0
LineWidth = 1
BrickGray1 = 100
BrickGray2 = 50
BackGroundGray = 0

Specks

Scale = 1.0
ForegroundGray = 100
BackgroundGray = 0

Psin

Rel. 12

Imports an EPS (encapsulated PostScript) file into the drawing (*short for PostScript INput; an external command in AcadPs.Exp*).

Command	Alias	Side Menu	Menu Bar	Tablet
psin	...	[FILE]	[File]	...
		[IMPORT]	[Import]	
			[PostScript]	
			[Import]	

Command: **psin**

Select filename from dialogue box.

Insertion point <0,0,0>: **[pick]**

Scale factor:

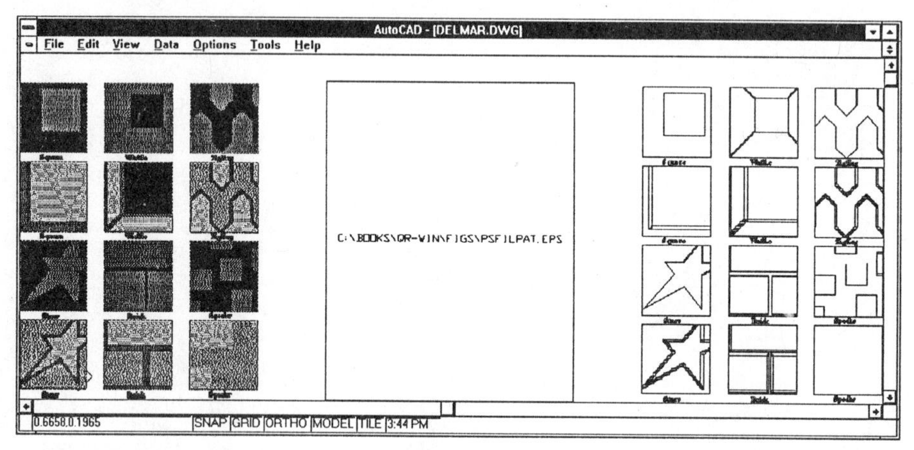

PsQuality = 75 **PsQuality = 0** **PsQuality = -75**

COMMAND OPTIONS

None

RELATED AUTOCAD COMMANDS

- **PsDrag** Toggles display of bounding box during placement.
- **GifIn** Imports a GIF raster file.
- **Import** Dialogue-box shell for the **PsIn** command.
- **PcxIn** Imports a PCX raster file.
- **PsOut** Exports an EPS file.
- **TiffIn** Imports a TIFF raster file.

RELATED SYSTEM VARIABLE

- **PsQuality** Display options for placing an EPS file (*see figure*):
  - **75** Display filled at 75dpi (*default*)
  - **0** Display bounding box and filename; no image.
  - **-75** Display image outline at 75dpi; no fill.

RELATED FILES

Found in \Acad13\Common\Support subdirectory:

- **Acad.Psf** PostScript font substitution mapping and fill definition.
- **AcadPsc.Ps** *An empty file.*
- **AcadPsd.Ps** Provides dummy 'statusdict', 'serverdict', and other LaserWriter operators.
- **AcadPsf.Ps** Font initialization for GhostScript.
- **AcadPsi.Ps** Ghostscript master initialisation file.
- **AcadPss.Ps** GhostScript symbol font encoding vector.

TIPS

- AutoCAD uses an in-house modified version of GhostScript, a freeware PostScript clone.
- The 'acadpsversion' operator returns local version number of AcadPs.Exp of '2.2-ACADPS:Q001-JW1'.
- The Acad.Psf file defines all fonts included with:
  - Adobe Type Manager for Windows.
  - Adobe Plus Pack.
  - Adobe Font Pack 1.
  - Linguist PostScript fonts distributed with AutoCAD.
- **PsIn** places the PostScript as an anonymous block '*U' in the drawing.
- When the EPS block is first placed in the drawing, it is of unit size.

PsOut

Rel. 12

Exports the current drawing as an encapsulated PostScript file (*an external command in Acadps.Exp*).

Command	Alt+	Side Menu	Menu Bar	Tablet
psout	F, E	[FILE]	[File]	W 25
		[Export]	[Export]	
		[PSout:]	[EPS]	

Command: **psout**
Specify filename in dialogue box.
What to export--Display, Extents, Limits, View or Windows <D>:
Include a screen preview image in the file?(None/EPSI/TIFF)<None>:
Screen preview image size (128x128 is standard)? (128/256/512)<128>:
Enter the Size or Width, Height (in Inches) <>:
Effective plotting area: *ww* by *hh* high

COMMAND OPTIONS

None

RELATED AUTOCAD COMMANDS

- **Export** Dialogue-box shell for the **PsOut** command.
- **Plot** Exports dawing in a variety of formats, including raster EPS.
- **PsIn** Imports EPS files.

RELATED SYSTEM VARIABLE

- **PsProlog** Specify the PostScript prologue information.

RELATED FILES

- ***.EPS** Extension of file produced by PsOut.
- **Acad.Psf** PostScript fonts substitution map found in \Acad13\Common\Support.

TIPS

- The 'screen preview image' is only used for screen display purposes since graphics software generally cannot display PostScript graphic files.

- Although Autodesk recommends using the smallest screen preview image size (128x128), even the largest preview image (512x512) has a minimal effect on file size and screen display time,

- The screen preview image size has no effect on the quality of the PostScript output.

- If you're not sure which screen preview format to use, select TIFF.

PSpace

Rel. 11

Switches from model space to paper space (*short for Paper SPACE*).

Command	Alias	Side Menu	Menu Bar	Tablet
pspace	ps	[VIEW] [Pspace:]	[View] [Paper Space]	V 13

Command: **pspace**

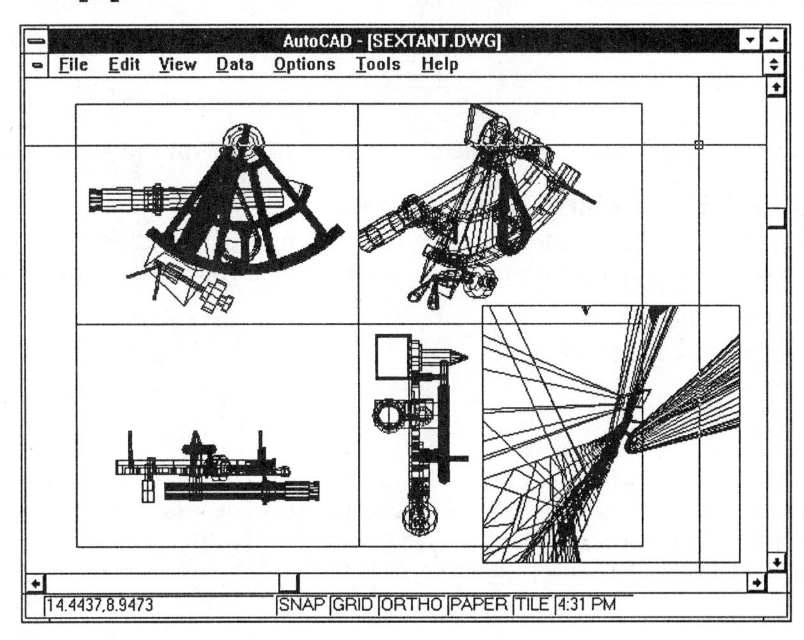

COMMAND OPTIONS

None

RELATED AUTOCAD COMMANDS

- **MSpace** Switches from paper space to model space.
- **MView** Creates viewports in paper space.
- **MvSetup** Creates paper space setup for a new drawing.
- **UcsIcon** Toggles display of paper space icon.
- **Zoom** The **XP** option scales paper space relative to model space.

RELATED SYSTEM VARIABLES

- **MaxActVp** Maximum number of viewports displaying an image.
- **TileMode** Must equal 0 for paper space to work.
- **PsLtScale** Linetype scale relative to paper space.

TIPS

■ Use paper space to layout mutiple views of a single drawing.

■ Paper space is known as 'drawing composition' in other CAD packages.

■ When a drawing is in paper space, AutoCAD displays '**P**' on the status line and the paper space icon:

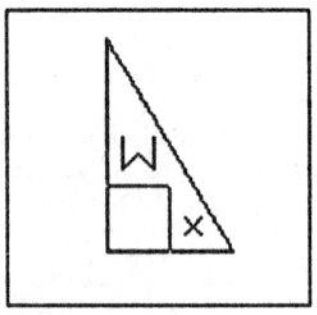

■ Switch to paper space by double-clicking on MODEL on the status bar.

QUICK START: Enabling paper space.

Entering paper space for the first time can be a mystifying experience, since your drawing literally disappears. Here are the steps you need to take:

1. Turn Tilemode off:

```
Command: tilemode 0
```

2. Enter paper space:

```
Command: pspace
```

3. Although the drawing area goes blank, don't worry: you drawing has not been erase. To see your drawing, you need to create at least one viewport:

```
Command mview fit
```

Your drawing reappears!

4. Now switch back to model space:

```
Command: mspace
```

5. Use the **Zoom** and **Pan** command to make the drawing smaller or larger within the paper space viewport.

6. Switch back to paper space with **PS**. Create a few more viewports by picking points with the **MView** command. Try overlapping a couple of viewports. Switch back to model space with **MS** and set different zoom levels for each viewport.

7. Switch back to paper space with **PS**. Now use the **Move** and **Stretch** commands to change the position and size of the paper space viewports. Draw a title border around all the viewports.

8. Some other paper space-related command to experiment with are: **VpLayer**, **Zoom XP**, **PsLtScale**, and **HpMPlot.**

Purge

Ver. 2.1

Removes unused named objects from the drawing: blocks, dimension styles, layers, linetypes, shapes, text styles, application id tables, and multiline styles.

Command	Alt+	Side Menu	Menu Bar	Tablet
purge	D,P	[DATA] [Purge:]	[Data] [Purge]	...

Command: **purge**
Purge unused Blocks/Dimstyles/LAyers/LTypes/SHapes/STyles /APpids/Mlinestyles/All: **A**

Sample response:
No unreferenced blocks found.
Purge layer DOORWINS? <N> **y**
Purge layer TEXT? <N> **y**
Purge linetype CENTER? <N> **y**
Purge linetype CENTER2? <N> **y**
No unreferenced text styles found.
No unreferenced shape files found.
No unreferenced dimension styles found.

COMMAND OPTIONS

Blocks Named but unused blocks.
Dimstyles Unused dimension styles.
LAyers Unused layers.
LTypes Unused linetypes.
SHapes Unused shape files.
STyles Unused text styles.
APpids Unused application id table of ADS and AutoLISP apps.
Mlinestyles Unused multiline styles.
All Purge drawing of all eight named objects, if necessary.

RELATED AUTOCAD COMMANDS

- **End** Two **End** commands in a row can remove spurious information from a drawing.
- **WBlock** Writes the current drawing to disk (with the * option) and removes spurious information from the drawing.

TIPS

- As of Release 13, **Purge** can be used at any time; it no longer must be used as the first command used after a drawing is loaded.

- It may be necessary to use the **Purge** command several times; follow each purge with the **End** command, then **Open** the drawing and **Purge** again. Repeat until **Purge** reports nothing to purge.

QSave

Rel. 12

Saves the current drawing without requesting a filename (*short for Quick SAVE*).

Command	Ctrl+	Side Menu	Menu Bar	Tablet
qsave	S	[FILE]	[File]	T 24
		[Save:]	[Save]	

```
Command: qsave
```

COMMAND OPTIONS

None

RELATED AUTOCAD COMMANDS

- **End** Saves the drawing, without requesting a filename, and ends AutoCAD.
- **Save** Saves drawing, after requesting the filename.
- **SaveAs** Saves the drawing with a different filename.

RELATED SYSTEM VARIABLES

- **DbMode** Indicates whether the drawing has changed since it was loaded.
- **DwgName** Current drawing filename (*default is "UNNAMED"*).
- **DwgTitled** Status of drawing's filename:
  - **0** Name is "UNNAMED".
  - **1** Name is other than UNNAMED.
- **DwgWrite** Drawing's read-write status:
  - **0** Read-only.
  - **1** Read-write.

TIPS

■ When the drawing is named, then the **QSave** command requests a file name.

■ When the drawing file, its subdirectory, or drive (such as a CD-ROM drive) are marked 'read-only,' use the **SaveAs** command to save the drawing to another filename, subdir, or drive.

QText

Ver. 2.0

Displays a line of text as a rectangular box (*short for Quick TEXT*).

Command	Alt+	Side Menu	Menu Bar	Tablet
qtext	O,D,T	[OPTIONS]	[Options]	Y 22
		[DISPLAY]	[Display]	
		[Qtext:]	[Text Frame Only]	

```
Command: qtext
ON/OFF <Off>: on
```

Normal text **Quick text**

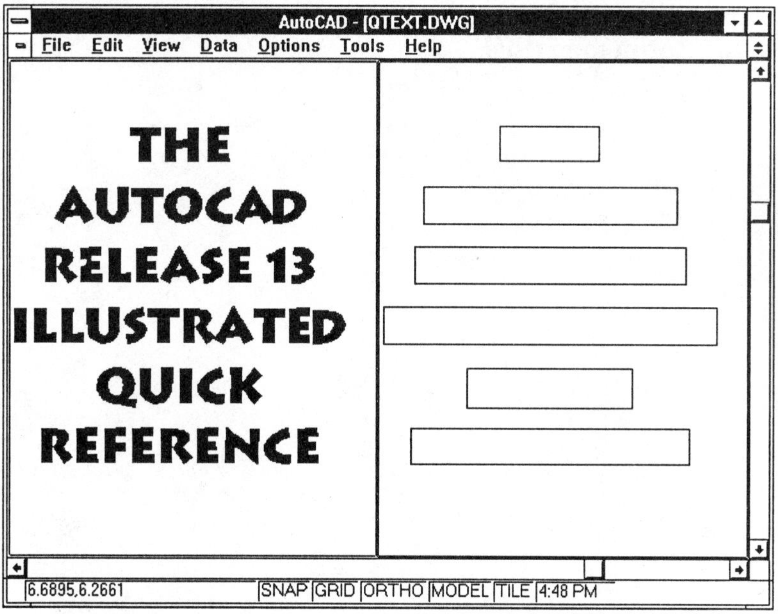

COMMAND OPTIONS

ON Turns on quick text, after the next **Regen** command.

OFF Turns off quick text, after the next **Regen** command.

RELATED AUTOCAD COMMANDS

- **DdRModes** Toggles **QText** via dialogue box.
- **Regen** Regenerates the screen; makes quick text take effect.

RELATED SYSTEM VARIABLE

■ **QTextMode** Holds the current state of quick text mode.

TIPS

■ To reduce the redraw and regen time of text, use **QText** to turn lines of text into rectangles, which redraw faster.

■ The length of a **QText** box does not necessarily match the actual length of text.

■ Turning on **QText** does affect text during plotting; qtext blocks are plotted as text.

■ To find invisible text (such as text made of spaces), turn on **QText**, thaw all layers, and **Zoom** to extents.

Quit

Ver. 1.0

Exits AutoCAD without saving changes to the drawing, from the most recent **Save** or **End** command.

Command	Alias	Side Menu	Menu Bar	Tablet
quit	exit	[FILE]	[File]	X 24
		[Exit:]	[Exit]	

Command: quit
Displays dialogue box.

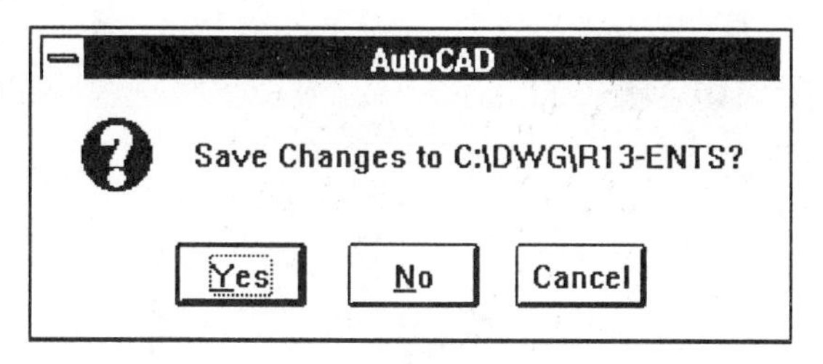

COMMAND OPTIONS

Yes Save changes before leaving AutoCAD.
No Don't save changes.
Cancel Don't quit AutoCAD.

RELATED AUTOCAD COMMANDS

- **End** Saves the drawing and exits AutoCAD.
- **SaveAs** Saves the drawing by another name or to another subdirectory or drive.

RELATED SYSTEM VARIABLE

- **DbMod** Indicates whether the drawing has changed since it was loaded.

RELATED FILES

- ***.DWG** AutoCAD drawing files.
- ***.BAK** Backup file.
- ***.BK1** Additional backup files.

TIPS

- You can make changes to a drawing, yet preserve its original format: first, use the **SaveAs** command to save the drawing by another name; then, use the **Quit** command to preserve the drawing in its original state.

- Even if you accidently save over a drawing, you can recover the previous version: first, use the DOS **Erase** or **Rename** command to rename the DWG file; then, use the DOS **Rename** command to rename the backup BAK file to DWG.

Ray

Creates a semi-infinite construction line.

Command	Alias	Side Menu	Toolbar	Tablet
ray	...	[DRAW 1] [Ray:]	[Draw] [Line] [Ray]	L 9

```
Command: ray
From point: [pick]
Through point: [pick]
```

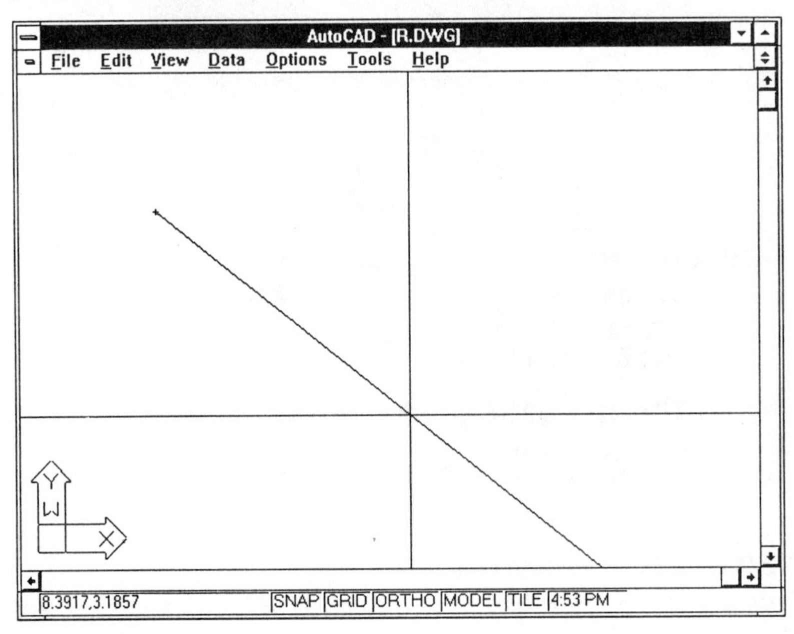

COMMAND OPTIONS

None

RELATED AUTOCAD COMMANDS

- **DdModify** Modifies the ray.
- **Line** Draws a line.
- **XLine** Creates an infinite construction line.

TIPS

- A ray is a 'contruction line' that displays but does not plot.
- The ray has all properties of a line: it can have color, layer, linetype, be used as a cutting edge, etc.

RConfig

Rel. 12

Configures output devices for the Render module (*short for Render CONFIG; an external command in Render.Arx*).

Command	Alt+	Side Menu	Menu Bar	Tablet
rconfig	O,R	[TOOLS] [RENDER] [Config:]	[Options] [Render Configure]	...

Command: **rconfig**

Switches to text screen:

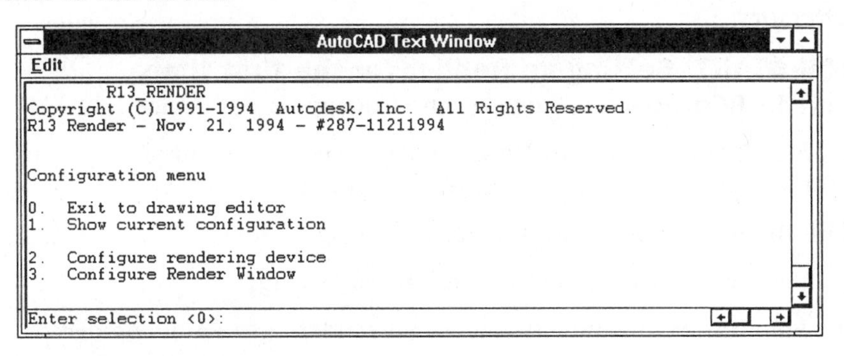

COMMAND OPTIONS

1	Show current **Render** configuration.
2	Configure rendering output device.
3	Configure **Render** window.
<0>	Exit to AutoCAD.

RELATED AUTOCAD COMMANDS

- **Config** Configures input and output devices for AutoCAD.
- **DlxConfig** Selects graphics board for Render.
- **HpMPlot** Mixes rendered and wireframe plot to HPGL/2 device.
- **Render** Performs the rendering.
- **RPref** Specifies rendering preferences.

RELATED DOS VARIABLES

It is not necessary to set any of these variables before using Render:

- **AveFaceDir** Points to working directory to store faces and triangles produced during rendering.
- **AveMaps** Points to subdirectory containing texture maps.
- **RenderCfg** Points to the location of the Render.Cfg configuration file.

RELATED FILE

- **Render.Cfg** The configuration file for Render in \Acad13\Win subdir.

TIPS

■ All Windows display drivers trade off higher resolution for fewer colors. For renderings, a larger number of colors is more important than a higher resolution. As a suitable tradeoff, select 256 colors and the highest associated resolution.

■ Use the **AveFaceDir** variable to point to a RAM drive to help speed up complex renderings.

■ The **RenderCfg** variable allows you to keep a number of different rendering configurations; place each Render.Cfg file in a different subdirectory.

QUICK START: Setting up Render for the first time.

1. With the **RConfig** command, select hardcopy output device (if any).

2. (Optional) Use the **HpConfig** command to configure AutoCAD for mixed wireframe/rendering output on an HPGL/2-compatible device.

3. With the **RPref** command, specify rendering options.

4. Use the **Render** command to create the rendering.

5. The **SaveImg** command lets you output renderings to a file on disk.

Recover

Rel. 12

Recovers a damaged drawing without user intervention.

Command	Alt+	Side Menu	Menu Bar	Tablet
recover	F,T,R	[FILE]	[File]	...
		[MANAGE]	[Management]	
		[Recover:]	[Recover]	

Command: **recover**

Sample output:

```
Drawing recovery.
Drawing recovery log.
Scanning for sentinels   99% done
Scanning completed.
Validating objects in the handle table.
Valid objects 1452   Invalid objects 0
Validating objects completed.
Used contingency data.
Salvaged database from drawing.
41       Blocks audited
Pass 1 956      objects audited
Pass 2 956      objects audited
Pass 3 1400     objects audited
Total errors found 0 fixed 0
Regenerating drawing.
```

COMMAND OPTIONS

None

RELATED AUTOCAD COMMAND

- **Audit** Checks a drawing for integrity.

TIPS

- The **Open** command automatically invokes the **Recover** command if AutoCAD detects that the drawing is damaged.
- **Recover** does not ask permission to repair damaged parts of the drawing file; use the **Audit** command if you want to control the repair process.
- The **Quit** command discards changes made by the **Recover** command.
- If the **Recover** and **Audit** commands do not fix the problem, try using the **DxfOut** command, followed by the **DxfIn** command.

Rectang

Rel. 12

Draws a rectangle out of a polyline.

Command	Alias	Side Menu	Menu Bar	Tablet
rectang	...	[DRAW 1] [Rectang:]	[Draw] [Polygon] [Rectang]	...

```
Command: rectangle
First corner: [pick]
Other corner: [pick]
```

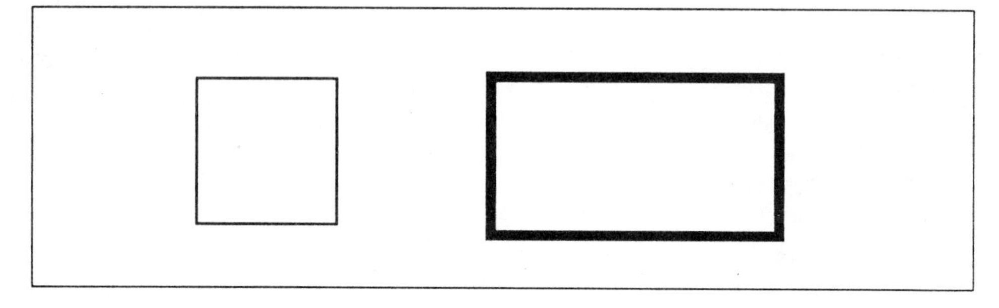

COMMAND OPTIONS

None

RELATED AUTOCAD COMMANDS

- **Donut** Draws solid-filled circles with a polyline.
- **Ellipse** Draws ellipsis with a polyline, when **PEllipse** = 1.
- **PEdit** Edits polylines, including rectangles.
- **PLine** Draws polylines and polyline arcs.
- **Polygon** Draws a polygon (*3 to 1,024 sides*) from a polyline.

TIPS

- Rectangles are drawn from polylines; use the **PEdit** command to change the rectangle, such as the width of the polyline.

- The pick point determines the location of the rectangle's first vertex; rectangles are drawn counterclockwise.

- Use the **Snap** command and object snap modes to precisely place the rectangle.

- Use object snap mode **INTersection** to snap to the rectangle's vertices.

Redefine

Rel. 9

Restores the meaning of an AutoCAD command after being disabled by the **Undefine** command.

Command	Alias	Side Menu	Menu Bar	Tablet
redefine	...	...	...	...

```
Command: redefine
Command name:
```

COMMAND OPTIONS

None

RELATED AUTOCAD COMMANDS

- *All commands* All AutoCAD commands can be redefined.
- **Undefine** Disables the meaning of an AutoCAD command.

TIPS

- Prefix any command with a . (*period*) to temporarily redefine the undefinition, as in:

```
Command: .line
```

- Prefix any command with an _ (*underscore*) to make an English-language command work in any lingual version of AutoCAD, as in:

```
Command: _line
```

 Redo

Ver. 2.5

Reverses the effect of the most recent **Undo** and **U** command.

Command	Alt+	Side Menu	Menu Bar	Tablet
redo	E,R	[ASSIST]	[Edit]	U 9-10
		[Redo:]	[Redo]	

Command: **redo**

COMMAND OPTIONS

None

RELATED AUTOCAD COMMAND

- **Undo** Undoes the most recent series of AutoCAD commands.

RELATED SYSTEM VARIABLE

- **UndoCtl** Determines the state of the **Undo** command.

TIP

- The **Redo** command is limited to undoing a single undo, while the **Undo** and **U** commands undo operations all the way back to the beginning of the editing session.

'Redraw

Ver. 1.0

Redraws the current viewport to clean up the screen.

Command	Alias	Side Menu	Menu Bar	Tablet
'redraw	r	[VIEW]	[View]	L11-P11
		[Redraw:]	[Redraw View]	

Command: **redraw**

Before redraw:

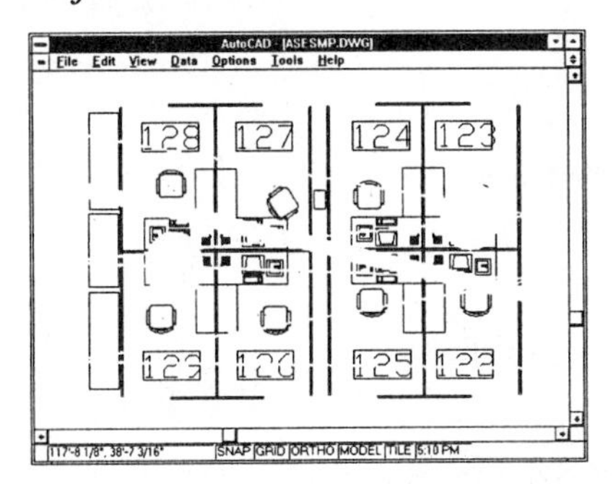

After redraw:

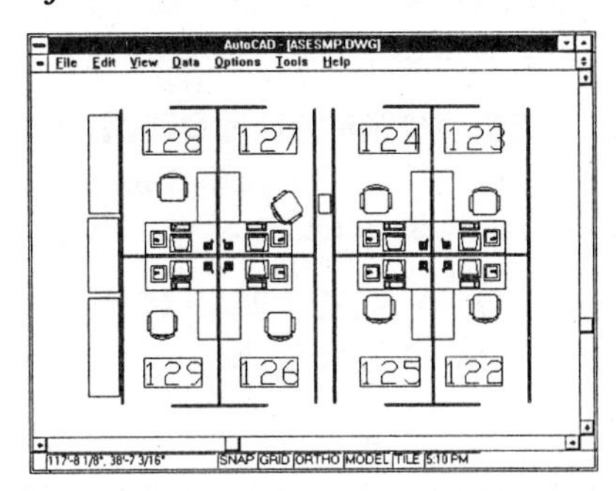

COMMAND OPTION

[Esc] Cancels the redraw.

RELATED AUTOCAD COMMANDS

- **RedrawAll** Redraws all viewports.
- **Regen** Regenerates the current viewport.
- **Zoom** The **VMax** option zooms the furthest out without causing a regeneration.

RELATED SYSTEM VARIABLE

- **SortEnts** Controls the order of redrawing objects:
  - **0** Sorts by order in the drawing database.
  - **1** Sorts for object selection.
  - **2** Sorts for object snap.
  - **4** Sorts for redraw.
  - **8** Sorts for creating slides.
  - **16** Sorts for regenerations.
  - **32** Sorts for plotting.
  - **64** Sorts for PostScript plotting.

TIPS

- Use **Redraw** to clean up the screen after a lot of editing; some commands automatically redraw the screen when they are done.

- **Redraw** does not affect objects on frozen layers.

- Use the **RedrawAll** command to redraw all viewports.

'RedrawAll

Rel. 10

Redraws all viewports to clean up the screen.

Command	Alt+	Side Menu	Menu Bar	Tablet
'redrawall	V,A	[VIEW]	[View]	Q11-R11
		[RedrwAl:]	[Redraw All]	

`Command: redrawall`

COMMAND OPTIONS

[Esc] Cancels the redraw.

RELATED AUTOCAD COMMANDS

- **Redraw** Redraws only the current viewport.
- **RegenAll** Regenerates all viewports.

RELATED SYSTEM VARIABLE

- **SortEnts** Controls the order of redrawing objects.

TIPS

- **RedrawAll** does not affect objects on frozen layers.
- Use the **Redraw** command to redraw a single viewport.

Regen

Ver. 1.0

Regenerates the current viewport to update the drawing.

Command	Alias	Side Menu	Menu Bar	Tablet
regen	...	...	...	J 11

```
Command: regen
Regenerating drawing.
```

COMMAND OPTION

[Esc] Cancels the regeneration.

RELATED AUTOCAD COMMANDS

- **Redraw** Quickly cleans up the current viewport.
- **RegenAll** Regenerates all viewports.
- **RegenAuto** Checks with you before doing most regenerations.
- **ViewRes** Controls whether zooms and pans are regens or redraws.

RELATED SYSTEM VARIABLE

- **RegenMode** Current setting of **RegenAuto**:
  - 0 Off.
  - 1 On (*default*).

TIPS

- Some commands automatically force a regeneration of the screen; other commands queue the regen.

- To save on regeneration time:
  - Freeze layers you are not working with.
  - Use **QText** to turn text into rectangles.
  - Place hatching last on its own layer.

- Use the **RegenAll** command to regenerate all viewports.

RegenAll

Rel. 10

Regenerates all viewports.

Command	Alias	Side Menu	Menu Bar	Tablet
`regenall`	...	...	...	K 11

```
Command: regenall
Regenerating drawing.
```

COMMAND OPTION

[Esc] Cancels the regeneration process.

RELATED AUTOCAD COMMANDS

- **RedrawAll** Redraws all viewports.
- **Regen** Regenerates the current viewport.
- **RegenAuto** Checks with you before doing most regenerations.
- **ViewRes** Controls whether zooms and pans are regens or redraws.

RELATED SYSTEM VARIABLE

- **RegenMode** Current setting of **RegenAuto**.

TIPS

- **RegenAll** does not regenerate objects on frozen layers.
- Use the **Regen** command to regenerate a single viewport.

'RegenAuto

Ver. 1.2

AutoCAD asks you before performing a regeneration, when turned off (*short for REGENeration AUTOmatic*).

Command	Alias	Side Menu	Menu Bar	Tablet
'regenauto	...	...	...	...

```
Command: regenauto
ON/OFF <On>: off
```

Example:

```
Command: regen
About to regen, proceed? <Y>:
```

COMMAND OPTIONS

OFF Turns on "About to regen, proceed?" message.
ON Turns off "About to regen, proceed?" message.

RELATED AUTOCAD COMMAND

■ **Regen** Forces a regeneration in the current viewport.

RELATED SYSTEM VARIABLES

■ **Expert** Suppresses the "About to regen, proceed?" message when value is greater than 0.

■ **RegenMode** Current setting of **RegenAuto** command.

TIPS

■ If a regeneration is caused by a transparent command, AutoCAD delays it with the message, "Regen queued."

■ Release 12 reduces the number of regenerations by expanding the virtual screen from 16 bits to 32 bits.

Region

Rel. 11

Creates a 2D region from closed objects *(formerly the **Solidify** command; an external command in Acis.Dll)*.

Command	Alias	Side Menu	Toolbar	Tablet
region	...	[CONSTRCT] [Region:]	[Draw] [Polygon] [Region]	J 8

```
Command: region
Select objects: [pick]
Select objects: [Enter]
1 loop extracted.
1 region created.
```

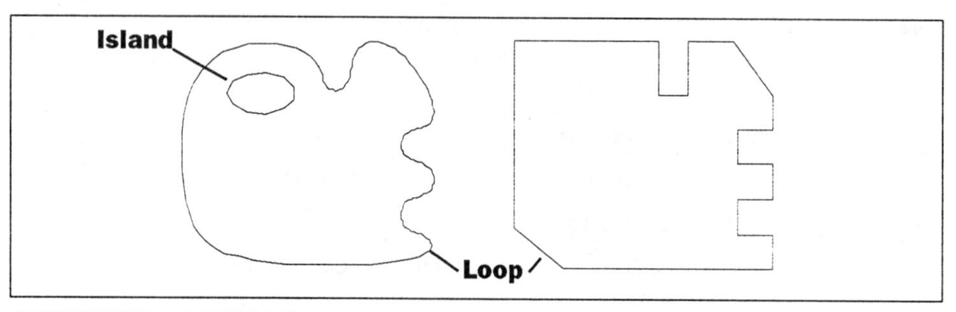

COMMAND OPTIONS

None

RELATED AUTOCAD COMMANDS

All

RELATED SYSTEM VARIABLE

- **DelObj** Toggles whether objects are deleted during the region conversion.

TIPS

- The **Region** command converts:
  - Closed line sets.
  - Closed 2D and planar 3D polylines.
  - Closed curves.
- The Region command rejects open objects, intersections, and self-intersecting curves.
- Splined and curve-fitted polylines are not converted into spline objects.

■ The resulting region is unpredictable when more than two curves share an endpoint.

■ Polylines with width lose their width when converted to a region.

DEFINITIONS

Curve

■ An object made of circles, ellipses, splines, and joined circular and elliptical arcs.

Island

■ A closed shape fully within (not touching or intersecting) another closed shape.

Loop

■ A closed shape made of closed polylines, closed lines, and curves.

Region

■ A 2D closed area defined as an ACIS object.

Reinit

Rel. 12

Reinitializes the digitizer, display, plotter and input-output ports, and reloads the Acad.Pgp file (*short for REINITialize*).

Command	Alt+	Side Menu	Menu Bar	Tablet
reinit	T,Z	[TOOLS [Reinit:]	[Tools] [Reinitialize]	...

Command: **reinit**

Displays dialogue box.

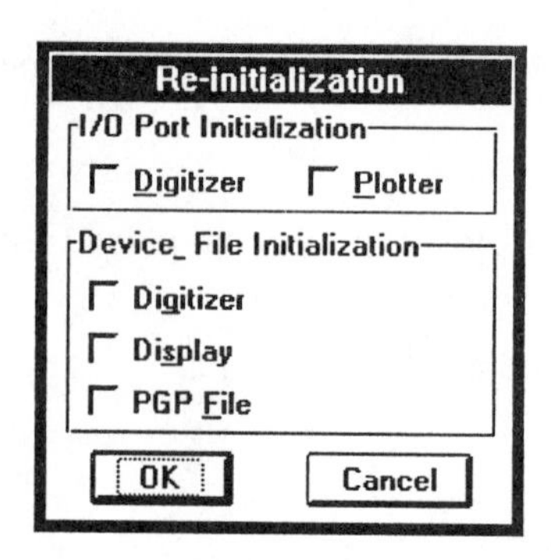

COMMAND OPTIONS

Digitizer Reinitializes port connected to digitizer.
Plotter Reinitializes port connected to plotter.
Digitizer Reinitializes digitizer driver.
Display Reinitializes display driver.
Pgp File Reloads Acad.Pgp file.

RELATED AUTOCAD COMMAND

- **Menu** Reloads menu file.

RELATED SYSTEM VARIABLE

- **Re-init** Reinitializes via system variable settings.

RELATED FILES

- **Acad.Pgp** The program parameters file in \Acad13\Common\Support subdir.
- ***.EXP** Device drivers in \Acad13\Dos\Drv subdirectory.

TIPS

- AutoCAD allows you to connect both the digitizer and the plotter to the same port since you do not need the digitizer during plotting; use the **Reinit** command to reinitialize the digitizer after plotting.

- AutoCAD reinitializes all ports and reloads the Acad.Pgp file each time another drawing is loaded.

Rename

Ver. 2.1

Allows you to change the names of blocks, dimension styles, layer, linetypes, text styles, UCS names, views, and viewports.

Command	Alias	Side Menu	Menu Bar	Tablet
rename	...	...	...	...

```
Command: rename
Block/Dimstyle/LAyer/LType/Style/Ucs/VIew/VPort:
```

Example:

```
Command: rename
Block/Dimstyle/LAyer/LType/Style/Ucs/VIew/VPort: B
Old block name: diode-20
New block name: diode-02
```

COMMAND OPTIONS

Block Changes the name of a block.
Dimstyle Changes the name of a dimension style.
LAyer Changes the name of a layer.
LType Changes the name of a linetype.
Style Changes the name of a text style.
Ucs Changes the name of a UCS configuration.
VIew Changes the name of a view configuration.
VPort Changes the name of a viewport configuration.

RELATED AUTOCAD COMMANDS

- **DdLModes** Changes layer names via a dialogue box.
- **DdRename** Dialogue box for renaming.
- **DdUcs** Changes UCS configuration names via a dialogue box.
- **Files** Changes the names of files on disk.

RELATED SYSTEM VARIABLES

- **CeLayer** Name of current layer.
- **CeLtype** Name of current linetype.
- **DimStyle** Name of current dimension style.
- **InsName** Name of current block.
- **TextStyle** Name of current text style.
- **UcsName** Name of current UCS view.

Render

Rel. 12

Creates a rendering of 3D objects (*an external command in Render.Arx*).

Command	Alias	Side Menu	Toolbar	Tablet
render	...	[TOOLS]	[Render]	L 1
		[RENDER]	[Render]	
		[Render:]		

Command: **render**

Displays dialogue box.

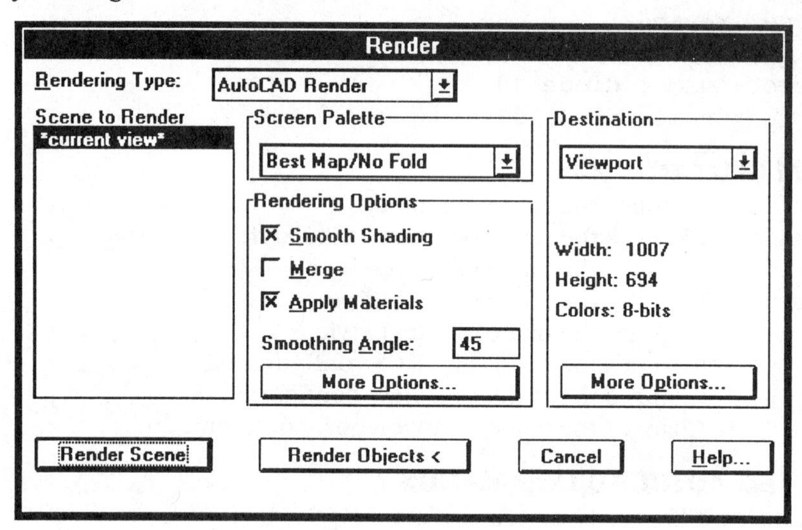


```
Using current view.
Default scene selected. | / - \
```

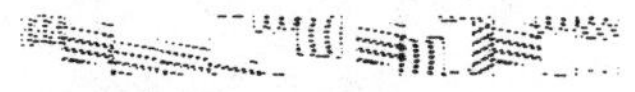

Rendering Type
Selects from AutoCAD Render, AutoVision, and third-party renderers, if loaded into system.

Scene to Render
Selects scene name, defined by the **Scene** command.

Screen Palette
Selects palette for 256-color graphics boards; not required for 16-bit or better graphics boards.

Smooth Shading
Toggles smooth or faceted shading.

Merge Combines multiple images.

Apply Materials
Applies texture mapping defined by the **RMat** command.

Smoothing Angle

Sets the threshold angle at which **Render** smooths facets.

More Options

Additional rendering options:

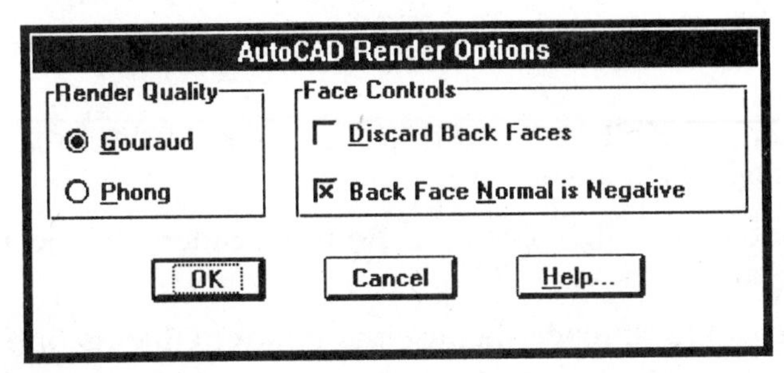

Destination Output rendering to viewport, window, or file.

More Options Additional output options:

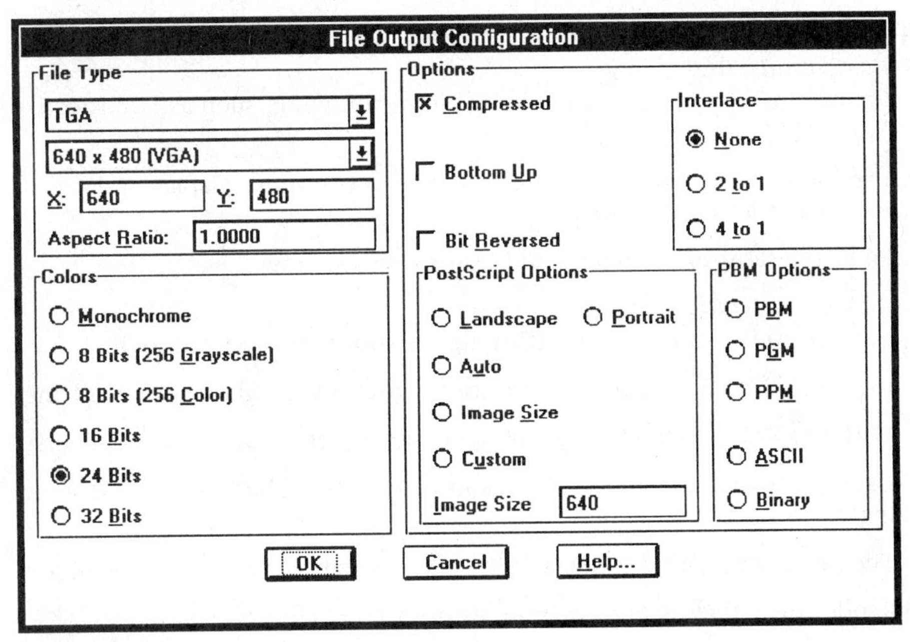

Render Scene Render entire drawing.

Render Objects

Render selected objects.

RELATED AUTOCAD COMMANDS

- *All rendering-related commands.*
- **DView** Create perspective view.
- **Hide** Removes hidden lines from wireframe view.
- **Shade** Simple flat shading of 3D objects.

TIPS

■ If you do not place a light or define a scene, **Render** uses the current view and places a single light at your eye.

■ If you do not select a light or scene, **Render** renders all objects using all lights and the current view.

■ To run a quick check rendering, use the **Render Objects** option.

■ Render outputs to the following file format: GIF, X11, PBM, BMP, TGA, PCX, Sun, FITS, PostScript, TIFF, Fax Group III, and IFF.

QUICK START: Your first rendering.

■ *Basic rendering:*

1. Create a 3D drawing or select a 3D sample drawing, such as Linkrods.Dwg in \Acad13\Common\Sample.

2. Use the **Config** and **RConfig** commands to configure display and hardcopy rendering devices.

3. Enter the **Render** command and wait a few seconds.

■ *Advanced Rendering:*

1. (*Optional*) Use **Config** and **RConfig**, as above; load a 3D drawing.

2. Use **MatLib** to load material definitions (*texture mapping*) into drawing.

3. With the **RMat** command, assign materials to colors, layers, and objects.

4. Use the **Light** command to place and aim lights: point, spot, and distant lights.

5. The **Scene** command collects lights and a viewpoint into a named object.

6. Render the named scene with the **Render** command.

7. Use the **SaveImg** command to save the rendering to a TIFF, Targa, or GIF file on disk.

8. View the save rendering file with the **Replay** command.

9. (*Optional*) Export the rendering to 3D Studio with the **3dsOut** command.

Output to viewport:

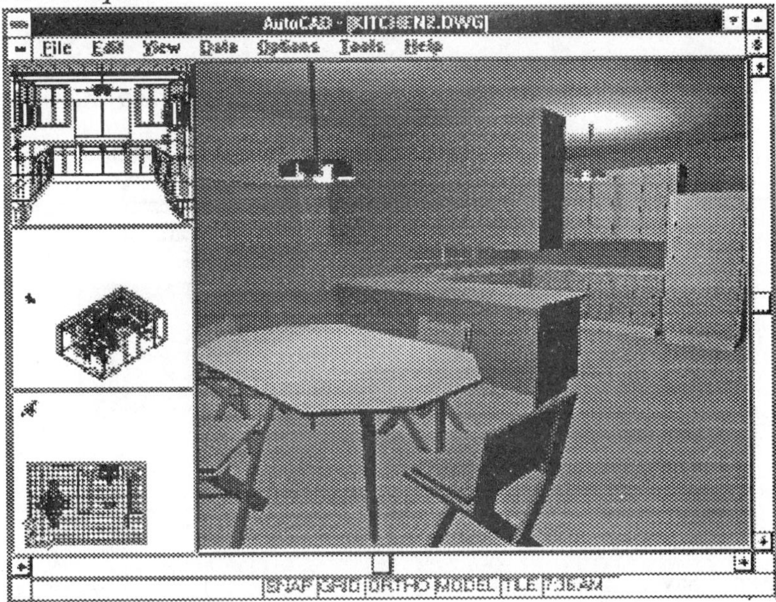

Output to Render window:

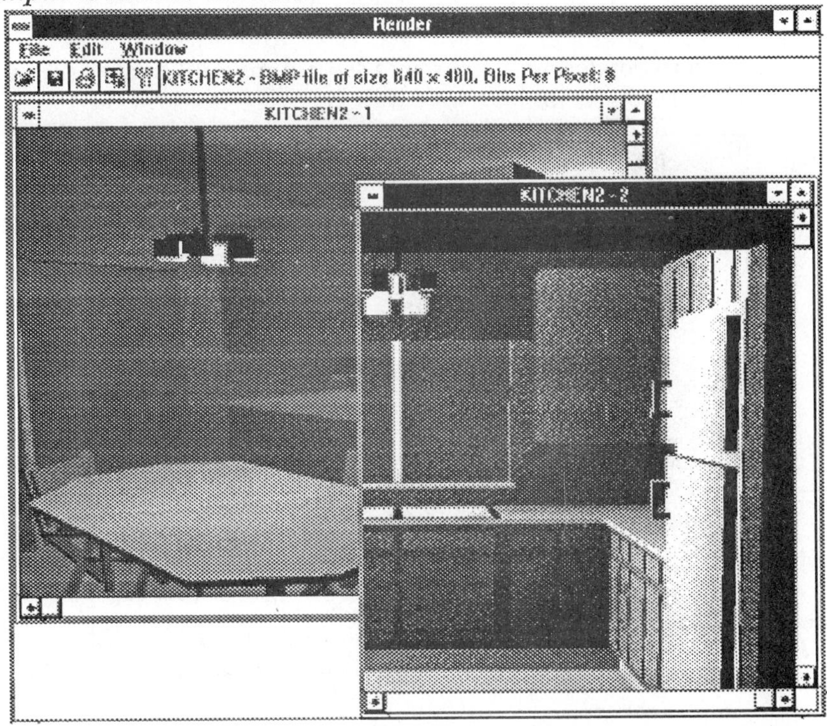

Open:

Opens a BMP (*bitmap*), DIB (*device-independent bitmap*), or RLE (*run-length encoded – compressed bitmap*) file:

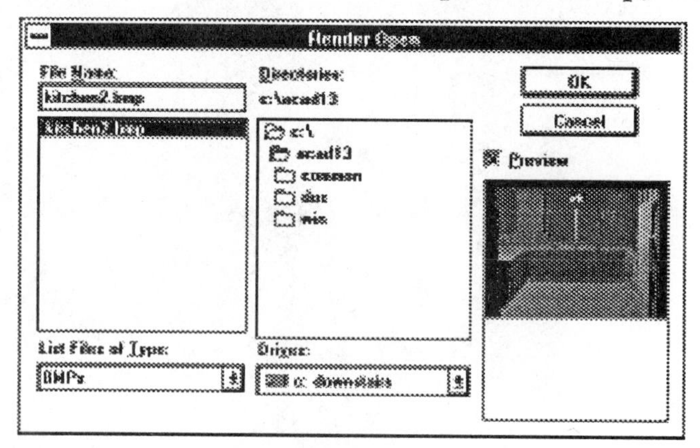

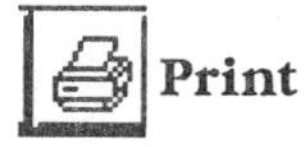

Save Saves rendering as a BMP file.

Print

Prints rendering; maximum tiled output is 130 pages (*13 across by 10 sheets down*):

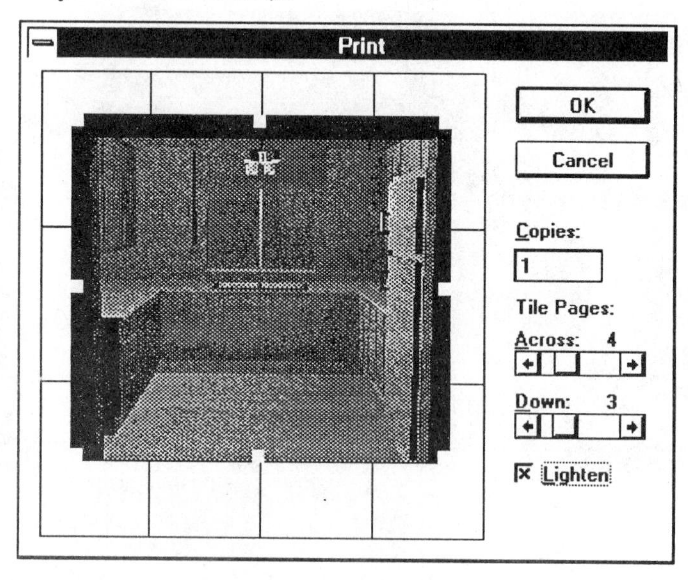

 **Copy** Copy rendering to the Windows Clipboard in DIB format.

 **Options** Render window options:

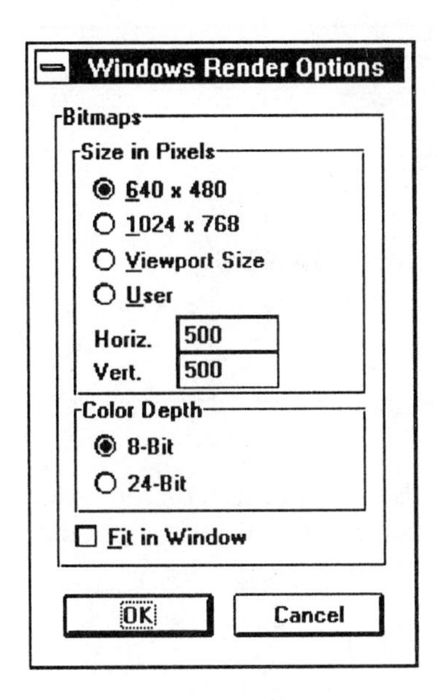

- **Size in Pixels**

  The larger the viewport, the better the rendering quality at a cost of longer rendering time and larger file sizes.

- **Color Depth**

  Select a color-depth of 24 bits only if your computer's graphics board and Windows display driver both support 24-bit output; otherwise, rendering quality is worse using 24-bit output with an 8-bit graphics board.

- **Fit in Window**

  Resize the rendering to fit the window; can degrade the quality of rendering.

'RenderUnload

Rel. 13

Unloads Render to free up memory for AutoCAD (*an external command in Render.Arx*).

Command	Alias	Side Menu	Menu Bar	Tablet
'renderunload...	...	...	...	...

```
Command: renderunload
AutoCAD Render has been unloaded from memory.
```

COMMAND OPTIONS

None

RELATED AUTOCAD COMMANDS

All rendering commands

TIPS

■ **RenderUnload** frees memory by removing the **Render** code from system RAM.

■ All **Render** commands are still immediately available since any rendering command automatically reloads the **Render** module.

NON-EXISTANT COMMAND: RendScr

The **RendScr** command does not work in the Windows version of AutoCAD Release 13.

Replay

Rel. 12

Displays a GIF, TIFF, or Targa file as a bitmap (*an external command in Render.Arx*).

Command	Alt+	Side Menu	Menu Bar	Tablet
replay	T,G,V	[TOOLS]	[Tools]	...
		[Replay:]	[Image]	
			[View]	

Command: **replay**

*Displays **Select File** dialogue box; select file.*
Displays dialogue box.

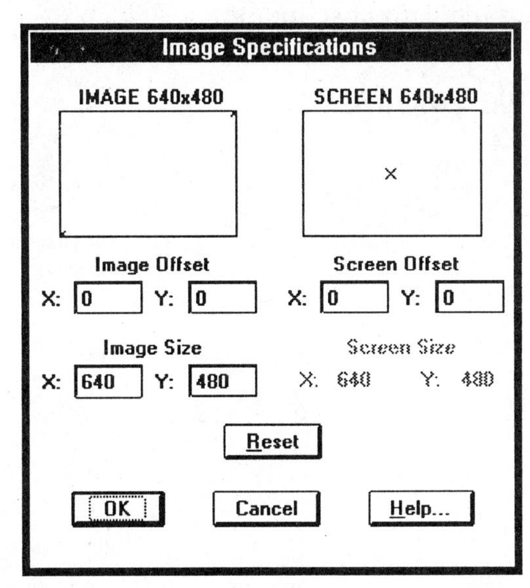

COMMAND OPTIONS

Image Selects displayed area by clicking on image tile.
Image Offset The x,y-coordinates of the image's lower-left corner.
Image Size Sizes of image in pixels.
Screen Fills entire viewport with image.
Screen Offset The x,y-coordinates of the image's lower-left corner.
Reset Restores values.

RELATED AUTOCAD COMMANDS

- **Import** Dialogue-box 'shell' for loading some raster and vector files.
- **SaveImg** Saves a rendering as a GIF, TIFF, or Targa raster file.
- **GifIn** Imports a GIF file as a vector file.
- **PcxIn** Imports a PCX file as a vector file.
- **TiffIn** Imports a TIFF file as a vector file.

'Resume

Ver. 2.0

Resumes a script file after pausing it by pressing the **[Backspace]** key.

Command	Alias	Side Menu	Menu Bar	Tablet
`'resume`	...	...	...	...

```
Command: resume
```

COMMAND OPTIONS

[Backspace] Pauses the script file.
[Esc] Stops the script file.

RELATED AUTOCAD COMMANDS

- **RScript** Reruns the current script file.
- **Script** Loads and runs a script file.

RELATED SYSTEM VARIABLES

- *None*

Revolve

Rel. 11

Creates a 3D solid object by revolving a closed object about an axis *(formerly the **SolRev** command; an external command in Acis.Dll)*.

Command	Alias	Side Menu	Toolbar	Tablet
revolve	...	[DRAW 2]	[Solids]	L 8
		[SOLIDS]	[Revolve]	
		[Revolve:]		

```
Command: revolve
Select objects: [pick]
Select objects: [Enter]
Axis of revolution - Object/X/Y/<Start point of axis>: [pick]
Angle of revolution <full circle>: [Enter]
```

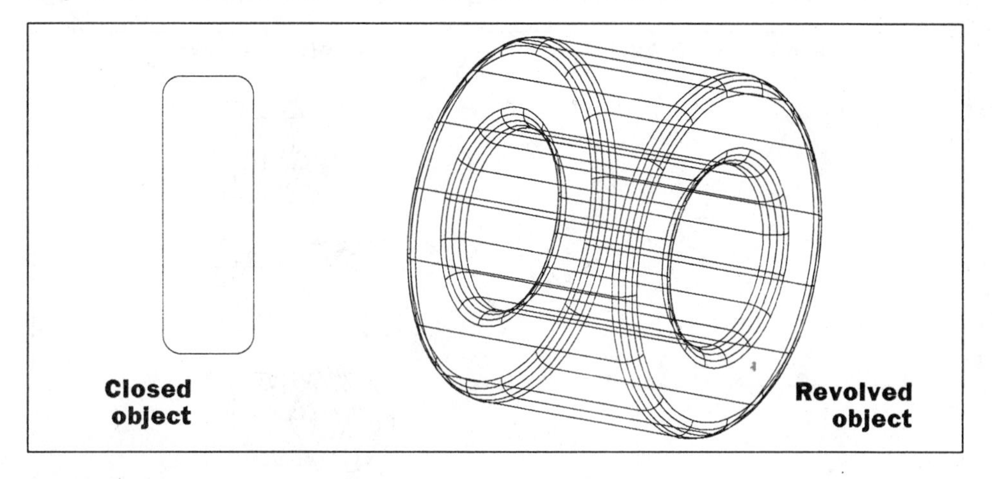

COMMAND OPTIONS

Object Selected object determines axis of revolution.
<Start point> Indicates the axis of revolution.
X Uses positive x-axis as axis of revolution.
Y Uses positive y-axis as axis of revolution.

RELATED AUTOCAD COMMANDS

- **Extrude** Extrudes a 2D object into a 3D solid model.
- **Rotate** Rotates open and closed objects.

TIPS

- **Revolve** works with just one object at a time.
- **Revolve** works with closed polylines, circles, ellipses, donuts, polygons, closed splines, and regions.
- **Revolve** does not work with open objects, crossing, or self-intersecting polylines.

RevSurf

Rel. 10

Generates a 3D surface of revolution defined by a path curve and an axis (*short for REVolved SURFace*).

Command	Alias	Side Menu	Toolbar	Tablet
revsurf	...	[DRAW 2] [SURFACES] [Revsurf:]	[Surfaces] [RevSurf]	N 8

```
Command: revsurf
Select path curve: [pick]
Select axis of revolution: [pick]
Start angle <0>: [Enter]
Included angle (+=ccw, -=cw) <Full circle>: [Enter]
```

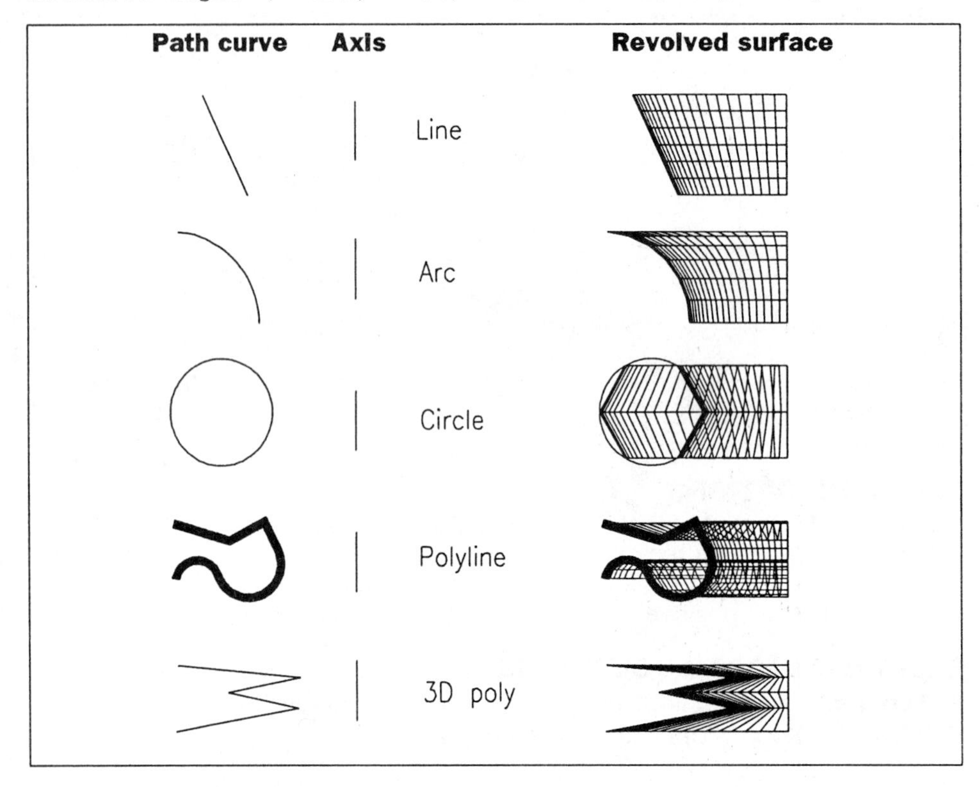

COMMAND OPTIONS

<Full circle> Revolves object through 360 degrees.

Included angle

Specifies angle of revolution.

RELATED AUTOCAD COMMANDS

- **EdgeSurf** Creates a 3D surface bounded by four edges.
- **PEdit** Edits revolved surfaces.
- **Revolve** Revolves a 2D closed object into a 3D solid.
- **RuleSurf** Creates a 3D ruled surface.
- **TabSurf** Creates a 3D tabulated surface.

RELATED SYSTEM VARIABLES

- **SurfTab1** Mesh density in m-direction.
- **SurfTab2** Mesh density in n-direction.

TIPS

- Unlike the **Revolve** command, **RevSurf** works with open and closed objects.

- If a multi-segment polyline is the axis of revolution, the rotation axis is defined as the vector pointing from the first vertex to the last vertex, ignoring the location of intermediate vertices.

DEFINITIONS

Axis of Revolution

- The axis about which the object is rotated.
- Defines the m-direction stored in system variable **SurfTab1**.

Path Curve

- The object being revolved.
- Defines the n-direction stored in system variable **SurfTab2**.

RMat

Rel. 13

Applies material definitions (texture maps) to colors, layers, and objects; used by the **Render** command (*short for Render MATerials; an external command in Render.Arx*).

Command	Alias	Side Menu	Toolbar	Tablet
rmat	...	[TOOLS]	[Render]	...
		[RENDER]	[RMat]	
		[Mater'1:]		

Command: **rmat**

Displays dialogue box.

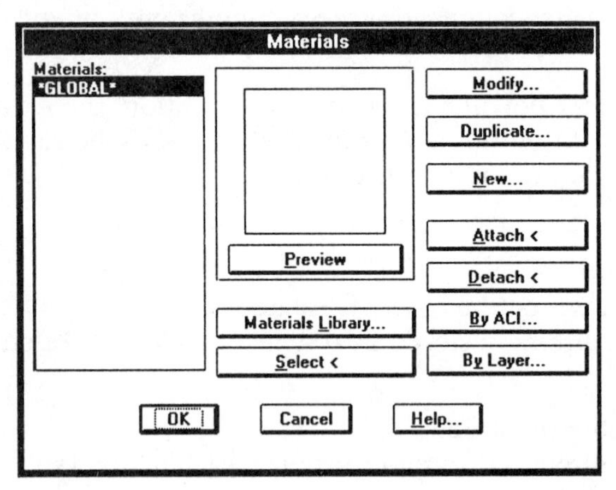

COMMAND OPTIONS

Materials Lists names of materials loaded into drawing by the **MatLib** command.

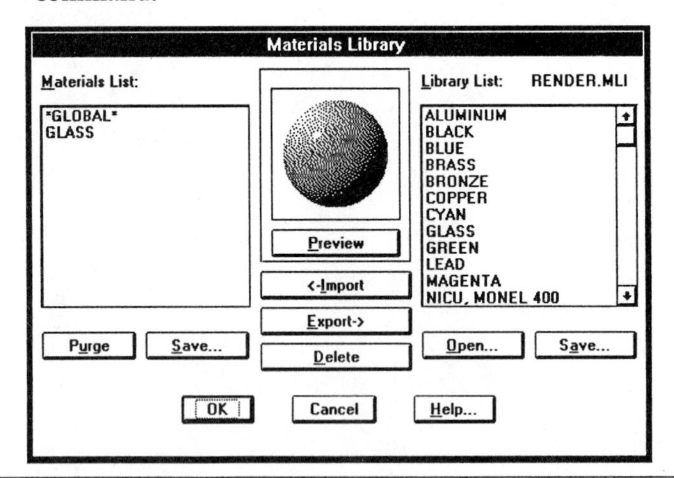

Preview Previews the material mapped to a sphere.

Materials Library

Displays the **MatLib** command's dialogue box.

Select Selects the objects to attach the material definition.

Modify Edits a material definition; displays dialogue box:

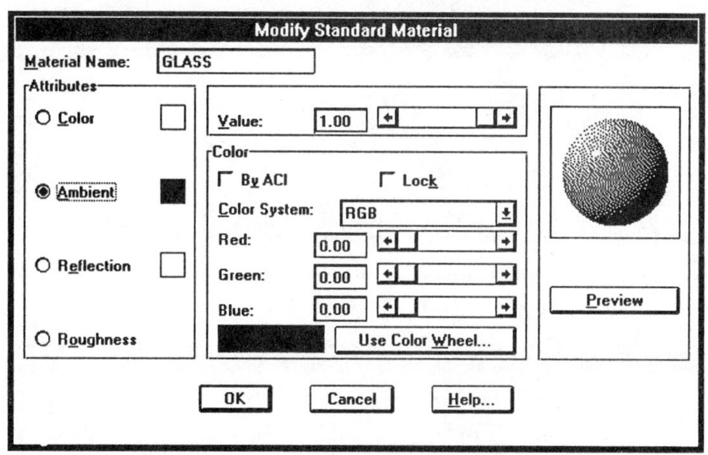

Duplicate Duplicates a material definition so that you can edit it.

New Creates a new material definition.

Attach Selecst the objects to attach the material definition.

Detach Selects the objects to dettach a material definition.

By ACI Attaches material to ACI number; displays dialogue box:

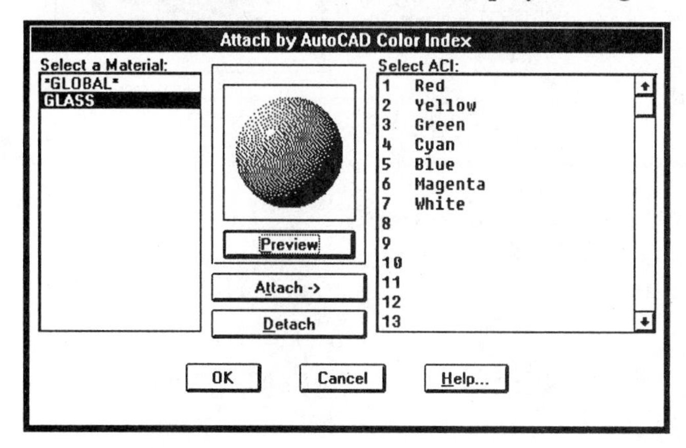

By Layer Attaches material to layer name.

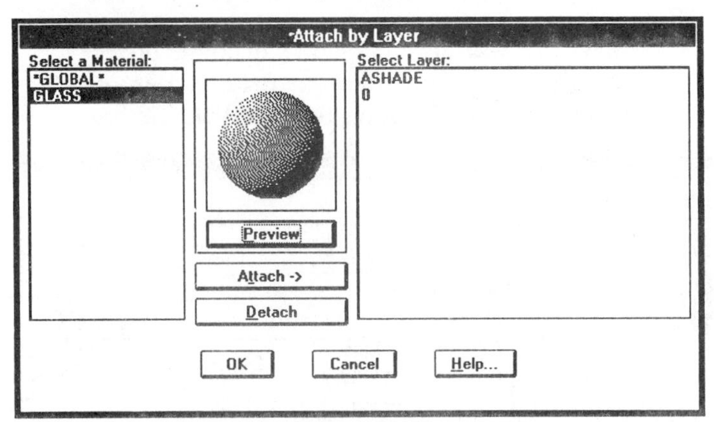

Color Selects color by RGB, HLS, or ACI specifications.

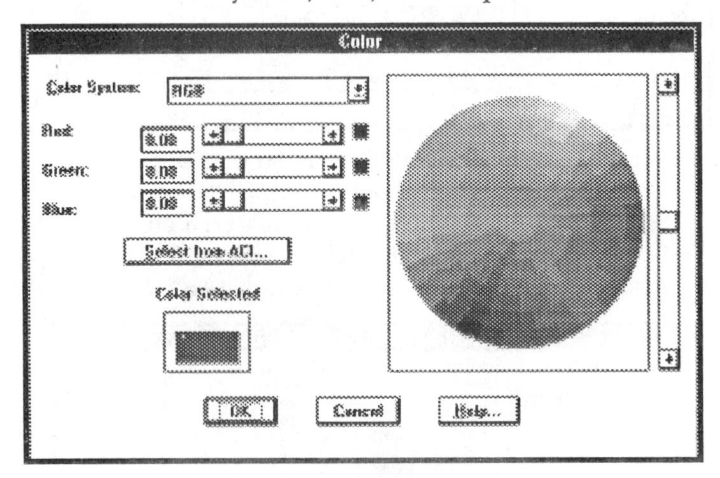

RELATED AUTOCAD COMMANDS

- **MatLib** Loads material definitions into drawing.
- **Render** Renders drawing using material definitions.

TIPS

- The **By ACI** option lets you attach a material definition to all objects of one color.

- The **By Layer** option lets you attach a material definition to all objects on one layer.

- One set of material definitions is in file \Acad13\Common\Support\Render.Mli. A more extensive set is in \Autovis\Av_supt\Autovis\Mli on the CD-ROM distribution disc.

DEFINITIONS

ACI

■ AutoCAD Color Index, the formal name for Autodesk's unique color numbering system (*see **Layer** command for list*).

GLOBAL

■ The default material definition in all drawings:

- ■ **Color** ACI #18
- ■ **Ambient** 0.10
- ■ **Reflection** 0.20
- ■ **Roughness** 0.50

HLS

■ The Hue-Lightness-Saturation method of defining colors.

RGB

■ The Red-Geen-Blue method of defining colors.

Material Definition

■ Defines a rendered surface texture by four parameters: color, ambient light, reflection, and roughness.

Rotate

Ver. 2.5

Rotates objects about a base point in a 2D plane.

Command	Alias	Side Menu	Toolbar	Tablet
rotate	...	[MODIFY] [Rotate:]	[Modify] [Rotate]	W 13

Command: **rotate**
Select objects: **[pick]**
Select objects: **[Enter]**
Base point: **[pick]**
<Rotation angle>/Reference: **R**
Reference angle <0>:
New angle:

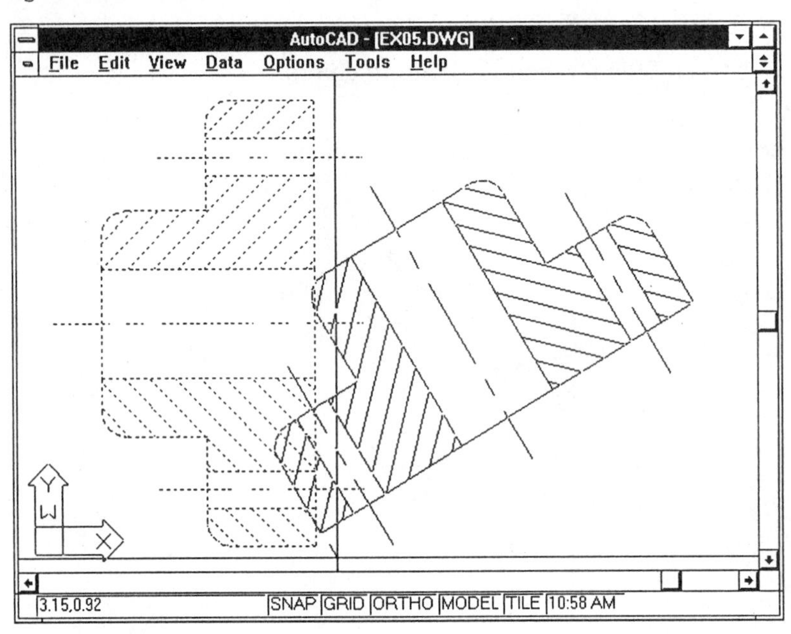

COMMAND OPTIONS

<Rotation angle>
Specifies the angle of rotation.

Reference Indicates a starting reference angle and an ending reference angle.

RELATED AUTOCAD COMMANDS

- **Change** Rotates text entities.
- **Rotate3D** Rotate objects in 3D spcae.

Rotate3D

Rel. 11

Rotates objects about an axis in 3D space (*an external command in Geom3d.Exp*).

Command	Alias	Side Menu	Toolbar	Tablet
rotate3d	...	[MODIFY]	[Modify]	...
		[Rotate:]	[Rotate]	
		[Rotat3D:]	[Rotate3D	

```
Command: rotate3d
Select objects: [pick]
Select objects: [Enter]
Axis by Object/Last/View/Xaxis/Yaxis/Zaxis/<2 points>:
<Rotation angle>/Reference: R
Reference angle <0>:
New angle:
```

COMMAND OPTIONS

Object Select object to specify rotation axis.
Last Select last-picked axis.
View Current view direction is the rotation axis.
Xaxis/Yaxis/Zaxis
X-, y- or z-axis is the rotation axis.
<2 points> Define two points on rotation axis.

RELATED AUTOCAD COMMANDS

- **Mirror3d** Mirrors objects in 3D space.
- **Rotate** Rotates objects in 2D space

RELATED SYSTEM VARIABLES

- *None*

RPref

Specify options for the **Render** command (*short for Render PREFerences; an external command in Render.Arx*).

Command	Alias	Screen menu	Toolbar	Tablet
rpref	...	[TOOLS] [RENDER] [Prefs:]	[Render] [RPref]	O 1

Command: **rpref**

Displays dialogue box.

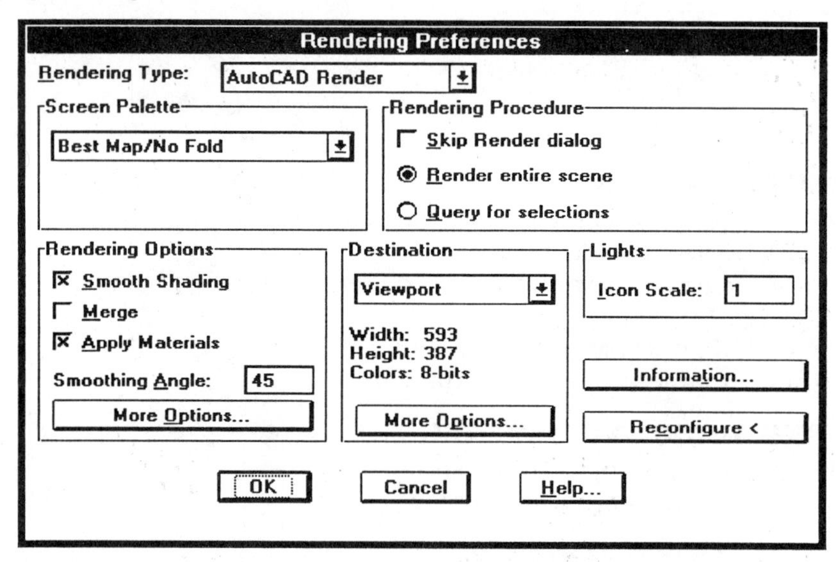

COMMAND OPTIONS

Rendering Type
Select from AutoCAD Render, AutoVision, and third-party renderers, if loaded into system.

Screen Palette
Select palette for 256-color graphics boards; not required for 16-bit or better graphics boards.

Smooth Shading
Toggle smooth or faceted shading.

Merge Combine multiple images.

Apply Materials
Apply texture mapping defined by the **RMat** command.

Smoothing Angle
Set the threshold angle at which **Render** smooths facets.

More Options Additional render options:

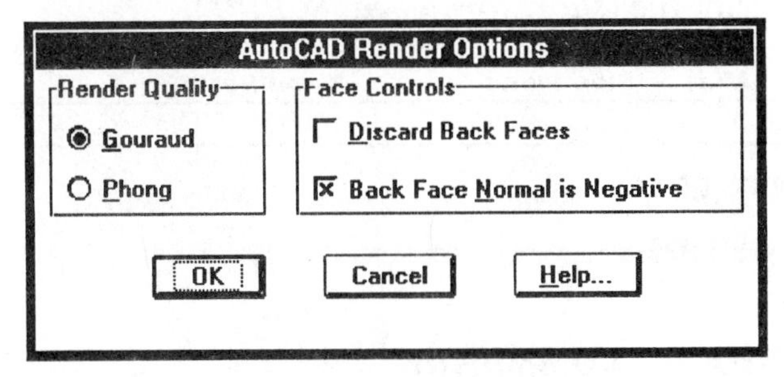

Skip Render Dialog
Create rendering without displaying **Render**'s dialogue box.

Render entire screen
Renders all objects in scene.

Query for selections
Select objects to render.

Destination Output rendering to viewport, window, or file.

More Options Additional file output options:

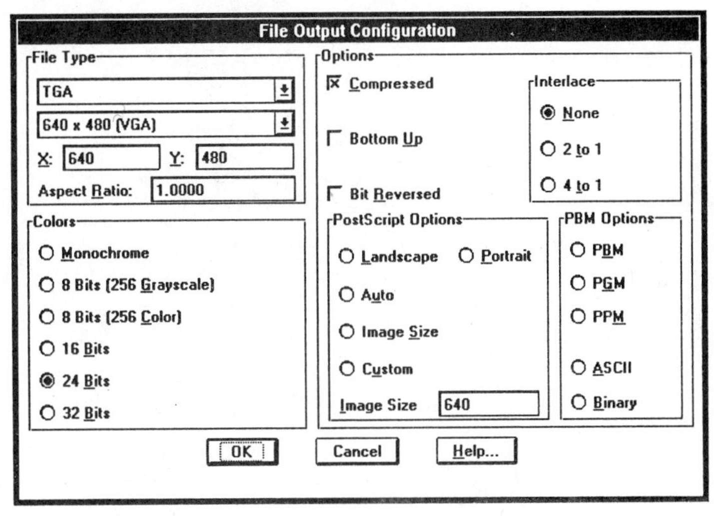

Lights Icon Scale
Size of light blocks relative to current scale factor.

Information Starts **Stats** command.

Reconfigure Starts **RConfig** command.

RELATED AUTOCAD COMMANDS

- *All rendering-related commands*

'RScript

Ver. 2.0

Repeats the script file (*short for Repeat SCRIPT*).

Command	Alias	Side Menu	Menu Bar	Tablet
'rscript	...	...	...	...

Command: **rscript**

COMMAND OPTIONS

None

RELATED AUTOCAD COMMANDS

- **Resume** Resumes a script file after being interupted.
- **Script** Loads and runs a script file.

RuleSurf

Rel. 10

Draws a 3D ruled surface between two objects (*short for RULEd SURFace*).

Command	Alias	Side Menu	Toolbar	Tablet
`rulesurf`	...	[DRAW 2] [SURFACES] [Rulsurf:]	[Surfaces] [RuleSurf]	...

```
Command: rulesurf
Select first defining curve: [pick]
Select second defining curve: [pick]
```

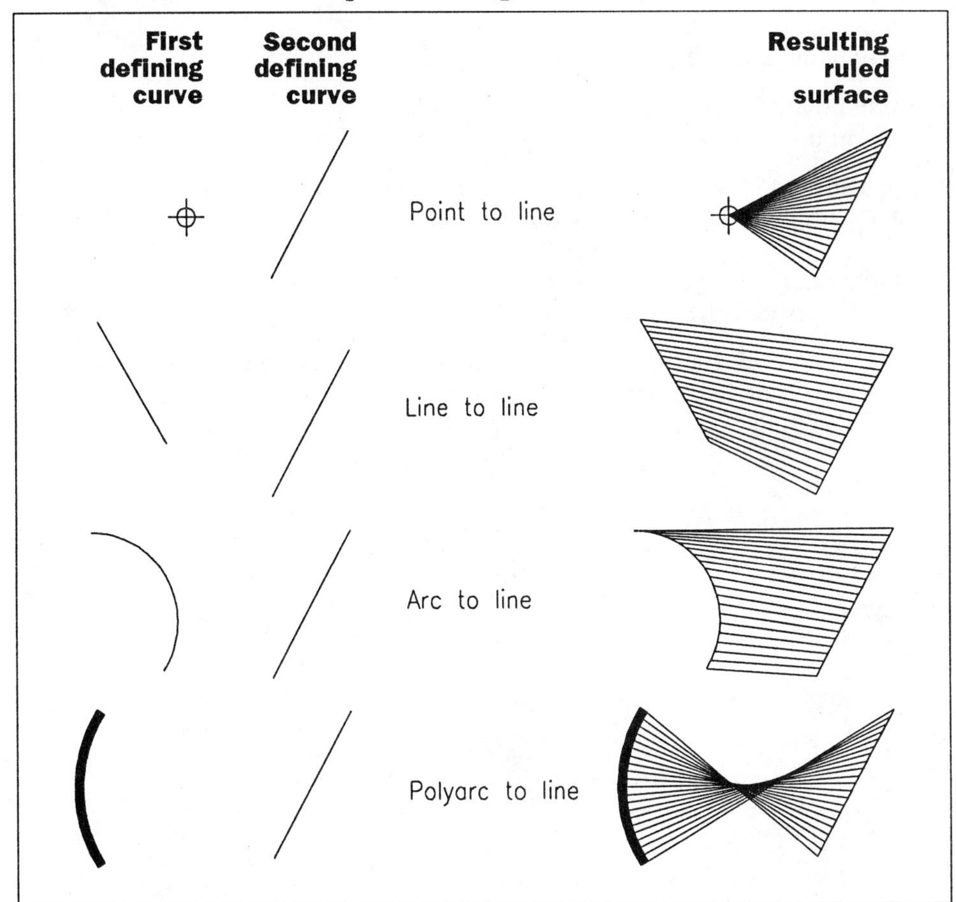

COMMAND OPTIONS

None

RELATED AUTOCAD COMMANDS

- **Edgesurf** Draws a 3D surface bounded by four edges.
- **Revsurf** Draws a 3D surface of revolution.
- **Tabsurf** Draws a 3D tabulated surface.
- **3D** Draws primitive 3D objects.

RELATED SYSTEM VARIABLE

- **Surftab1** Determines the number of rules drawn.

TIPS

- The **RuleSurf** command uses these objects as the boundary curve:
  - Point.
  - Line, arc, or circle.
  - Polyline and 3D polyline.

- If one boundary is closed, then the other boundary must also be closed; the exception is using a point as a boundary.

- The **RuleSurf** command begins drawing its mesh as follows:
  - **Open objects** From the object's endpoint closest to your pick point.
  - **Circles** From the zero-degree quadrant.
  - **Closed polylines** From the last vertex.

- Since **RuleSurf** draws its mesh with a circle in the opposite direction from a closed polylines, use a donut in place of the circle.

Ver. 2.0

Saves the drawing to disk, after always prompting for a filename.

Command	Alias	Side Menu	Menu Bar	Tablet
save	...	...	...	...

```
Command: save
```

Displays dialogue box.

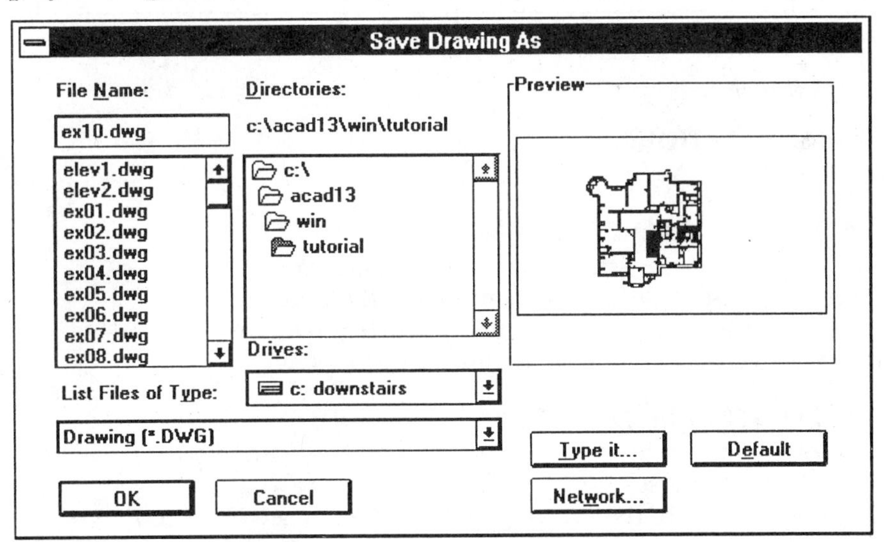

COMMAND OPTIONS

None

RELATED AUTOCAD COMMANDS

- **End** Saves the drawing and exits AutoCAD.
- **Quit** Exits AutoCAD without saving the drawing.
- **QSave** Saves drawing without prompting for name.
- **SaveAsR12** Saves drawing in Release 12 format.

RELATED SYSTEM VARIABLES

- **DbMod** Indicates that the drawing was modified.
- **DwgName** Name of the drawing; 'UNNAMED' when unnamed.

TIPS

- The **Save** command always displays the **Save Drawing** dialogue box, unlike other software applications. To avoid the dialogue box, use the **QSave** command.

- When the drawing is unnamed, the **Save** command mimics the **SaveAs** command and displays the **Save Drawing As** dialogue box.

SaveAs

Rel. 12

Saves the current drawing to disk as a Release 13-format DWG drawing file; when you save the drawing with different filename, the drawing takes on the new name.

Command	Alt+	Side Menu	Menu Bar	Tablet
saveas	F,A	[FILE] [SaveAs:]	[File] [Save As]	T 25

Command: **saveas**

Displays dialogue box.

COMMAND OPTIONS

None

RELATED AUTOCAD COMMANDS

- **End** Saves the drawing and exits AutoCAD.
- **Quit** Exits AutoCAD without saving the drawing.
- **Save** Saves the drawing with the current name.
- **SaveAsR12** Saves drawing in Release 12-format DWG.

RELATED SYSTEM VARIABLES

- **DbMod** Indicates whether the drawing was modified during the current editing session.
- **DwgName** Name of the drawing; 'UNNAMED' when unnamed.

SaveAsR12

Rel. 13

Saves the current drawing to disk as a Release 12-format DWG drawing file.

Command	Alias	Side Menu	Menu Bar	Tablet
saveas	...	[FILE] [EXPORT] [SaveR12:]	...	...

Command: **saveasr12**

Example log output:

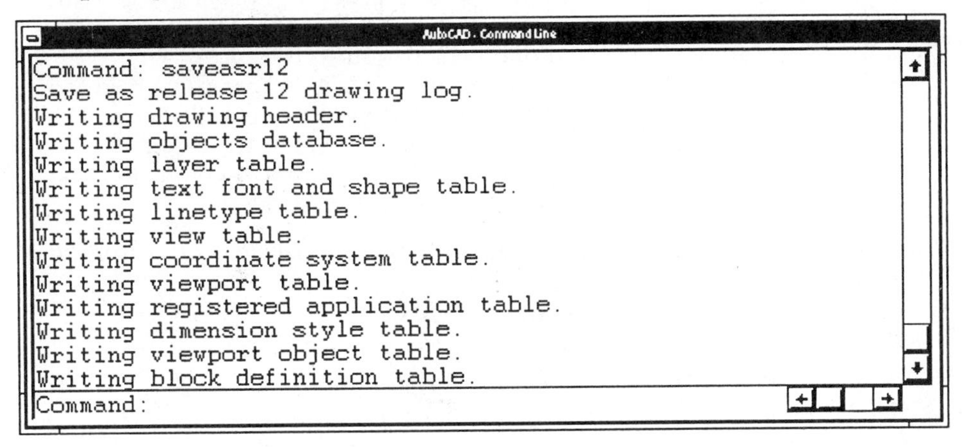

COMMAND OPTIONS

None

RELATED AUTOCAD COMMANDS

- **Quit** Exits AutoCAD without saving the drawing.
- **Save** Saves the drawing with the current name in Release 13 format.
- **SaveAs** Saves drawing by another name in Release 13-format.

RELATED SYSTEM VARIABLES

- **DwgName** Name of the drawing; 'UNNAMED' when unnamed.
- **HpBound** Draws boundary as a polyline boundary (*R12-compatible*)
- **PEllipse** Draws ellipsis as a polyarc (*R12-compatible*).

TIPS

- The **SaveAsR12** command translates the Release 13 drawing to a DWG file compatible with Release 12 by:
  - Converting R13-specific objects to R12 equivalents.
  - Stripping out objects that cannot be translated into R12.

- The following Release 13 objects are converted by the **SaveAsR12** command *(this list is more accurate than Autodesk's documentation)*:
  - Ellipse — Polyarc.
  - Multiline — Parallel polylines, arcs, filled arcs, and filled polygons.
  - Spline — Splined polyline.
  - Ray and Xline: Converted to lines, cut off at the drawing extents.
  - Hatch pattern: Associativity is dropped.
  - Dimension: Remains as an associative dimension with text intact.
  - Leader: Leader line becomes a polyline; arrowhead becomes a solid; MText becomes text.
  - Tolerance: Polylines and text; tolerance symbols are gibberish.
  - MText: Paragraph text becomes lines of text.
  - TrueType fonts: Converted to the TXT font.
  - ACIS solids: Bodies, 3D solids, and 2D regions are converted to a loose collection of polylines and arcs. To convert these objects to 3D polyfaces, use the **3dsOut** command, re-import with the **3dsIn** command, and then use the **SaveAsR12** command.

- These R13 objects are deleted *(this list is more accurate than Autodesk's documentation)*:
  - Text formatting codes specific to **MText**.
  - Shapes in linetypes; global linetype scale in **CeLtScale**.
  - **Rays** and **Xlines** outside of the drawing extents.
  - User-defined objects.
  - Groups and multiline styles.
  - OLE objects (from Windows version).
  - **XRef** overlays and ASE link information.
  - Preview BMP image.
  - **Render**'s material assignments.
  - Dictionary group codes, ADE lock bit, and object visibility flag and
  - All R13-specific system variables.

- Exert some control over SaveAsR12 with the **Explode** and **Xplode** commands, which convert complex objects – such as multilines and 3D solids – into polylines, arcs, and other simpler objects.

- During conversion, the **SaveAsR12** command displays a list changed and deleted objects (*called 'the drawing log'*); unfortunately, AutoCAD does not save the log to disk, unless you first type the **LogFileOn** command to capture the text screen to the Acad.Log file.

- There is *no* equivalent **SaveAsR12DXF** command for creating Release 12-compatible DXF files; the **Dxfix** utility provided with earlier releases of AutoCAD does not work.

SaveImg

Saves a rendered image as a GIF, TIFF, or Targa file on disk (*an external command in Render.Arx*).

Command	Alt+	Side Menu	Menu Bar	Tablet
`saveimg`	`T,G,S`	`[TOOLS]` `[SaveImg:]`	`[Tools]` `[Image]` `[Save]`	`...`

`Command:` **`saveimg`**

Displays dialogue box.

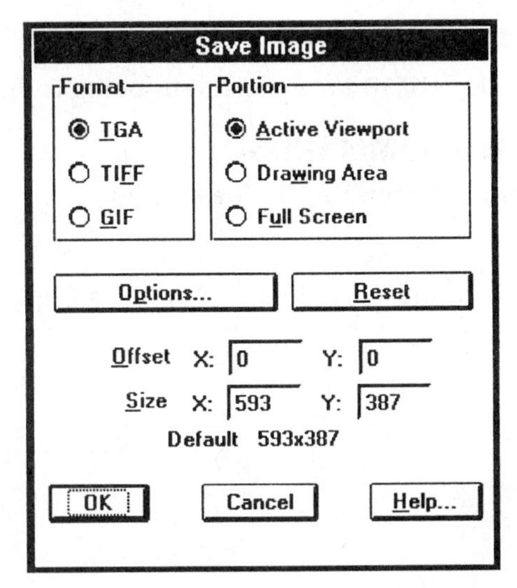

COMMAND OPTIONS

- **Image Name** Name of file.
- **Directory** Subdirectory name.
- **Portion** Selects the portion of image to be saved.
- **Format** TGA (Targa), TIFF, or GIF format.
- **Options** Options for TGA and TIFF output; displays dialogue box:

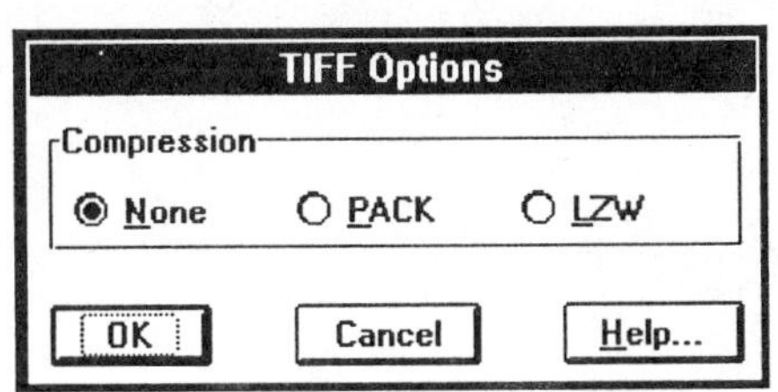

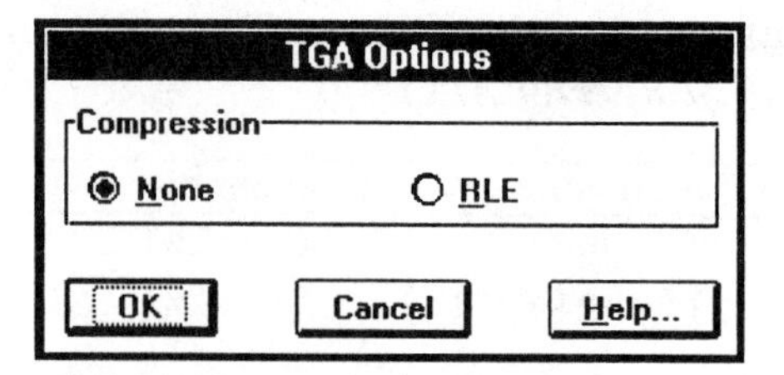

Reset Reset values to their original settings.
Offset Offset distance in pixels, where 0,0 is lower-left corner.
Size Upper-right distance in pixels.

RELATED AUTOCAD COMMAND

- **SaveImage** Saves a thumbnail image of the drawing to BMP file.

RELATED WINDOWS COMMANDS

- **[Prt Scr]** Save entire screen to Windows Clipboard.
- **[Alt]+[Prt Scr]** Save the topmost window to Windows Clipboard.

Scale

Ver. 2.5

Changes the size of selected objects, making them smaller or larger.

Command	Alias	Side Menu	Toolbar	Tablet
scale	...	[MODIFY] [Scale:]	[Modify] [Resize] [Scale]	W 12

```
Command: scale
Select objects: [pick]
Select objects: [Enter]
Base point: [pick]
<Scale factor>/Reference: r
Reference length <1>:
```

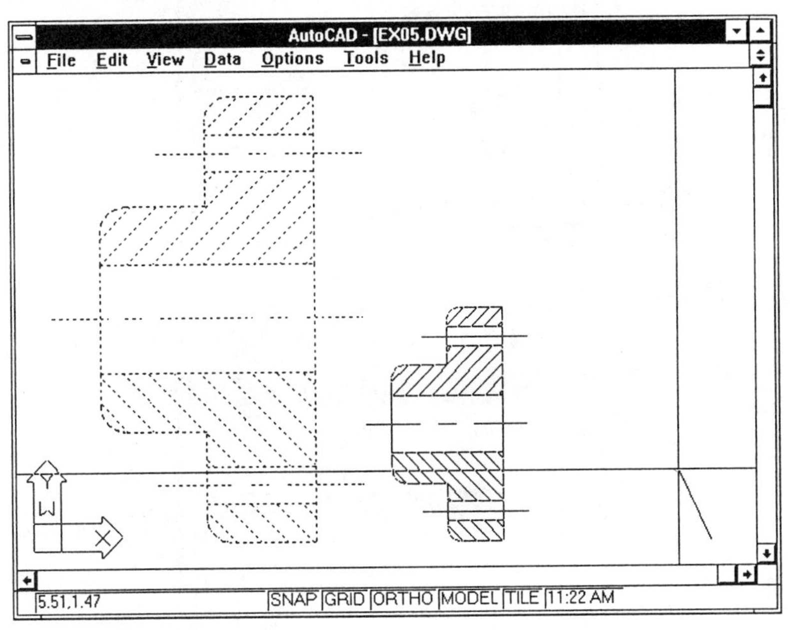

COMMAND OPTIONS

Reference Supplies a reference value.

<Scale factor> Indicates scale factor, which applies equally in the x-, y- and z-directions.

RELATED AUTOCAD COMMANDS

- **Insert** Allows a block to be scaled independently in the x-, y-, and z-directions.
- **Plot** Allows a drawing to be plotted at any scale.

Scene

Rel. 12

Collects lights and a viewpoint into a named scene (*an external command in Render.Arx*).

Command	Alias	Side Menu	Toolbar	Tablet
scene	...	[TOOLS] [RENDER] [Scene:]	[Render] [Scene]	L 2

Command: **scene**

Displays dialogue box.

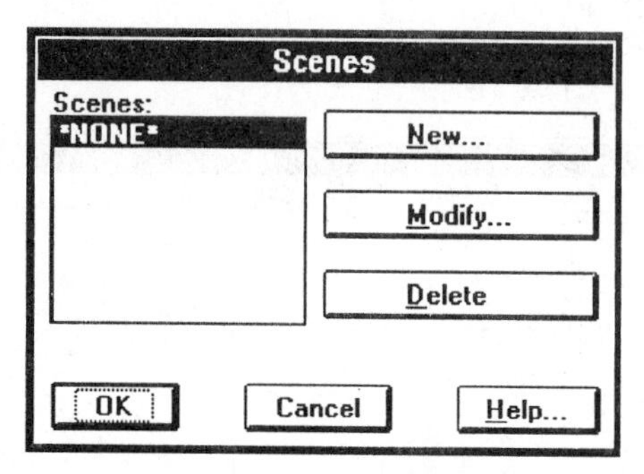

COMMAND OPTIONS

New Creates a new named scene; displays dialogue box.

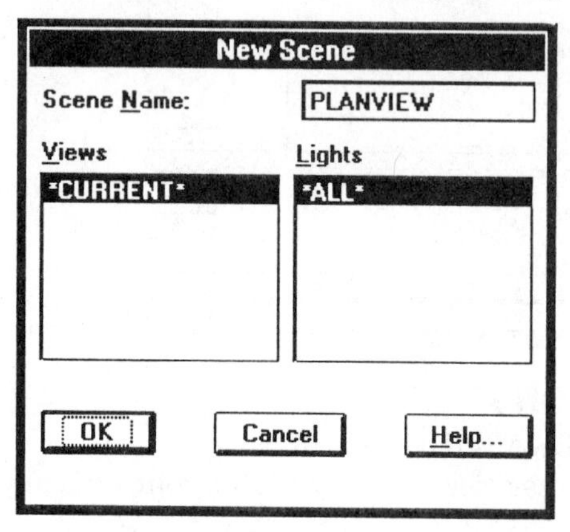

Scene name Enters a name for the scene.
Views Selects a named view.
Lights Selects one or more lights.

Modify Changes an existing scene definition.

Delete Deletes a scene from drawing.

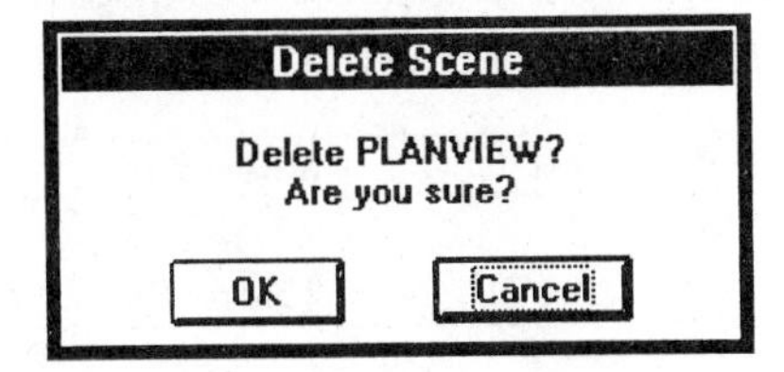

RELATED COMMANDS

- **Light** Places lights in the drawing for **Scene** command.
- **Render** Uses scenes to create renderings.
- **View** Creates named views for the **Scene** command.

TIPS

- Before you can use the **Scene** command, you need to create:
    - At least one named view with the **View** command; or
    - Place at least one light with the **Light** command.

Otherwise, there is no need to use the **Scene** command.

- If you select no lights for a scene, **Render** uses ambient light.
- Scene parameters are stored as attributes definitions in a block.

'Script

Ver. 1.4

Runs an ASCII file containing a sequence of AutoCAD instructions to execute a series of commands as if typed at the keyboard.

Command	Alt+	Side Menu	Menu Bar	Tablet
'script	T,R	[TOOLS]	[Tools]	...
		[Script:]	[Run Script]	

```
Command: script
Script file <>:
```

COMMAND OPTIONS

[Backspace] Interrupts the script.
[Esc] Stops the script.
~ Displays the file dialogue box.

RELATED AUTOCAD COMMANDS

- **Delay** Pauses, in milliseconds, before executing the next command.
- **Resume** Resumes a script after a script has been interrupted.
- **RScript** Repeats a script file.

TIPS

- Since the **Script** command is a transparent command, it can be used during another command.

- Prefix the **VSlide** command to preload it into memory; this results in a faster slide show:

```
*vslide
```

- AutoCAD can start with a script file on the command line:

```
C:\ acad13 dwgname scrname
```

Since the script filename must follow the drawing filename, use a dummy drawing filename, such as 'X'.

- You can make a script file more flexible (*pause for user input, branch with conditionals, and so on*) by inserting AutoLISP functions.

QUICK START: Writing a script file.

1. A script file consist of the exact keystrokes you type for a series of command. The script file must be plain ASCII text, so write the script file using a text editor, such as Notepad, rather than a word processor.

2. Here is an example script that places a door symbol in the drawing:

```
; Inserts DOOR2436 symbol at x,y = (76,100)
; x-scale = 0.5, y-scale = 1.0, rotation = 90 degrees
insert door2436 76,100 0.5 1.0 90
```

3. In the script, these characters have special meaning:

- (*Space or end-of-line*) Equivalent to pressing the spacebar or **[Enter]** key.
- ; (*Semicolon*) Allows a comment in the script file.
- * (*Asterisk*) Prefixes the **VSlide** command to preload the SLD file.

4. Save script file with any 8-character file name and the .SCR extension. For this example, use 'InsDoor.Scr'.

5. Return to AutoCAD and run the script with the **Script** command:

```
Command: script
Script file: insdoor.scr
Command: insert
Block name (or ?): door2436
Insertion point: 76,100
X scale factor <1>/Corner/XYZ: 0.5
Y scale factor (default = X):1.0
Rotation angle <0>:90
```

6. Rerun the script with the **RScript** command.

Section

Rel. 11

Creates a 2D region object from the intersection of a plane and a 3D solid (*formerly the **SolSect** command; an external command in Acis.Dll*).

Command	Alias	Side Menu	Toolbar	Tablet
section	...	[DRAW 2] [SOLIDS] [Section:]	[Solids] [Section]	Y 16

```
Command: section
Select objects: [pick]
Select objects: [Enter]
Section plane by Object/Zaxis/View/XY/YZ/ZX/<3 points>:
```

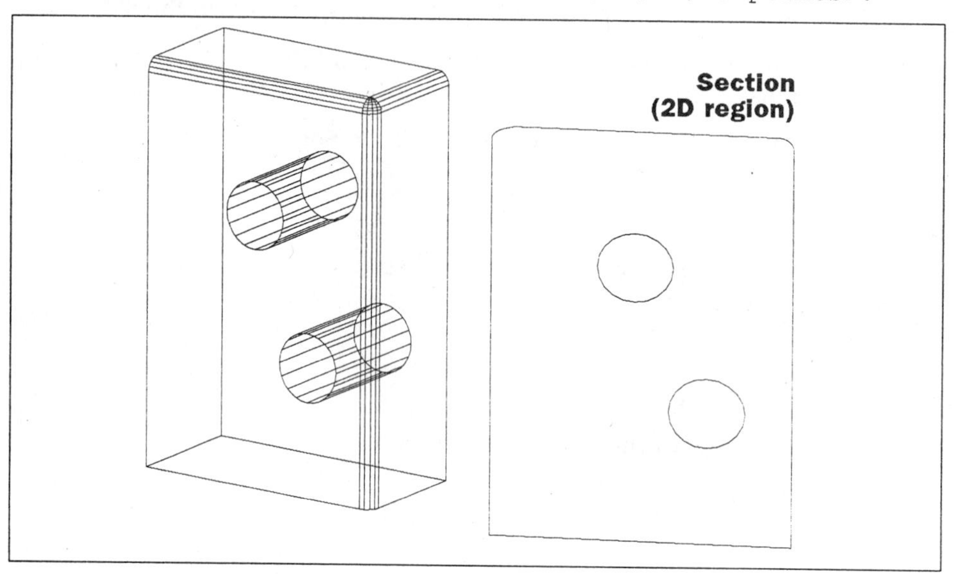

COMMAND OPTIONS

Object Align section plane with an object:
- Circle or ellipse.
- Arc or elliptical arc.
- 2D spline or 2D polyline.

Zaxis Specify the normal (*z-axis*) to the section plane.
View Use the current view plane as the section plane.
XY Use the x,y-plane of the current view.
YZ Use the y,z-plane of the current view.
ZX Use the z,x-plane of the current view.
<3 points> Pick three points to specify the section plane.

RELATED AUTOCAD COMMAND

- **Slice** Cuts a slice out of a solid model.

TIPS

- Section blocks are placed on the current layer, not the object's layer.
- Regions are ignored.
- One cutting plane is required for each selected solid.
- The **Last** option was removed from Release 13.

Select

Ver. 2.5

Creates a selection set of objects before executing a command.

Command	Alias	Side Menu	Toolbar	Tablet
select	...	[ASSIST] [SERVICE]	[Standard] [Select Objects]	U14-U22

```
Command: select
Select objects: [pick]
```

COMMAND OPTIONS

A Continues to add objects after using the **R** option (*short for Add*).

AU Switches from **[pick]** to **C** or **W** modes, depending on whether an object is found at the initial pick point (*short for AUtomatic*).

ALL Selects all objects in the drawing.

BOX Goes into **C** or **W** mode, depending on how the cursor moves.

C Selects objects in and crossing the selection box (*Crossing*).

CP Selects all objects inside and crossing the selection polygon.

F Selects all objects crossing a polyline (*short for Fence*).

G Selects objects contained in a named group (*short for Group; new to Release 13*).

L Selects the last-drawn object still visible on the screen (*Last*).

M Makes multiple selections before AutoCAD scans the drawing; saves time in a large drawing (*short for Multiple*).

P Selects the previously selected objects (*short for Previous*).

R Removes objects from the selection set (*short for Remove*).

SI Selects only a single set of objects before terminating **Select** command (*short for SIngle*).

U	Removes the most-recently added selected objects (*short for Undo*).
W	Selects all objects inside the selection box (*short for Window*).
WP	Selects a objects inside the selection polygon (*short for windowed polygon)*.
[pick]	Selects a single object.
[Enter]	Exits the **Select** command.
[Esc]	Aborts the **Select** command.

RELATED AUTOCAD COMMAND

- **Filter** Specifies objects that are added to the selection set.

RELATED SYSTEM VARIABLES

- **PickAdd** Controls how objects are added to a selection set.
- **PickAuto** Controls automatic windowing at the 'Select objects:' prompt.
- **PickDrag** Controls method of creating a selection box.
- **PickFirst** Controls command-object selection order.

TIP

- The selection set is lost when you switch AutoCAD between model and paper space.

'SetVar

Ver. 2.5

Lists the settings of system variables; allows you to change variables that are not read-only (*short for SET VARiable*).

Command	Alias	Side Menu	Menu Bar	Tablet
'setvar	...	[OPTIONS] [Sys Var:]	...	...

```
Command: setvar
Variable name or ?:
```

Example usage:

```
Command: setvar
Variable name or ?: visretain
New value for VISRETAIN <0>: 1
```

COMMAND OPTIONS

Variable name Indicates the system variable name you want to access.

? Lists the names and settings of system variables.

TIPS

■ See Appendix A for the complete list of all system variables found in AutoCAD Release 13.

■ Almost all system variables can be entered without the **SetVar** command. For example,

```
Command: visretain
New value for VISRETAIN <0>: 1
```

■ The following system variables are not listed by the **SetVar** command (*these system variables are used by third-party programmers, for debugging, or are obsolete*):

- **_LInfo**, **_PkSer**, and **_Server**
- **AuxStat**, **AxisMode**, and **AxisUnit**
- **DbglInstall**
- **EntExts**, **EntMods**, and **ErrNo**
- **Flatland** and **Force_Paging**
- **GlobCheck**
- **LazyLoad**
- **MaxObjMem** and **MacroTrace**
- **NodeName**
- **PHandle**
- **QaFlags**
- **Re-Init**
- **UserI1** through **UserI5**, **UserR1** through **UserR5**, and **UserS1** through **UserS5**

Shade

Rel. 11

Performs 16- and 256-color shaded renderings, and quick hidden-line removal of 3D drawings.

Command	Alias	Side Menu	Toolbar	Tablet
shade	...	[TOOLS]	[Render]	O 2
		[SHADE]	[Shade]	
		[Shade]		N 2

```
Command: shade
Regenerating drawing.
Shading 50% done.
Shading complete.
```

ShadEdge = 0

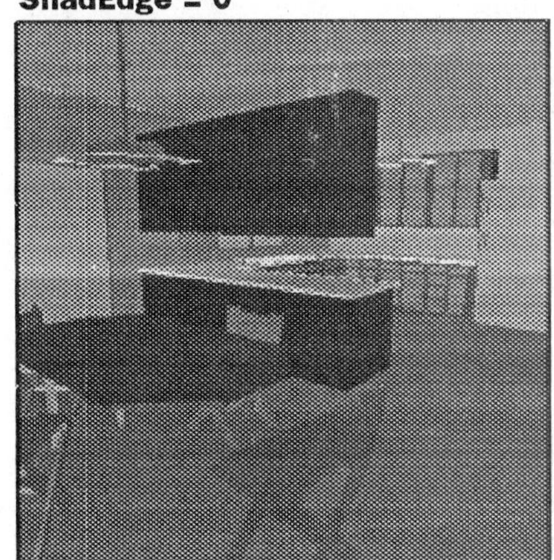

ShadEdge = 1

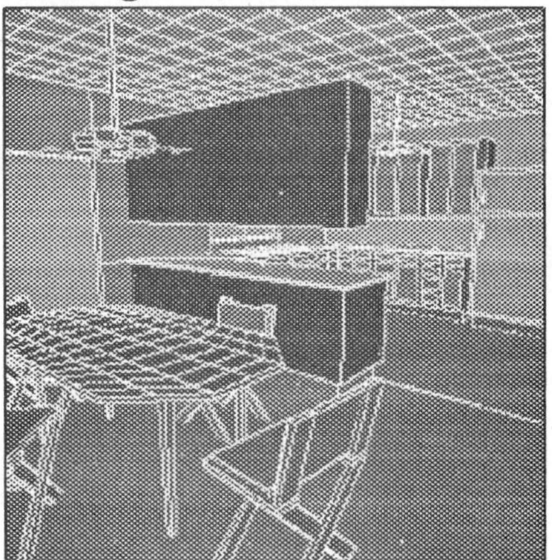

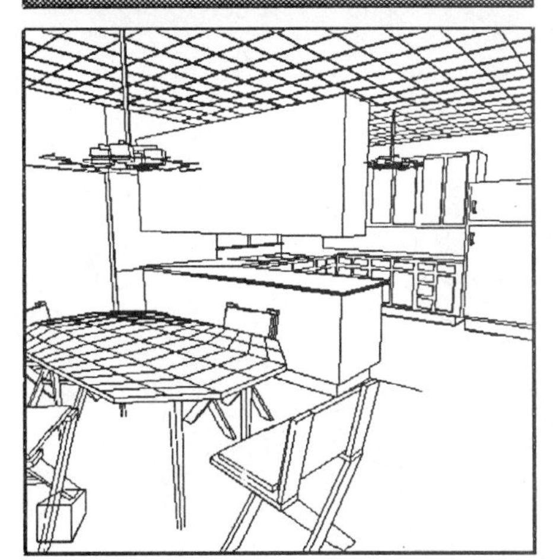

ShadEdge = 2

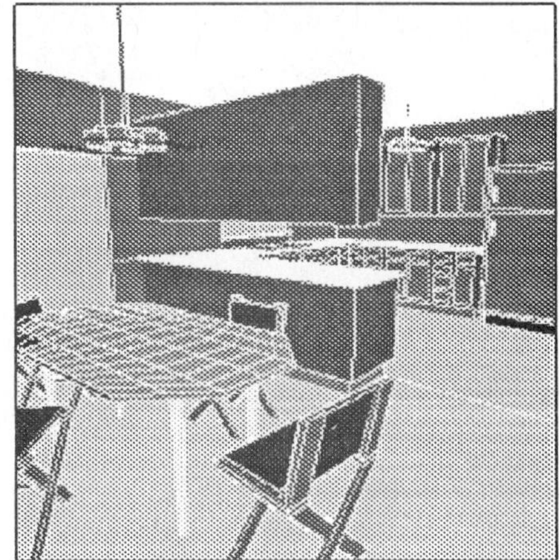

ShadEdge = 3

COMMAND OPTIONS

None

RELATED AUTOCAD COMMANDS

- **DView** Does hidden-line removal of perspective views.
- **Hide** Does true hidden-line removal of 3D drawings.
- **MSlide** Saves a rendered view as an SLD-format slide file.
- **MView** Does a hidden-line view of individual viewports during plots and prints.
- **Plot** Does a hidden-line view during plotting.
- **Render** Performs a more realistic rendering.

RELATED SYSTEM VARIABLES

- **ShadEdge** Determines the style of shading:
  - 0 256-color shading.
  - 1 256-color shading with outlined polygons.
  - 2 Hidden-line removal.
  - 3 16-color shading (*default*).
- **ShadeDif** Determines the shading contrast (*default* = *70*).

TIPS

- As an alternative to the **Shade** command, the **Render** module does high-quality renderings of 3D drawings but takes longer to complete the rendering.

- The smaller the viewport, the faster the rendering.

- The more accurate hidden-line algorithm (*known as the "Release 11" algorithm in Release 12*) was removed from Release 13.

Shape

Ver. 1.0

Inserts a predefined shape in the current drawing; shapes are more compact than blocks but are more difficult to create.

Command	Alias	Side Menu	Toolbar	Tablet
shape	...	[DRAW 2] [Shape]	[Miscellaneous] [Shape]	...

```
Command: shape
Shape name (or ?):
Starting point: [pick]
Height <>:
Rotation angle <0>:
```

Some shapes included in Es.Shp.

COMMAND OPTIONS

Shape name Indicates the name of the shape to insert.

? Lists the names of currently loaded shapes.

RELATED AUTOCAD COMMANDS

- **Load** Loads an SHX-format shape file into the drawing.
- **Insert** Inserts a block into the drawing.
- **Style** Loads SHX font files into the drawing.

RELATED SYSTEM VARIABLE

- **ShpName** Current SHP filename.

TIPS

- Shapes are defined by SHP files, which must first be compiled into SHX files before they can be loaded by the **Load** command.
- Compile an SHP file into an SHX file with the **Compile** command.
- AutoCAD comes with three SHX shape files, located in the \Acad13\Common\Sample subdirectory:

  Es.Shx Electronic component shapes (*see figure*).

  Pc.Shx Printed circuit board shapes.

  St.Shx Surface texture shapes for mechanical parts drawings.

Shell

Ver. 2.5

Temporarily exit AutoCAD to the DOS operating system (*an external command defined in Acad.Pgp*).

Command	**Alias**	**Side Menu**	**Menu Bar**	**Tablet**
shell	sh	...	[Tools]	Y24 - 25

```
Command: shell
OS Command:
```

COMMAND OPTIONS

[Enter] Stays in DOS for more than one command.

Exit Returns to AutoCAD from DOS.

RELATED COMMANDS

- **End** Exits AutoCAD back to the Windows desktop.

The following commands are defined by Acad.Pgp:

- **Catalog** Equivalent to the DOS command DIR /W.
- **Del** Executes the DOS command DEL.
- **Dir** Executes the DOS command DIR.
- **Edit** Executes the DOS program EDIT.
- **Type** Executes the DOS command TYPE.

RELATED FILE

- **Acad.Pgp** The external command definition file, in subdirectory \Acad13\Common\Support.

QUICK START: Adding a command to the Acad.Pgp file.

1. Load the Acad.Pgp file into the Notepad text editor.

2. The PGP file uses this format to add a command:

```
CommandName, [DOS request], MemoryReserve, [*]Prompt, ReturnCode
```

Meaning of the format:

■ **CommandName**	The name you type at AutoCAD's 'Command:' prompt.
■ **DOS Request**	The command AutoCAD feeds to DOS.
■ **MemoryReserve**	Always 0 (*a holdover from older versions*).
■ **Prompt**	A phrase to prompt user action.
■ ***Prompt**	User reponse to prompt may contain spaces.
■ **ReturnCode 0**	Return to AutoCAD's text sreen.
1	Load $Cmd.Dxb file into drawing upon return.
2	Load $Cmd.Dxb as a block into the drawing.
4	Return to AutoCAD's previous screen mode (*usually the graphics screen*).

3. For our example, we want to add fast access to the WordPerfect for DOS word processor. Add the this line anywhere in the Acad.Pgp file:

```
WP, WP, 0, File to edit: ,4
```

4. Save the file and return to AutoCAD.

5. Use the **ReInit** command to reload the Acad.Pgp file.

6. Enter **WP** at the Command: prompt:

```
Command: wp
```

7. AutoCAD shells out to DOS and prompt you:

```
File to edit:
```

Enter the name of a text file; AutoCAD launches WordPerfect with the file.

7. To return to AutoCAD, hold down the **[Alt]** key and press the **[Tab]** key until AutoCAD shows up.

ShowMat

Rel. 13

Lists the material attached to an object (*short for SHOW MATerial; an undocumented command in Render.Arx*).

Command	Alias	Side Menu	Menu Bar	Tablet
showmat	...	...	...	...

```
Command: showmat
Select object: [pick]
```

Example output:

```
Material BRONZE is explicitly attached to the object.
```

COMMAND OPTIONS

None

RELATED AUTOCAD COMMANDS

- **MatLib** Loads material definitions into the drawing.
- **RMat** Attaches materials to objects, colors, and layers.

Sketch

Ver. 1.4

Allows freehand drawing as lines or polylines.

Command	Alias	Side Menu	Toolbar	Tablet
sketch	..	[DRAW 1] [Sketch:]	[Miscellaneous] [Sketch]	...

```
Command: sketch
Record increment <0.1000>: [Enter]
Sketch. Pen eXit Quit Record Erase Connect .
```

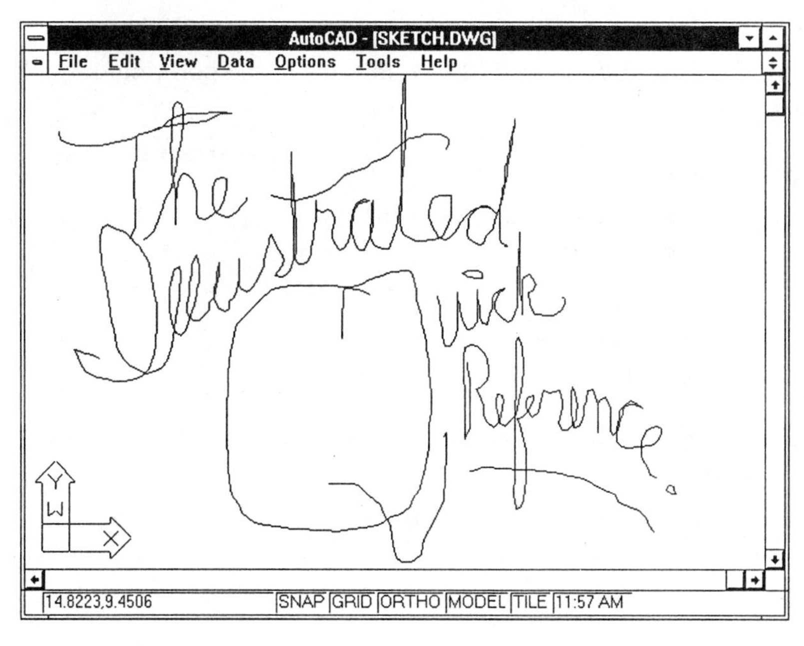

COMMAND OPTIONS

Commands can be invoked by digitizer buttons:

Connect Connect to the last drawing segment (*button #6*).

Erase Erase temporary segments as the cursor moves over them (*button #6*).

eXit Record the temporary segments and exit the **Sketch** command (*button #3*).

Pen Lifts and lowers the pen (*pick button #1*).

Quit Discard temporary segments and exit the **Sketch** command (*button #4*).

Record Record the temporary segments as permanent (*button #2*).

. (Period) Connects the last segment to the current point (*button #1*).

RELATED AUTOCAD COMMANDS

- **Line** Draws line segments.
- **PLine** Draws polyline and polyline arc segments.

RELATED SYSTEM VARIABLES

- **SketchInc** The current recording increment for **Sketch**.
- **SkPoly** Controls the type of sketches recorded:
  - **0** Record sketches as lines.
  - **1** Record sketches as polylines.

TIPS

■ During the **Sketch** command, the definitions of the pointing device's buttons change to:

Button Number	Meaning	Equivalent Keystroke
0	Raise and lower the ***p***en	P
1	Draw line to current ***point***	.
2	***R***ecord sketch	R
3	Record sketch and e***X***it	X
4	Discard sketch and ***Q***uit	Q
5	***E***rase sketch	E
6	***C***onnect to last-drawn segment	C

■ Only the first three (*or two*) button-commands are available on three- (*or two-*) button mice.

■ Pull-down menus are unavailable during the **Sketch** command.

Slice

Rel. 11

Cuts a 3D solid with a plane, creating two 3D solids *(formerly the SolCut command; an external command in Acis.Dll)*.

Command	Alias	Side Menu	Toolbar	Tablet
solcut	...	[DRAW 2]	[Solids]	Y17
		[SOLIDS]	[Slics]	
		[Slice:]		

```
Command: slice
Select objects: [pick]
Select objects: [Enter]
Slicing plane by Object/Zaxis/View/XY/YZ/ZX/<3 points>:
Both sides/<Point on desired side of the plane>:
```

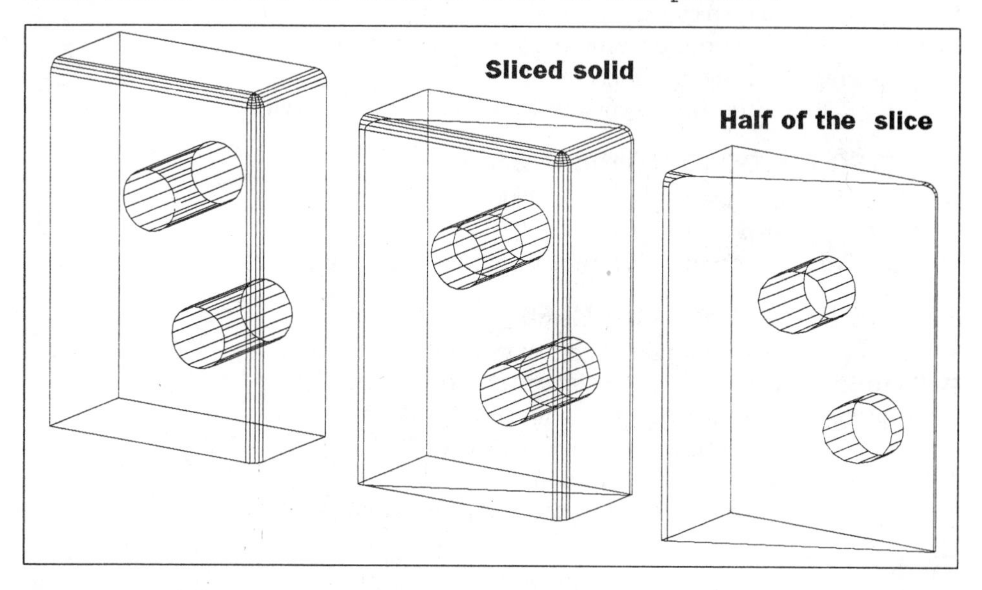

COMMAND OPTIONS

Object	Aligns cutting plane with a circle, ellipse, arc, elliptical arc, 2D spline, or 2D polyline.
View	Aligns cutting plane with viewing plane
XY	Aligns cutting plane with x,y-plane of current UCS
YZ	Aligns cutting plane with y,x-plane of current UCS
Zaxis	Aligns cutting plane with two normal points
ZX	Aligns cutting plane with z,x-plane of current UCS
<3 points>	Aligns cutting plane with three points

Both sides Retains both halves of cut solid model

<Point on desired side of the plane>
Retains either half of cut solid model

'Snap

Ver. 1.0

Sets the drawing resolution, grid origin, isometric mode, and angle.

Command	Status line	Side Menu	Menu Bar	Tablet
'snap	SNAP	[ASSIST]	[Assist]	V 21
		[Snap:]	[Snap]	
	[F9]			

```
Command: snap
Snap spacing or ON/OFF/Aspect/Rotate/Style <1.0000>:
```

COMMAND OPTIONS

Aspect Set separate x- and y-increments.
OFF Turn snap off.
ON Turn snap on.
Rotate Rotate the crosshairs for snap and grid.
Snap spacing Set the snap increment.
Style Switch between standard and isometric style.

RELATED AUTOCAD COMMANDS

- **DdRModes** Set snap values via a dialogue box.
- **Grid** Turn on the grid.
- **Isoplane** Switch to a different isometric drawing plane.

RELATED SYSTEM VARIABLES

- **SnapAng** Current angle of the snap rotation.
- **SnapBase** Base point of the snap rotation.
- **SnapIsopair** Current isometric plane setting.
- **SnapMode** Determines whether snap is on.
- **SnapStyl** Determines style of snap.
- **SnapUnit** The current snap increment in x- and y-directions.

TIPS

- You toggle snap mode by double-clicking **SNAP** on the status bar.
- The **Snap** command's **Style** option toggles isometric mode.
- The **Snap** command's **Rotate** option lets you change the origin point for hatching.

Solid

Ver. 1.0

Draws solid filled triangles and quadrilaterals; does *not* create a 3D ACIS solid.

Command	Alias	Side Menu	Menu Bar	Tablet
solid	...	[DRAW 1] [Solid:]	[Draw] [Polygon] [Solid]	...

```
Command: solid
First point: [pick]
Second point: [pick]
Third point: [pick]
Fourth point: [pick]
```

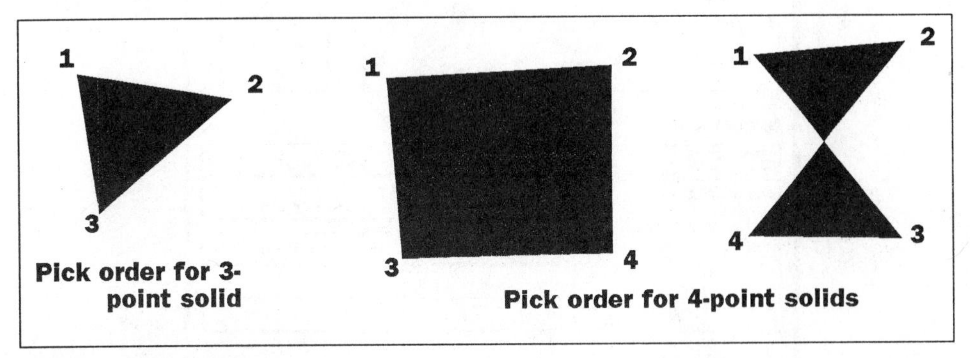

COMMAND OPTIONS

None

RELATED AUTOCAD COMMANDS

- **Fill** Turns object fill off and on
- **Trace** Draws lines with width.
- **PLine** Draws polylines and polyline arcs with width.

RELATED SYSTEM VARIABLE

- **FillMode** Determines whether solids are displayed filled or outlined.

Checks the spelling of text in the drawing (*an external command in AcSpell.Dll*).

Command	Alt+	Side Menu	Menu Bar	Tablet
spell	T,S	[TOOLS] [Spell:]	[Tools] [Spelling]	T 4

```
Command: spell
Select objects:
```

If misspelled text is found, displays dialogue box:

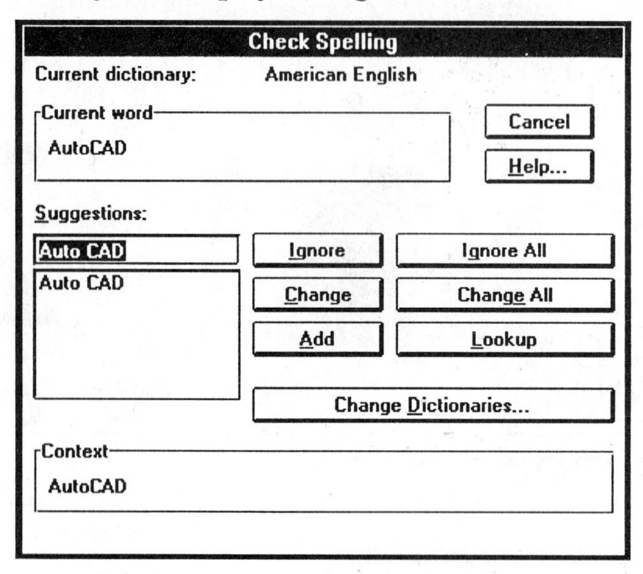

If selected text is spelled correctly:

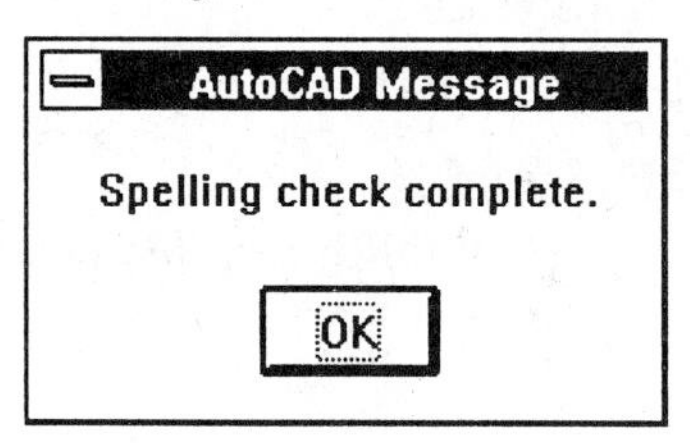

COMMAND OPTIONS

Ignore Ignores the spelling and go on to next word.
Ignore all Ignores all words with this spelling.
Change Changes to suggested spelling.
Change all Changes all words with this spelling.
Add Adds word to user dictionary.
Lookup Checks spelling of work in **Suggestions** box.
Change dictionaries
Selects a different dictionary.

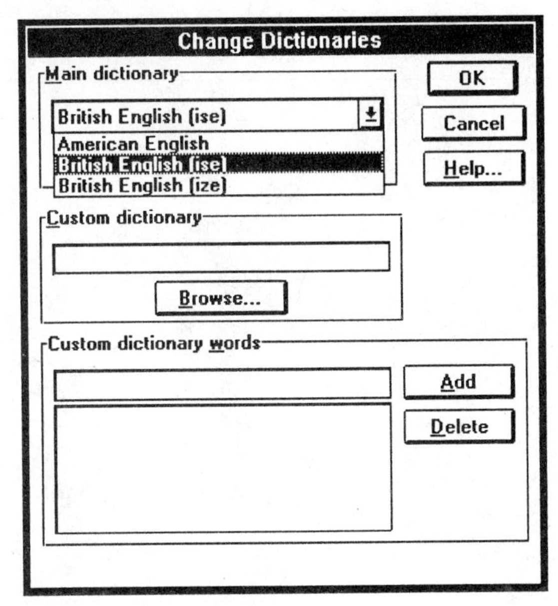

RELATED AUTOCAD COMMAND

- **DdEdit** Edits text.

RELATED SYSTEM VARIABLES

- **DctMain** Name of main spelling dictionary.
- **DctCust** Name of custom spelling dictionary.

RELATED FILES

- **Enu.Dct** Dictionary word file in \Acad13\Common\Support.
- ***.Cus** Custom dictionary files.

Sphere

Rel. 11

Draws a 3D sphere as a solid model *(formerly the **SolSphere** command; an external command in Acis.Dll)*.

Command	Alias	Side Menu	Toolbar	Tablet
sphere	...	[DRAW 2] [SOLIDS] [Sphere:]	[Solids] [Sphere]	K 7

```
Command: sphere
Center of sphere <0,0,0>: [pick]
Diameter/<Radius> of sphere: [pick]
```

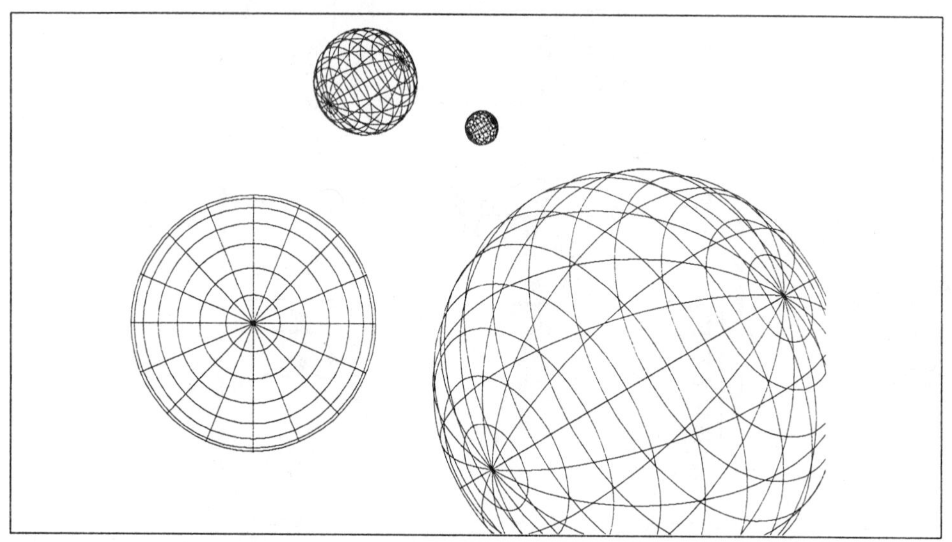

COMMAND OPTIONS

Center of sphere

Locates the center point of the sphere.

Diameter Specifies diamter of the sphere.

Radius Specifies radius of the sphere

RELATED AUTOCAD COMMANDS

- **Ai_Sphere** Draw surface meshed sphere.
- **Box** Draws solid boxes.
- **Cone** Draws solid cones.
- **Cylinder** Draws solid cylinders.
- **Torus** Draws solid tori.
- **Wedge** Draws solid wedges.
- **3D** Draws a surface meshed objects.

Spline

Rel. 13

Draws NURBS – non-uniform rational Bezier spline – curves (*an external command in Acis.Dll*).

Command	Alias	Side Menu	Toolbar	Tablet
spline	...	[DRAW 1]	[Draw]	K 4
		[Spline:]	[Spline]	

```
Command: spline
Object/<First point>: [pick]
Enter point: [pick]
Close/Fit tolerance/<Enter point>: [pick]
Close/Fit tolerance/<Enter point>: [Enter]
Enter start tangent: [pick]
Enter end tangent: [pick]
```

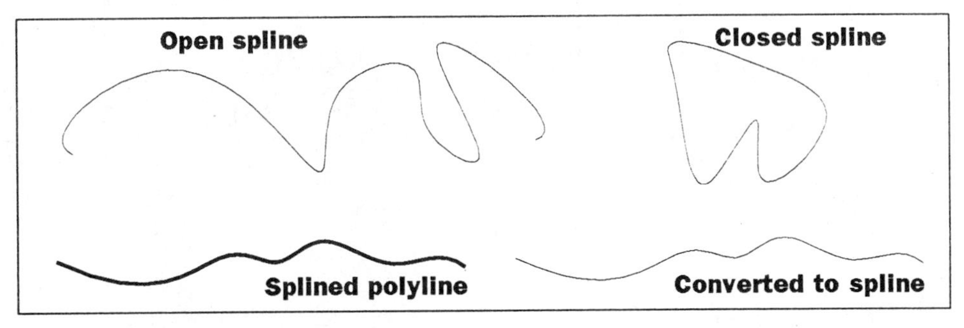

COMMAND OPTIONS

Close Closes spline at the start point.

Fit Changes spline tolerance; 0 = curve passes through fit points.

Object Converts 2D and 3D splined polylines into a NUBS spline.

RELATED AUTOCAD COMMANDS

- **PLine** Draws splined polyline.
- **SplinEdit** Edits a NURBS spline.

RELATED SYSTEM VARIABLE

- **DelObj** Toggles whether the original polyline is deleted with the Object option.

SplinEdit

Rel. 13

Edits a NURBS spline (*an external command in Acis.Dll*).

Command	Alias	Side Menu	Toolbar	Tablet
splinedit	...	[MODIFY] [SplinEd:]	[Modify] [Special Edit] [SplinEdit]	W 18

```
Command: splinedit
Select spline: [pick]
Fit data/Close/Move vertex/Refine/rEverse/Undo/eXit <X>: F
Add/Close/Delete/Move/Purge/Tangents/toLerance/eXit <X>:
```

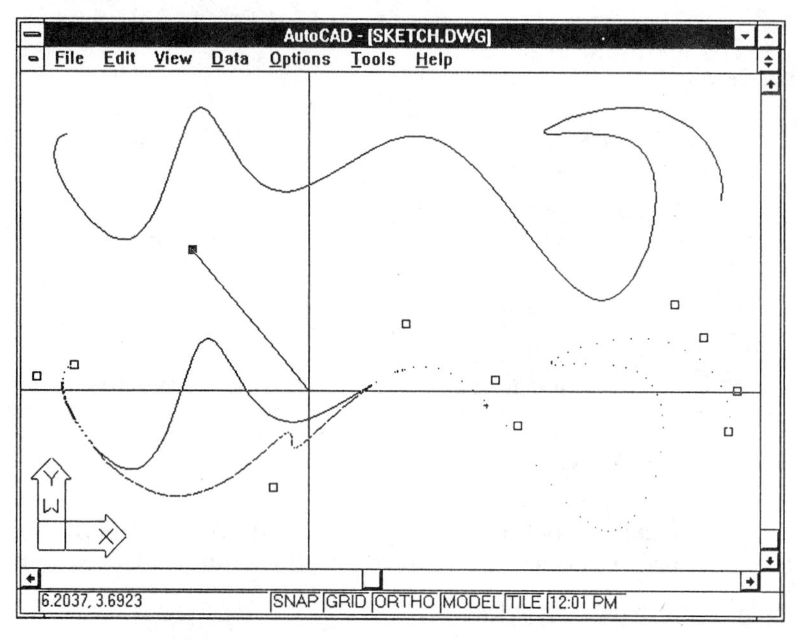

COMMAND OPTIONS

Fit data Edits the spline's fit points:

- **Add** Adds fit points.
- **Close** Closes the spline, if open.
- **Delete** Removes fit points.
- **Move** Moves fit points.
- **Open** Opens the spline, if closed.
- **Purge** Removes fit point data from drawing.
- **Tangents** Edits the start and end tangents.
- **toLerance** Refits spline with new tolerance value.
- **<eXit>** Exits suboptions.

Close	Closes the spline, if open.
Move vertex	Moves a control vertex.
Open	Opens the spline, if closed.
Refine	Adds a control point, change the spline's order or weight.
rEverse	Reverses the spline's direction.
Undo	Undoes the most-recent edit change.
<eXit>	Exits the **SplinEdit** command.

RELATED AUTOCAD COMMANDS

- **PEdit** Edits a splined polyline.
- **Spline** Draws a NURBS spline.

TIPS

- The spline looses its fit data when you use the following **SplinEdit** command options:
  - **Refine.**
  - **Fit Purge.**
  - **Fit Tolerance** followed by **Fit Move.**
  - **Fit Tolerance** followed by **Fit Open** or **Fit Close**.

- The maximum order for a spline is 26; once the order has been elevated, it cannot be reduced.

- The larger the 'weight,' the closer the spline is to the control point.

Stats

Rel. 12

Lists statistics of the most-recent rendering (*short for STATisticS; an external command in Render.Arx*).

Command	Alias	Side Menu	Toolbar	Tablet
stats	...	[TOOLS] [RENDER] [Stats:]	[Render] [Stats]	...

Command: **stats**
Displays dialogue box.

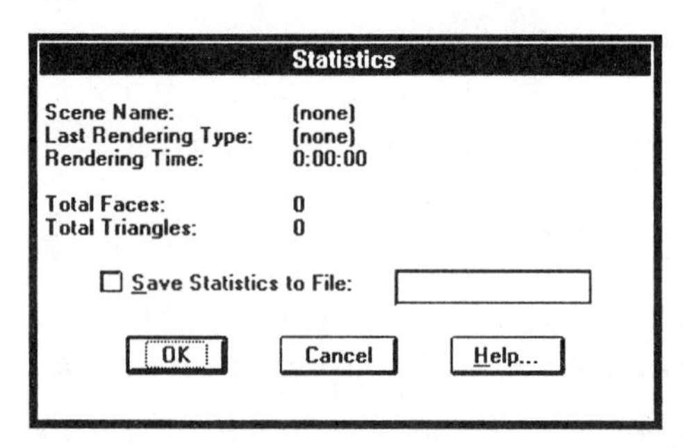

COMMAND OPTION

Save Statistics to File
Saves rendering statistics to file.

RELATED AUTOCAD COMMAND

- **Render** Creates renderings.

DEFINITIONS

Scene name
- Name of the currently selected scene.
- When no scene is current, displays '(none)'.

Last Rendering Type
- Name of currently selected renderer.
- Default is AutoCAD Render.

Rendering Time
- Time required to create most-recent rendering.
- Reported in HH:MM:SS (*hours, minutes, seconds*) format.

Total Faces
- Number of faces processed in most-recent rendering.
- A single 3D objects consists of many faces.

Total Triangles
- Number of triangles processed in most-recent rendering.
- A rectangular face is typically divided into two triangles.

'Status

Ver. 1.0

Displays information about the current drawing and environment.

Command	Alt+	Side Menu	Menu Bar	Tablet
'status	D,U	[DATA]	[Data]	...
		[STATUS:]	[Status]	

Command: **status**

Example output for the Acad.Dwg prototype drawing:

Command: **status**

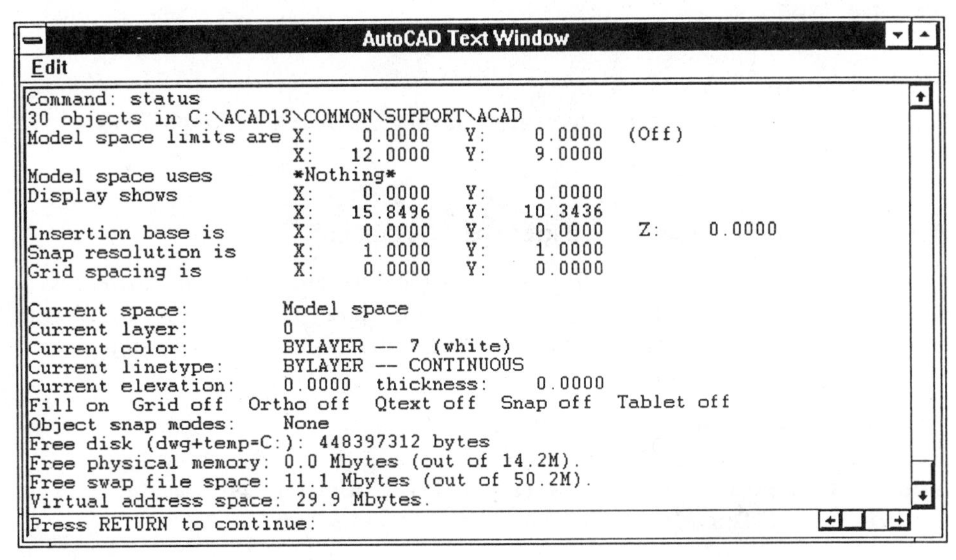

AutoCAD Text Window
Edit
Command: status
30 objects in C:\ACAD13\COMMON\SUPPORT\ACAD
Model space limits are X: 0.0000 Y: 0.0000 (Off)
X: 12.0000 Y: 9.0000
Model space uses *Nothing*
Display shows X: 0.0000 Y: 0.0000
X: 15.8496 Y: 10.3436
Insertion base is X: 0.0000 Y: 0.0000 Z: 0.0000
Snap resolution is X: 1.0000 Y: 1.0000
Grid spacing is X: 0.0000 Y: 0.0000
Current space: Model space
Current layer: 0
Current color: BYLAYER -- 7 (white)
Current linetype: BYLAYER -- CONTINUOUS
Current elevation: 0.0000 thickness: 0.0000
Fill on Grid off Ortho off Qtext off Snap off Tablet off
Object snap modes: None
Free disk (dwg+temp=C:): 448397312 bytes
Free physical memory: 0.0 Mbytes (out of 14.2M).
Free swap file space: 11.1 Mbytes (out of 50.2M).
Virtual address space: 29.9 Mbytes.
Press RETURN to continue:

COMMAND OPTION

[F2] Returns to the graphics screen.

RELATED AUTOCAD COMMANDS

- **DbList** Lists information about all objects in the drawing.
- **List** Lists information about selected objects.
- **Stats** Lists information of the most-recent rendering.

DEFINITIONS

Model space limits, paper space limits

- The x,y-coordinates stored in the **LimMin** and **LimMax** system variables.
- 'Off' indicates limits checking is turned off (*variable **LimCheck***).

Model space uses, paper space use
■ The x,y-coordinates fo the lower-left and upp-right extends of objects in the drawing.
■ 'Over' indicates drawing extents exceeds the drawing limits (*listed above*).

Display shows
■ The x,y-coordinates of the lower-left and upper-right corners of the current display.

Insertion base is
■ The x,y,z-coordinates stored in system variable InsBase.

Snap resolution is, grid spacing is
■ The snap and grid settings, as stored in the **SnapUnit** and **GridUnit** system variables.

Current space
■ Indicates whether model or paper space is current.

Current layer, current color, current linetype, current elevation, thickness
■ The current values for the layer name, color, linetype name, elevation, and thickness, as stored in system variables **CeLayer, CeColor, CeLType, Elevation**, and **Thickness**.

Fill, grid, ortho, qtext, snap, tablet
■ The current settings for the fill, grid, ortho, qtext, snap, and tablet modes, as stored in the system variables **FillMode, GridMode, OrthoMode, QTextMode, SnapMode**, and **TabMode**.

Object snap modes
■ The currently set object modes, as stored in system variable **OsMode**.

Free disk (dwg + temp = C)
■ Amount of free disk space on the drive storing AutoCAD's temporary files, as pointed to by system variable **TempPrefix**.

Free physical memory
■ Amount of free RAM.

Free swap file space
■ Amount of free space in the Windows's permament swap file on disk.

Virtual address space (Windows only)
■ Amount of the Windows permament swap file used by AutoCAD.

StlOut

Rel. 12

Exports 3D solids and bodies in binary or ASCII SLA format (*short for STereoLithography OUTput; formerly the **SolStlOut** command; an external command in Acis.Dll*).

Command	Alias	Side Menu	Menu Bar	Tablet
stlout	...	[FILE] [EXPORT] [STLout:]	...	...

```
Command: stlout
Select a single solid for STL output.
Select objects: [pick]
Create binary STL file? <Y>:
```

Example of a small portion of an STL file in ASCII format:

```
solid AutoCAD
   facet normal -7.0203459e-016 -9.8078528e-001 1.9509032e-001
      outer loop
         vertex 1.0000000e+001 1.1903397e+001 1.9011325e+001
         vertex 1.0000000e+001 1.2095498e+001 1.9977082e+001
         vertex 1.2726627e-001 1.2095498e+001 1.9977082e+001
      endloop
   endfacet
...
endsolid AutoCAD
```

COMMAND OPTIONS

Y — Creates binary-format SLA file.
N — Creates ASCII-format SLA file.

RELATED AUTOCAD COMMANDS

- *All solid modelling commands.*
- **AcisOut** Exports 3D solid models to an ASCII SAT-format ACIS file.
- **AmeConvert** Converts AME v2.x solid models into ACIS models.

RELATED SYSTEM VARIABLE

- **FaceTRes** Determines the 'resolution' of triangulating solid models.

RELATED FILE

- *.STL Command creates SLA-compatible file with STL extension.

TIPS

■ The solid model must lie in the postive x,y,z-octant of the WCS.

■ The **StlOut** command exports a single 3D ACIS solid; it does not export ACIS regions or any other AutoCAD object.

DEFINITIONS

STL

■ Stereolithography data file, which consists of a faceted representation of the ACIS model.

SLA

■ Stereolithography Apparatus.

ACIS

■ Andrew, Charles, Ian's Solids, the solid modelling engine used by AutoCAD Release 13.

Stretch

Ver. 2.5

Stretches objects to lengthen, shorten, or distort them.

Command	Alias	Side Menu	Menu Bar	Tablet
stretch	...	[MODIFY] [Stretch:]	[Modify] [Stretch]	X 17

```
Command: stretch
Select objects: c
First corner: [pick]
Other corner: [pick]
Select objects: [Enter]
Base point: [pick]
New point: [pick]
```

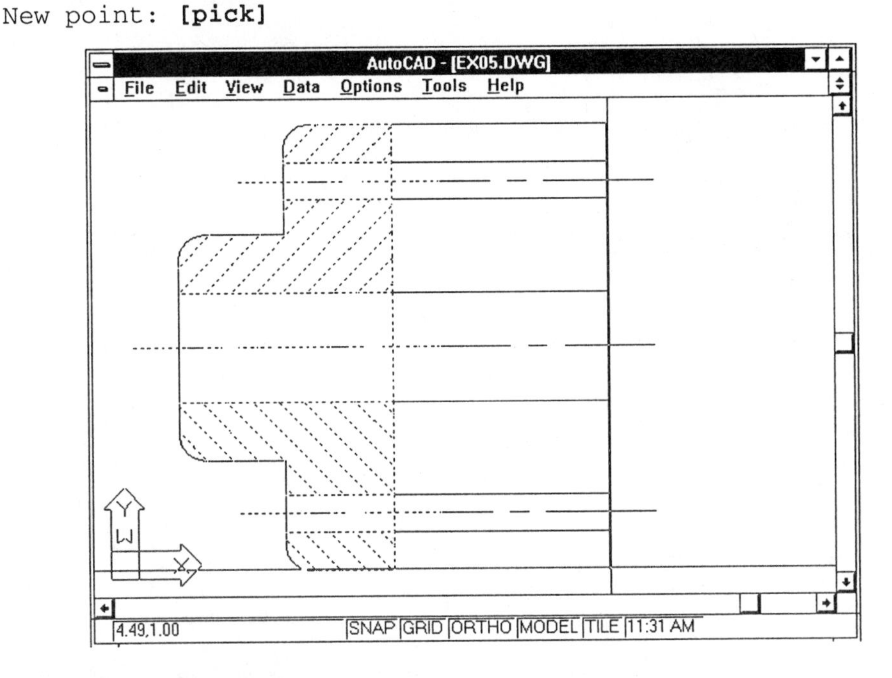

COMMAND OPTIONS

None

RELATED AUTOCAD COMMANDS

- **Change** Changes the size of lines, circles, text, blocks, and arcs.
- **Scale** Increases or decreases the size of any object.

TIPS

- The effect of the **Stretch** command is not always obvious; be prepared the use the **Undo** command.

■ The first time you select objects for the **Stretch** command, you must use **Crossing** object selection.

■ Objects entirely within the selection window are moved, rather than stretched.

■ The **Stretch** command will not move a hatch pattern unless the hatch's origin is included in the selection set.

■ Use the **Stretch** command to automatically update associative dimensions by including the dimension's endpoint in the selection set.

'Style

Ver. 2.0

Creates and modifies text styles based on a font file.

Command	Alt+	Side Menu	Menu Bar	Tablet
'style	D,T	[DATA]	[Data]	U 4
		[Style:]	[Text Style]	

```
Command: style
Text style name (or ?) <STANDARD>:
Font file <ROMANS>:
 Height <0.0000>: [Enter]
 Width factor <1.00>: [Enter]
Obliquing angle <0>: [Enter]
Backwards? <N> [Enter]
Upside-down? <N> [Enter]
Vertical? <N> [Enter]
STANDARD is now the current text style.
```

COMMAND OPTIONS

None

RELATED AUTOCAD COMMANDS

- **Change** Changes the style assigned to selected text.
- **Compile** Compiles SHP and PFB source files into SHX font files.
- **DText** Places text using current style; allows change of style.
- **MText** Places paragraph text.
- **Purge** Removes unused text style definitions.
- **Rename** Renames a text style name.
- **Text** Enters text using current style; allows change of style.

RELATED SYSTEM VARIABLES

- **TextStyle** The current text style.
- **TextSize** The current text height.

RELATED FILES

Font files are found in subdirectory \Acad13\Common\Support:

- ***.SHP** Autodesk's format for vector source fonts.
- ***.SHX** Autodesk's format for compile vector fonts.
- ***.PFB** PostScript font definition files.
- ***.PFM** PostScript font metrication files.
- ***.TTF** TrueType font files.

TIPS

- TTF TrueType fonts cannot be compiled.

■ Source files need not be compiled (*with the **Compile** command*) into SHX files before being loaded into the drawing; as of Release 12, the **Style** command compiles on the fly.

■ AutoCAD Release 12 added support for PostScript fonts; Release 13 added support for TrueType fonts.

■ AutoCAD's SHX fonts display the faster than TrueType fonts, which display faster than PostScript fonts.

■ The **TextFill** and **TextQlty** system variables greatly affect the speed of text display in a drawing. Here are the results from timing a drawing containing all fonts supplied with Release 13. The default settings is the baseline (= *1*):

- Default settings (*TextFill = 0, TextQlty = 50, QText = 0*): **1.0**
- Raise text quality to 100: **0.70** (*70% as fast as the default settings, or about 1.4 times slower*).
- Lower text quality to 0: **2.0** (*Twice as fast as the default settings*).
- Turn text fill on: **0.34** (*Three times as slow as the default settings*).
- Turn quick text on: **8.7** (*Nearly nine times faster than the default settings*).

■ AutoCAD includes the fonts shown below.

PostScript PFB fonts:

cibt.pfb ABC abc 123 !@#
cobt.pfb ABC abc 123 !@#
eur.pfb ABC abc 123 !@#
euro.pfb ABC abc 123 !@#
par.pfb ABC abc 123 !@#
rom.pfb ABC abc 123 !@#
romb.pfb ABC abc 123 !@#
romi.pfb ABC abc 123 !@#
sas.pfb ABC abc 123 !@#
sasb.pfb ABC abc 123 !@#
sasbo.pfb ABC abc 123 !@#
saso.pfb ABC abc 123 !@#
te.pfb ABC ABC 123 !@#
teb.pfb ABC ABC 123 !@#
tel.pfb ABC ABC 123 !@#

Vector SHX fonts:

complex.shx	ABC abc 123 !@#
gothice.shx	ABC abc 123 !@#
gothicg.shx	ABC abc 123 !@#
gothici.shx	ABC abc 123 !@#
greekc.shx	??? ??? 123 !@#
greeks.shx	??? ??? 123 !@#
isocp.shx	ABC abc 123 !@#
isocp2.shx	ABC abc 123 !@#
isocp3.shx	ABC abc 123 !@#
isoct.shx	A B C a b c 1 2 3 ! @ #
isoct2.shx	A B C a b c 1 2 3 ! @ #
isoct3.shx	A B C a b c 1 2 3 ! @ #
italic.shx	*ABC abc 123 !@#*
italicc.shx	*ABC abc 123 !@#*
italict.shx	*ABC abc 123 !@#*
monotxt.shx	ABC abc 123 !@#
romanc.shx	ABC abc 123 !@#
romand.shx	ABC abc 123 !@#
romans.shx	ABC abc 123 !@#
romant.shx	ABC abc 123 !@#
scriptc.shx	*ABC abc 123 !@#*
scripts.shx	*ABC abc 123 !@#*
simplex.shx	ABC abc 123 !@#
syastro.shx	⊙☿♀ ⚹ 123 !@#
symap.shx	123 !@#
symath.shx	ℵ′| ←↓∂ 123 !@#
symeteo.shx	123 !@#
symusic.shx	123 !@#
txt.shx	ABC abc 123 !@#

TrueType TTF fonts:

bgothl.ttf	ABC ABC 123 !@#
bgothm.ttf	ABC ABC 123 !@#
compi.ttf	±°′ ©®© ○○▫ □
comsc.ttf	ABC abc 123 !@#
dutch.ttf	ABC abc 123 !@#
dutchb.ttf	ABC abc 123 !@#
dutchbi.ttf	ABC abc 123 !@#
dutcheb.ttf	ABC abc 123 !@#
dutchi.ttf	ABC abc 123 !@#
monos.ttf	ABC abc 123 !@#
monosb.ttf	ABC abc 123 !@#
monosbi.ttf	ABC abc 123 !@#
monosi.ttf	ABC abc 123 !@#
swiss.ttf	ABC abc 123 !@#
swissb.tff	ABC abc 123 !@#
swissbi.ttf	ABC abc 123 !@#
swissbo.ttf	ABC abc 123 !@#
swissc.ttf	ABC abc 123 !@#
swisscb.ttf	ABC abc 123 !@#
swisscbi.ttf	ABC abc 123 !@#
swisscbo.ttf	ABC abc 123 !@#
swissci.ttf	ABC abc 123 !@#
swissck.ttf	ABC abc 123 !@#
swisscki.ttf	ABC abc 123 !@#
swisscl.ttf	ABC abc 123 !@#
swisscli.ttf	ABC abc 123 !@#
swisse.ttf	ABC abc 123 !@#
swisseb.ttf	ABC abc 123 !@#
swissek.ttf	ABC abc 123 !@#
swissel.ttf	ABC abc 123 !@#
swissi.ttf	ABC abc 123 !@#
stylu.ttf	ABC abc 123 !@#
swissk.ttf	ABC abc 123 !@#
swisski.ttf	ABC abc 123 !@#
swissko.ttf	ABC abc 123 !@#
swissl.ttf	ABC abc 123 !@#
swissli.ttf	ABC abc 123 !@#
umath.ttf	ABΨ αβψ + − × /
vinet.ttf	ABC abc 123 !@#

Subtract

Removes the volume of one 3D model or 2D region from another *(formerly the SolSub command; an external command in Acis.Dll)*.

Command	Alias	Side Menu	Toolbar	Tablet
subtract	...	[DRAW 2]	[Modify]	Y 14
		[SOLIDS]	[Explode]	
		[Subtrac:]	[Subtract]	

```
Command: subtract
Select objects: [pick]
Select objects: [Enter]
1 solid selected.
Objects to subtract from them..
Select objects: [pick]
Select objects: [Enter]
1 solid selected.
```

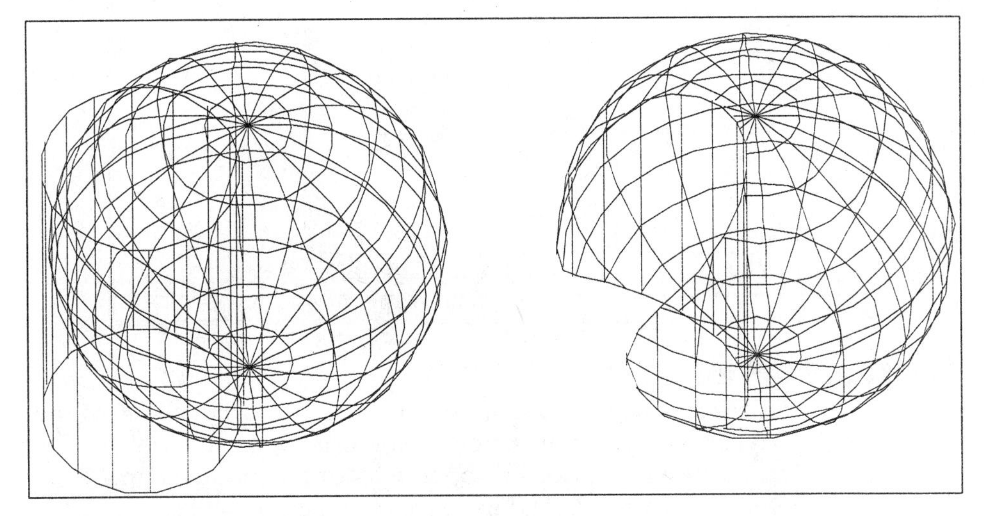

COMMAND OPTIONS

None

RELATED ACIS COMMANDS

- **Intersect** Removes all but the intersection of two solid volumes
- **Union** Joins two solids together

SysWindows

Rel. 13

Controls multiple windows; has no practical use in Windows v3.1

Command	Alias	Side Menu	Menu Bar	Tablet
syswindows	...	...	...	...

```
Command: syswindows
Cascade/tileHorz/tileVert/Arrangeicons:
```

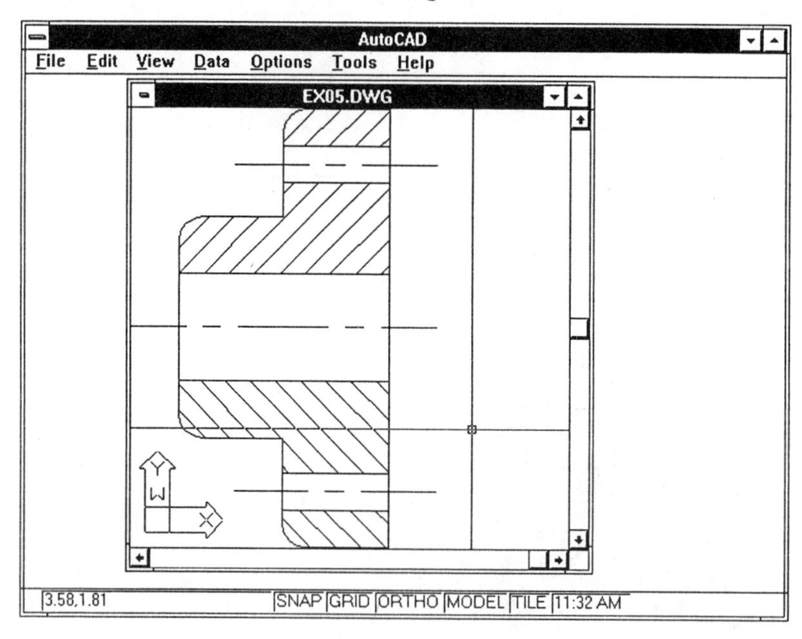

COMMAND OPTIONS

Cascade Resizes and repositions all open windows to overlap so that the title bar is visible; the current window is topmost.

tileHorz Resizes and repositions all open windows to not overlap; windows are positioned horizontally when possible.

tileVert Resizes and repositions all open windows to not overlap; windows are positioned vertically when possible.

Arrangeicons
Arranges iconized windows into a neat and orderly line at the bottom of the AutoCAD window.

TIPS

■ The **SysWindows** command operates under Windows but has no practical effect, since AutoCAD for Windows v3.1 supports a single drawing/window.

■ Under NT and Windows 95, AutoCAD can load more than one drawing and display more than one window.

Tablet

Ver. 1.0

Configures, calibrates, and toggles the digitizing tablet for menus and pointing area.

Command	Ctl+	Side Menu	Menu Bar	Tablet
tablet	O,T	[Settings]	[Options]	S 19-22
		[next]	[Tablet]	
		[TABLET:]		
	[F4]			X 25

```
Command: tablet
Options (ON/OFF/CAL/CFG):
```

COMMAND OPTIONS

CAL Calibrates the coordinates for the tablet.
CFG Configures the menu areas on the tablet.
OFF Turns off the tablet's digitizing mode.
ON Turns on the tablet's digitizing mode.

RELATED SYSTEM VARIABLE

■ **TabMode** Toggles use of the tablet:
0 Tablet mode disabled.
1 Tablet mode enabled.

RELATED FILES

■ **Acad.Mnu** Menu source code that defines functions of tablet menu areas, in \Acad13\Dos\Support.

■ **Mc.Exe** Menu compiler; semi-automates the creation of a tablet menu file; in \Acad13\Common\Sample.

■ **Tablet.Dwg** AutoCAD drawing of the printed template overlay; in the \Acad13\Common\Sample.

TIPS

■ Since version 2.5, AutoCAD includes a tablet overlay in the package (*see figure*).

■ To customize or change the size of the AutoCAD tablet overlay, edit the Tablet.Dwg file, then plot it out to fit your digitizer.

■ The **Tablet** command does not work if a digitizing tablet has not first been configured with the **Config** command.

■ AutoCAD supports up to four independent menu areas; macros are specified by the ***TABLET1 through ***TABLET4 sections of the Acad.Mnu menu file.

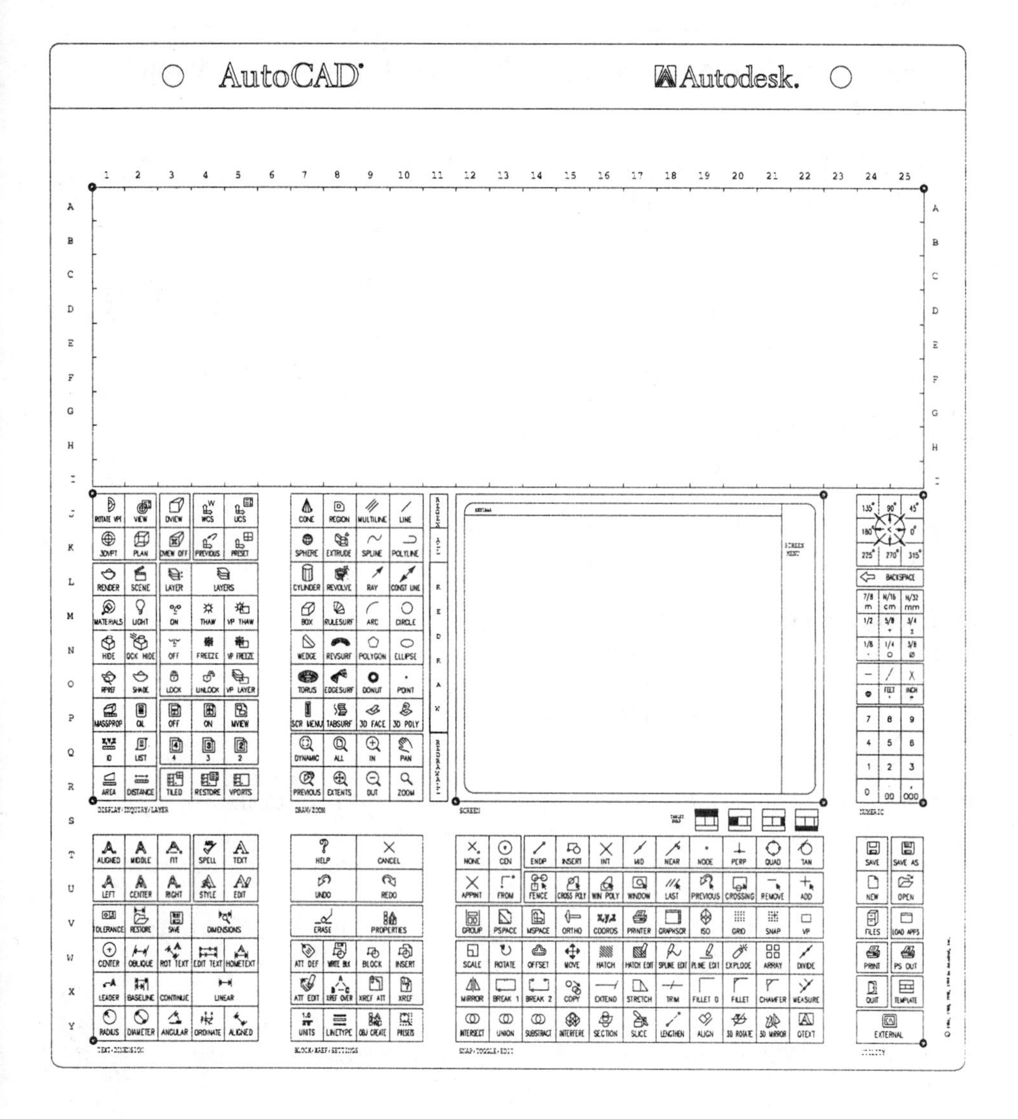

- Menu areas may be skewed but corners must form a right angle.
- As of Release 13, AutoCAD for Windows no longer supports 'mole' mode; instead, you must use a WinTab-compatible digitizing tablet.
- AutoCAD supports any size of digizing tablet; however, tablets smaller than A-size (8"x11") are – in most cases – too small for useful digitizing.

DEFINITIONS

Affine transformation

■ 3 pick points.

■ Sets an arbitrary linear 2D transformation with independent x,y-scaling and skewing.

Orthagonal transformation

■ 2 pick points.

■ Sets the translation; scaling and rotation angle remain uniform.

Outcome of fit

■ A report on the results of the 3 transformation types: affine, orthagonal, projective.

■ AutoCAD reports 5 types of outcomes:

- ■ **Exact** Enough points to transform data.
- ■ **Success** More than enough points to transform data.
- ■ **Impossible** Not enough points to transform data.
- ■ **Failure** Too many colinear and coincident points.
- ■ **Cancelled** Fitting cancelled during projective transform.

Projective transformation

■ 4 pick points.

■ Maps a perspective projection from one plane to another plane.

■ A limited form of 'rubber sheeting:' straight lines remain straight but not necessarily parallel.

Residual

■ Largest: where mapping is least accurate.

■ Second largest: second-least accurate.

RMS error

■ Root mean square error

■ Smaller is better: measures closeness of fit.

Standard deviation

■ Measures standard deviation of residuals.

■ Near zero: residual at each point is roughly the same.

TabSurf

Rel. 10

Draws a tabulated surface as a 3D mesh; defined by a path curve, and a direction vector (*short for TABulated SURFace*).

Command	Alias	Side Menu	Toolbar	Tablet
tabsurf	...	[DRAW 2] [SURFACES] [Tabsurf:]	[Surfaces] [TabSurf]	P 8

```
Command: tabsurf
Select path curve: [pick]
Select direction vector: [pick]
```

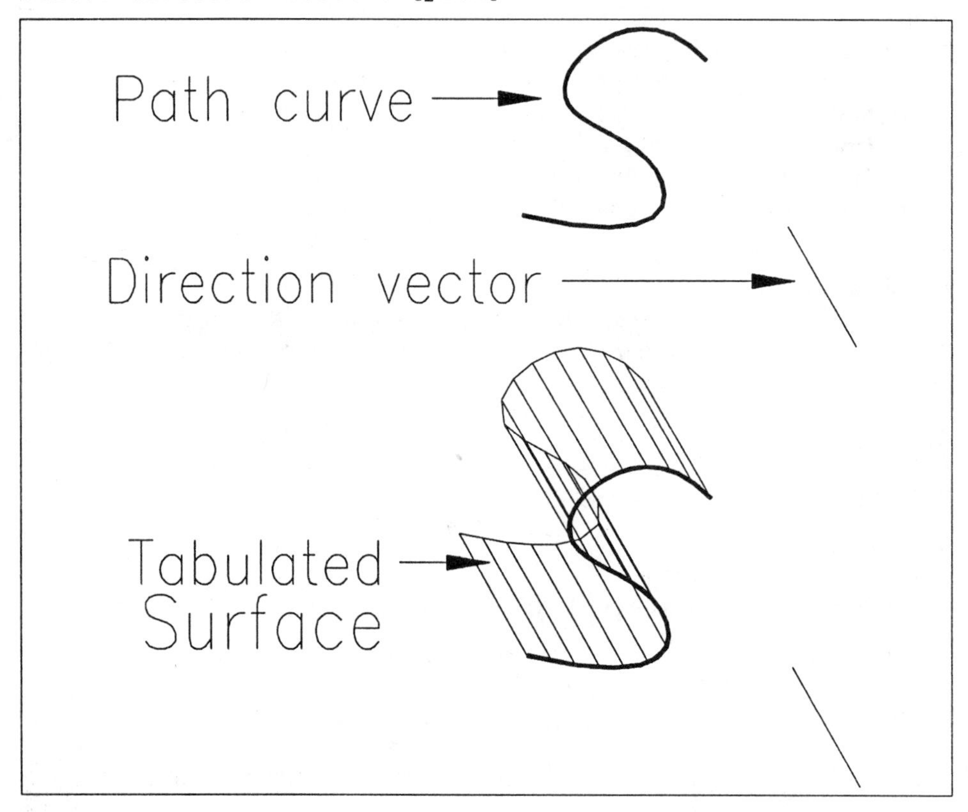

COMMAND OPTIONS

None

RELATED AUTOCAD COMMANDS

- **Edge** Changes the visibility of 3D face edges.
- **Explode** Reduces a tabulated surface into 3D faces.

- **PEdit** Edits a 3D mesh, such as tabulated surfaces.
- **EdgeSurf** Draws a 3D mesh surface between boundaries.
- **RevSurf** Draws a revolved 3D mesh surface around an axis.
- **RuleSurf** Draws a 3D mesh surface between open or closed boundaries.
- **3D** Creates 3D objects out of surface meshes.

RELATED SYSTEM VARIABLE

- **SurfTab1** Defines the number of tabulations drawn by **TabSurf** in n-direction.

TIPS

- The path curve can be open or closed:
  - Line, 2D polyline, and 3D polyline.
  - Arc, circle, and ellipse.

- The direction vector defines the direction and length of extrusion.

- The number of m-direction tabulations is always 2 and lies along direction vector.

- The number of n-direction tabulations is determined by system variable **SurfTab1** (*default* = *6*) along curves only.

TbConfig

Rel. 13

Creates, customizes, changse, and deletes icon toolbars (*short for ToolBar CONFIGuration*).

Command	Alt+	Side Menu	Menu Bar	Tablet
tbconfig	T,B	...	[Tools] [Customize Toolbars]	...

Command: **tbconfig**

Displays dialogue box.

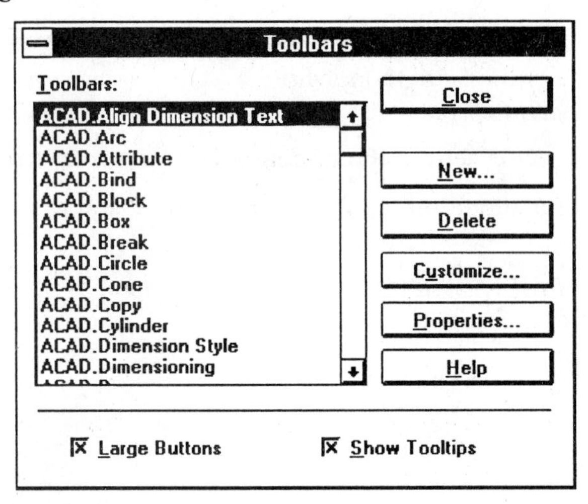

COMMAND OPTIONS

Toolbars Lists defined toolbars.

Close Exits dialogue box.

New Creates a new toolbar:

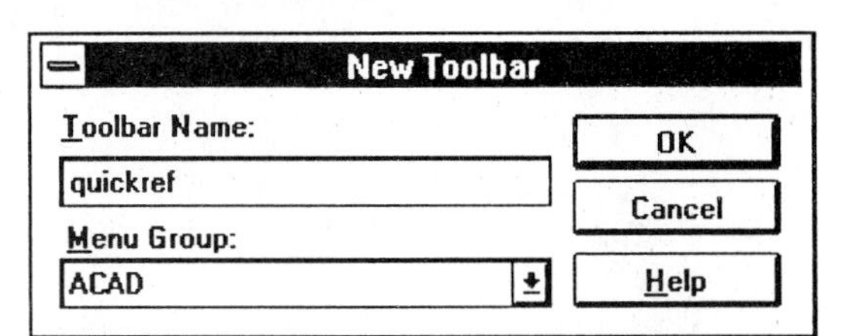

Delete Deletes a toolbar definition:

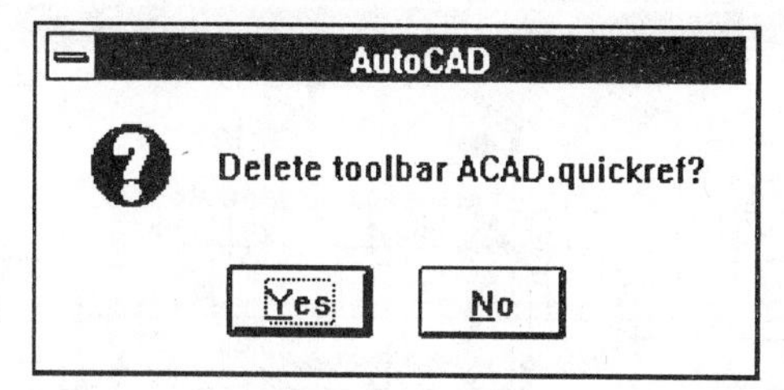

Customize Customizes a toolbar:

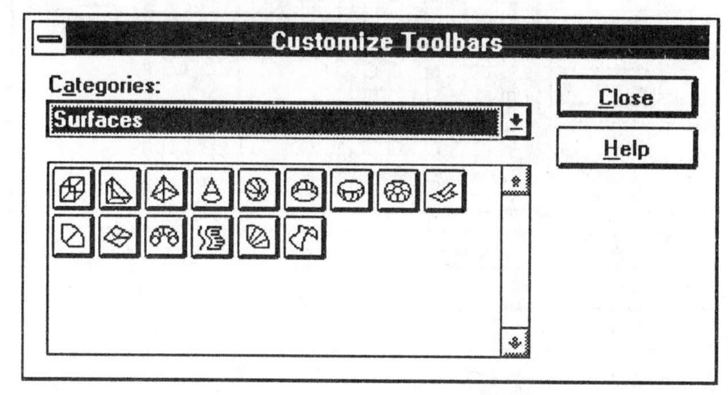

Properties Changes the meaning and icons of a toolbar:

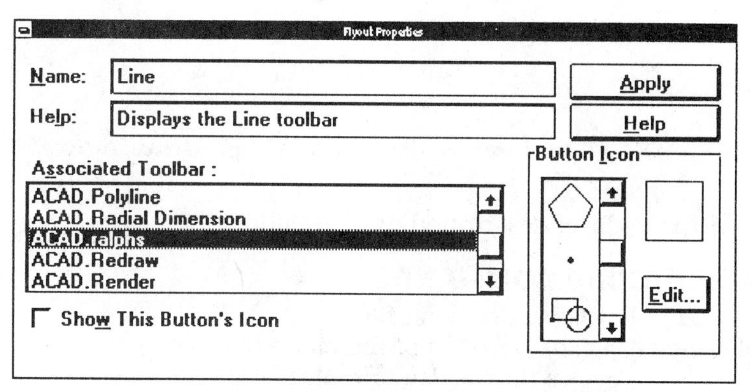

Edit Edits the icon:

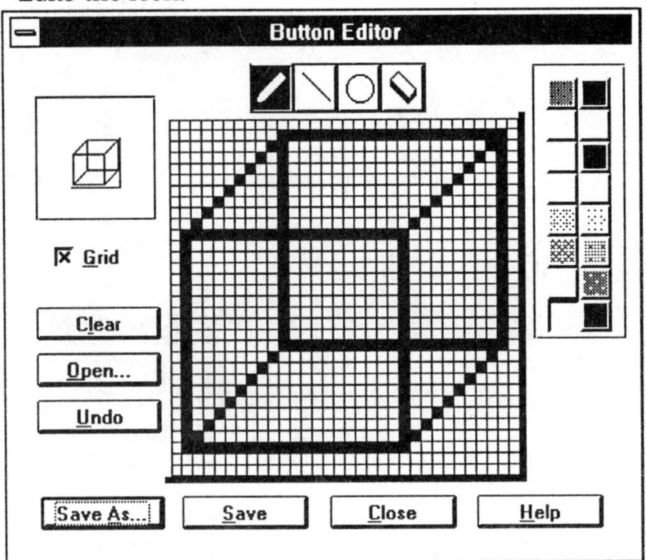

Large Buttons

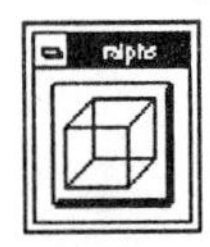

On All buttons are displayed larger (*32 x 32 pixels*).

Off All buttons are displayed smaller (*16 x 16 pixels*).

Show Tooltips Brief description of icon button.

RELATED AUTOCAD COMMANDS

- **MenuLoad** Loads partial menu files.
- **MenuUnload** Unloads portions of the menu file.
- **Toolbar** Toggles the display of toolbars.

RELATED SYSTEM VARIABLE

- **ToolTips** Toggles display of tooltops.

RELATED FILES

- **Acad.Ini** Stores toolbar definitions.
- **Acad.Mnc** Compiled menu file.
- **Acad.Mnt** Toolbar definitions.
- **Acad.Mnu** AutoCAD source menu file.
- ***.BMP** BMP bitmap files define custom icon buttons.

Text

Ver. 1.0

Places one line of text in the drawing.

Command	Alias	Side Menu	Toolbar	Tablet
text	...	...	[Draw] [Text]	...

```
Command: text
Justify/Style/<Start point>: j
Align/Fit/Center/Middle/Right/TL/TC/TR/ML/MC/MR/BL/BC/BR: r
Height <0.2000>: [Enter]
Rotation angle <0>: [Enter]
Text:
```

COMMAND OPTIONS

[Enter] Continues text one line below previously-placed text line.

Justify Displays the justification submenu:

- **Align** Aligns the text between two points with adjusted text height.
- **Fit** Fits the text between two points with fixed text height.
- **Center** Centers the text along the baseline.
- **Middle** Centers the text horizontally and vertically.
- **Right** Right-justifies the text.
- **TL** Top-left justification.
- **TC** Top-center justification.
- **TR** Top-right justification.
- **ML** Middle-left justification.
- **MC** Middle-center justification.
- **MR** Middle-right justification.
- **BL** Bottom-left justification.
- **BC** Bottom-center justification.
- **BR** Bottom-right justification.

<Start point> Left-justifies the text.

Style Displays the style submenu:

- **Style name** Indicates a different style name.
- **?** Lists the currently loaded styles.

RELATED AUTOCAD COMMANDS

- **Change** Changes the text height, rotation, style, and content.
- **DText** Places new text to the drawing interactively.
- **MText** Places paragraph text.
- **Style** Creates new text styles.

RELATED SYSTEM VARIABLES

- **TextSize** The current height of text.
- **TextStyle** The current style.

'TextScr

Ver. 2.1

Switches from the graphics screen to the text screen in single-screen systems (*short for TEXT SCReen*).

Command	Alias	Side Menu	Function Key	Tablet
'textscr	...	...	[F2]	...

Command: **textscr**

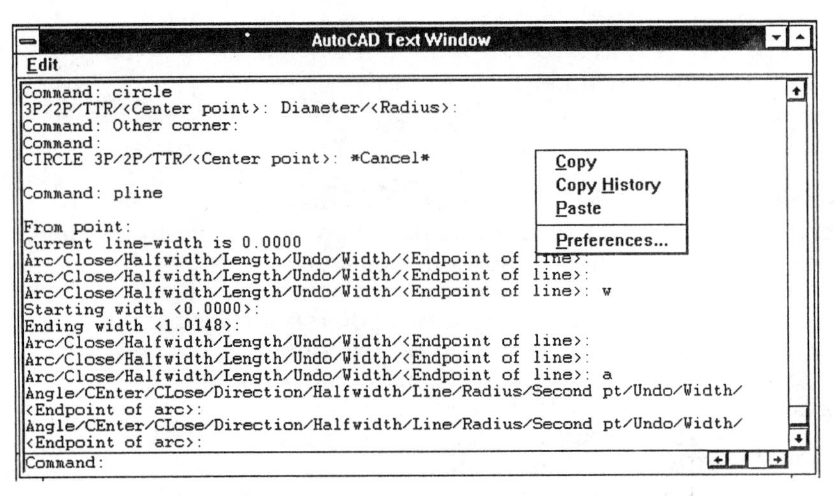

COMMAND OPTIONS

Copy Copies selected text to the Windows Clipboard.
Copy History Copies all text to the Clipboard.
Paste Pastes text from Clipboard into text window.
Preferences Displays **Preferences** dialogue box.

RELATED AUTOCAD COMMAND

- **GraphScr** Switches from text screen to graphics screen.

RELATED SYSTEM VARIABLE

- **ScreenMode** Reports whether screen is in text or graphics mode.
  - **0** Text screen.
  - **1** Graphics screen.
  - **2** Dual screen displaying both text and graphics.

TIPS

- **TextScr** does not work with dual-screen systems.
- Right-click on the text screen to display cursor menu.

'Tiffin

Rel. 12

Imports TIF raster files into the drawing as a block (*an external file in Raster.Exp*).

Command	Alt+	Side Menu	Menu Bar	Tablet
'tifin	F, I	[FILE]	[File]	...
		[IMPORT]	[Import]	
		[TIFFin:]	[TIF]	

```
Command: tifin
TIFF file name:
Insertion point <0,0,0>:
Scale factor:
```

COMMAND OPTIONS

None

RELATED AUTOCAD COMMANDS

- **PsIn** Imports EPS files.
- **GifIn** Imports GIF raster files.
- **Import** Dialogue-box shell for the **TiffIn** command.
- **PcxIn** Imports PCX raster files.
- **Replay** Displays GIF, TIFF, and TGA files as raster images.

RELATED SYSTEM VARIABLES

- **RiAspect** Adjusts image's aspect ratio.
- **RiBackG** Changes the image's background color.
- **RiEdge** Outlines edges.
- **RiGamut** Specifies number of colors.
- **RiGrey** Imports as a grey scale image.
- **RiThresh** Controls brightness threshold.

TIPS

- The TIFF format (*short for tagged image file format*) was invented by Aldus and Microsoft for desktop publishing software.

- **TiffIn** is limited to displaying a maximum of 256 colors.

- Exploding an imported TIFF block doubles the drawing file size.

- Each run of similarly-colored pixel is defined as a solid.

- Turn off system variable **GripBlock** (*set it to 0*) to avoid highlighting all the solid entities making up the block.

'Time

Ver. 2.5

Displays timely information about the current drawing.

Command	Alt+	Side Menu	Menu Bar	Tablet
'time	D,E	[DATA]	[Data]	...
		[TIME:]	[Time]	

Command: **time**

Sample output:

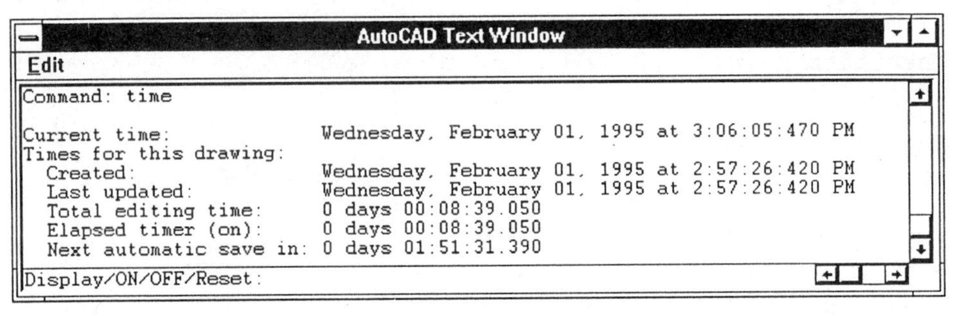

```
Command: time

Current time:                Wednesday, February 01, 1995 at 3:06:05:470 PM
Times for this drawing:
  Created:                   Wednesday, February 01, 1995 at 2:57:26:420 PM
  Last updated:              Wednesday, February 01, 1995 at 2:57:26:420 PM
  Total editing time:        0 days 00:08:39.050
  Elapsed timer (on):        0 days 00:08:39.050
  Next automatic save in:    0 days 01:51:31.390
Display/ON/OFF/Reset:
```

COMMAND OPTIONS

Display Displays the current time information.
OFF Turns off the user timer.
ON Turns on the user timer.
Reset Resets the user timer.
[F2] Returns to graphics screen.

RELATED AUTOCAD COMMAND

- **Status** Displays information about the current drawing and environment.

RELATED SYSTEM VARIABLES

- **CDate** The current date and time.
- **Date** The current date and time in Julian format.
- **TdCreate** Date and time the drawing was created.
- **TdInDwg** The time the drawing spent in AutoCAD.
- **TdUpdate** The last date and time the drawing was changed.
- **TdUsrTimer** The current user timer setting.
- **SaveTime** The automatic drawing save interval.

TIP

- The time displayed by the **Time** command is only accurate when the computer's clock is accurate; unfortunately, the clock in most personal computers strays by many minutes per week.

Tolerance

Rel. 13

Places geometric tolerancing symbols and text.

Command	Alias	Side Menu	Toolbar	Tablet
tolerance	tol	[DRAW DIM] [Toleran:]	[Dimensioning] [Tolerance]	V 1

```
Command: tolerance
[select collection of tolerance symbols]
Enter tolerance location: [pick]
```

Displays dialogue box.

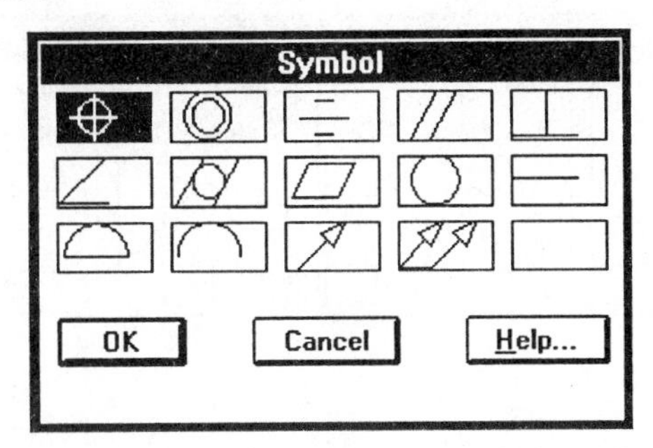

Select a tolerance symbol:

Location symbols:

⌖ Position.

◎ Concentricity and coaxiality.

⌯ Symmetry.

Orientation symbols:

// Parallelism.

⊥ Perpendicularity.

∠ Angularity.

Form symbols:

⌭ Cylindricity.

▱ Flatness.

⌭ Circularity and roundness.

■ Straightness.

Profile symbols:

Profile of the surface.

Profile of the line.

Circular runout.

Total runout.

After selecting a symbol, the following dialogue box appears:

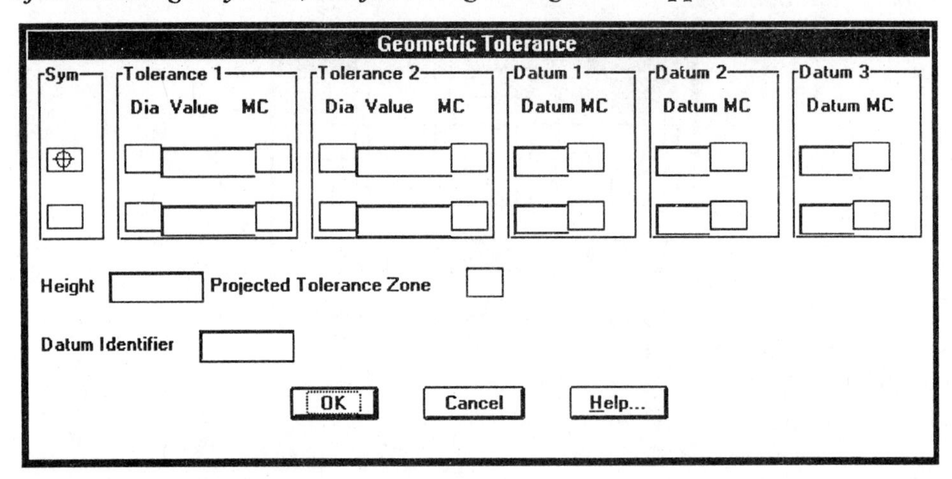

COMMAND OPTIONS

Sym Geometric characteristic symbol.

Tolerance First tolerance value:

- **Dia** Places optional Ø (*diameter*) symbol.
- **Value** Tolerance value.
- **MC** Material Condition: modifies tolerance symbol; displays dialogue box:

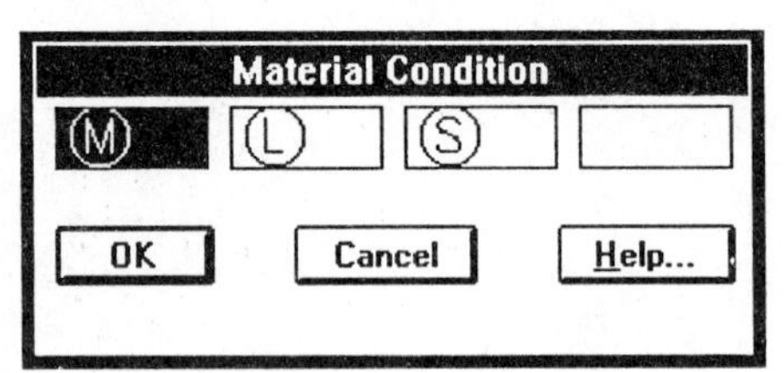

Material Condition symbols:

Ⓜ Maximum material condition.

Ⓛ Least material condition.

Ⓢ Regardless of feature size.

Datum Datum reference.
Height Projected tolerance zone value.
Projected Tolerance Zone
Places projected tolerance zone symbol.
Datum Identifier
Creates datum identifier symbol, such as: –A–

RELATED FILES

In \Acad13\Common\Support:

- **Gdt.Shp** Tolerance symbol definition source file.
- **Gdt.Shx** Compiled tolerance symbol file.

DEFINITIONS

Datum:

- A theoretically-exact geometric reference.
- Establishes the tolerance zone for the feature.
- These objects can be used as a datum:
  - Point, line, and plane.
  - Cylinder, and other geometry.

Material condition:

- These symbols modify the geometric charactertistics and tolerance values.
- Modifiers for features that vary in size.

Projected tolerance zone:

- Specifies the height of the fixed perpendicular part's extended portion.
- Changes the tolerance to positional tolerance.

Tolerance:

- Indicates amount of variance from perfect form.

Toolbar

Rel. 13

Displays or hides one or all toolbars.

Command	Alt+	Side Menu	Menu Bar	Tablet
toolbar	T,T	...	[Tools] [Toolbars]	...

```
Command: toolbar
Toolbar name or All: draw
Show/Hide/Left/Right/Top/Bottom/Float:  <Show>: f
Position <0,0>: [pick]
Rows <1>: [Enter]
```

*Effect of the **Toolbar All Show** command:*

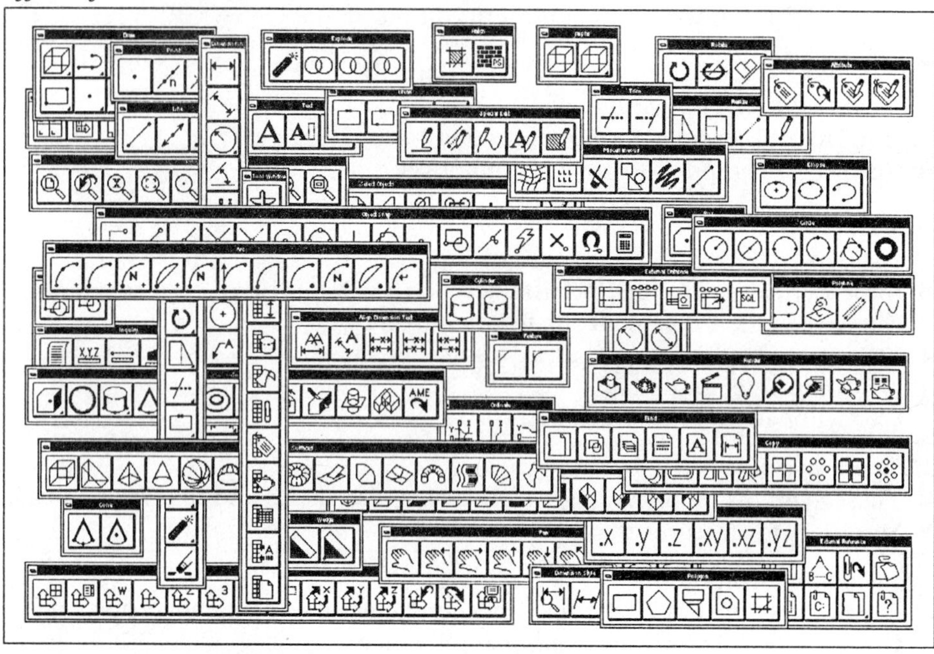

COMMAND OPTIONS

Show Displays the toolbar.
Hide Dismisses the toolbar.
Left Docks toolbar at left edge of AutoCAD drawing area.
Right Docks toolbar at right edge of AutoCAD drawing area.
Top Docks toolbar at top edge of AutoCAD drawing area.
Bottom Docks toolbar at bottom edge of AutoCAD drawing area.
Float Floats the toolbar anywhere on the Windows screen.
  Position Pixel coordinates of the upper-left corner of the floating toolbar.
  Rows Number of rows for floating toolbar.

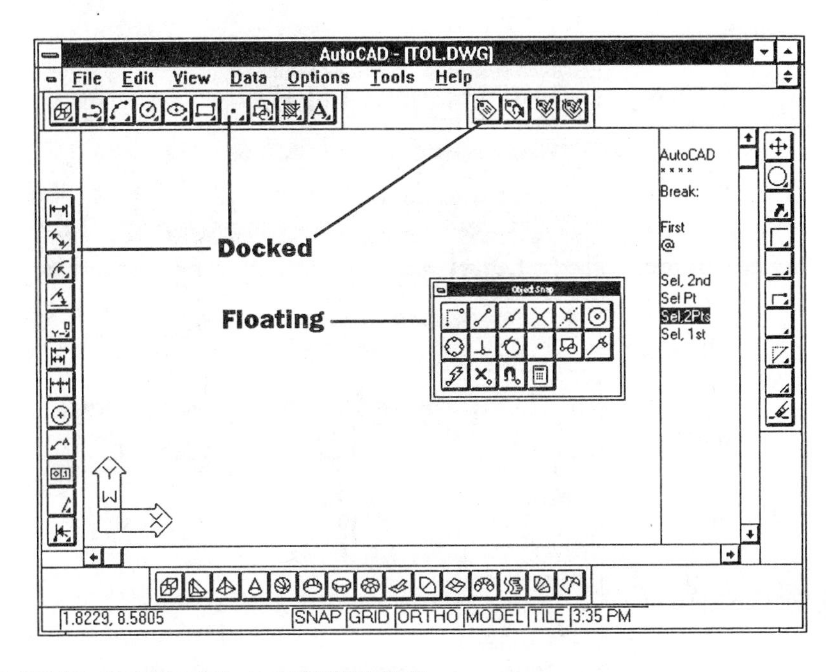

RELATED AUTOCAD COMMANDS

- **MenuLoad** Loads partial menu file.
- **MenuUnload** Unloads part of the menu file.
- **TbConfig** Creates and configures toolbars.

TIPS

- *Caution:* Opening all toolbars (*with the **Toolbar All Show** command*) consume one-tenth of GDI memory; when GDI memory falls to approximately 15%, Windows can crash.

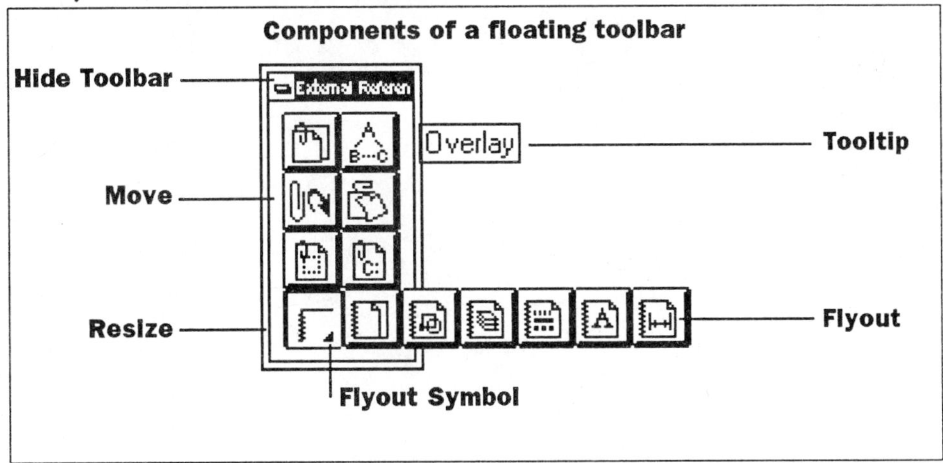

Default Toolbars

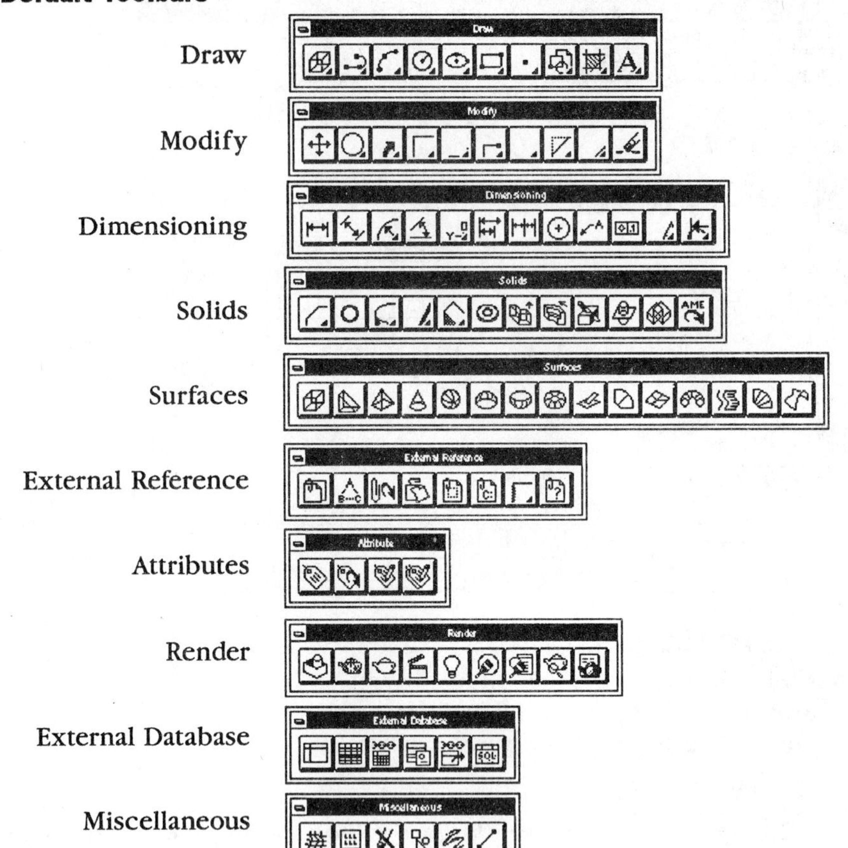

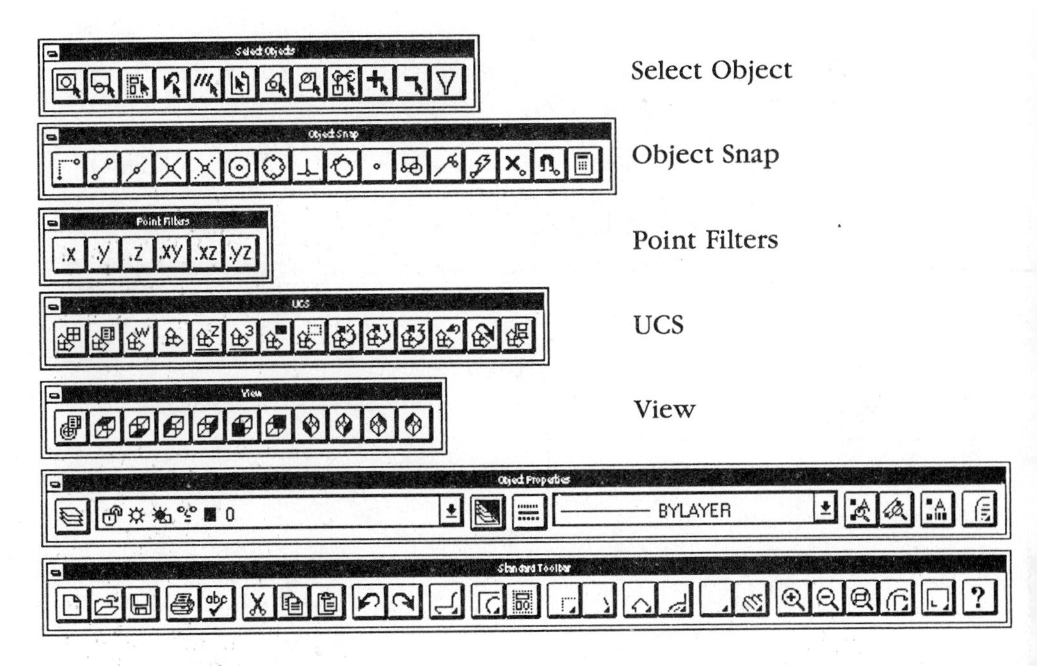

Above: Object Properties
Standard Toolbar

Torus

Rel. 11

Draws a 3D torus as a solid model *(formerly the **SolTorus** command; an external command in Acis.Dll)*.

Command	Alias	Side Menu	Toolbar	Tablet
torus	...	[DRAW 2] [SOLIDS] [Torus:]	[Solids] [Torus]	O 7

```
Command: torus
Center of torus <0,0,0>: [pick]
Diameter/<Radius> of torus: [pick]
Diameter/<Radius> of tube: [pick]
```

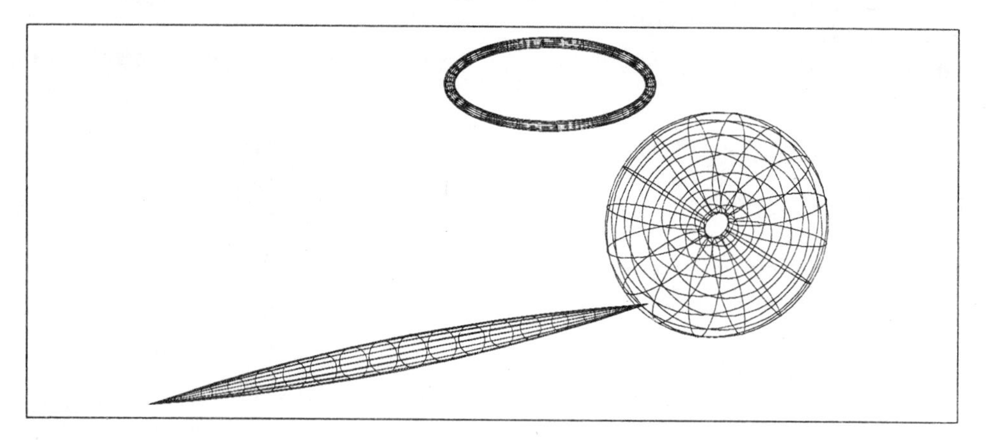

COMMAND OPTIONS

Diameter Indicates the diameter of the torus and the tube

<Radius> Indicates the radius of the torus and the tube

RELATED AUTOCAD COMMANDS

- **Ai_Torus** Creates a torus made from 3D polyfaces.
- **Box** Draws solid boxes.
- **Cone** Draws solid cones.
- **Cylinder** Draws solid cylinders.
- **Sphere** Draws solid spheres.
- **Wedge** Draws solid wedges.

TIPS

- **Torus** allows self-intersecting tori.
- A negative torus radius creates the football shape.
- When the torus radius is negative, the tube radius must be a larger positive number; for example, with a torus radius of -1.99, the tube radius must be greater than +1.99.

Trace

Ver. 1.0

Draws lines with width.

Command	Alias	Side Menu	Toolbar	Tablet
trace	...	[DRAW 2]	[Miscellaneous]	...
		[Trace:]	[Trace]	

```
Command: trace
Trace width <0.0500>: [Enter]
From point: [pick]
To point: [pick]
To point: [Enter]
```

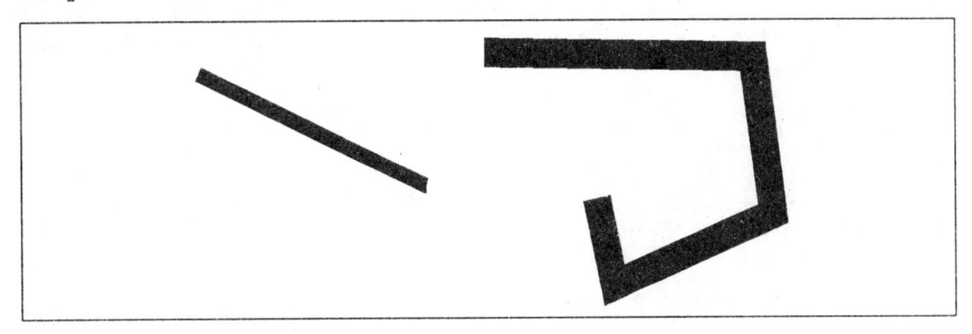

COMMAND OPTION

[Enter] Exits the **Trace** command.

RELATED AUTOCAD COMMANDS

- **Line** Draws lines with zero width.
- **MLine** Draws up to 16 parallel lines.
- **PLine** Draws polylines and polyline arcs with varying width.

RELATED SYSTEM VARIABLES

- **FillMode** Toggles display of fill or outline traces (default = 1, on).
- **TraceWid** The current width of the trace (default = 0.05).

TIPS

- Traces are drawn along the centerline of the pick points.

- Display of a trace segment is delayed by one pick point.

- During drawing of traces, you cannot backup since an **Undo** option is missing; if you require this feature, draw wide lines with the **PLine** command, setting the **Width** option.

- There is no option for controlling joints (*always beveled*) or endcapping (*always square*); if you require these features, draw wide lines with the **MLine** command, setting the solid fill, endcap, and joint options with the **MlStyle** command.

'TreeStat

Rel. 12

Displays the status of the drawing's spatial index, including the number and depth of nodes.

Command	Alias	Side Menu	Menu Bar	Tablet
'treestat	...	...	...	...

Command: **treestat**

Sample output for Acad.Dwg:

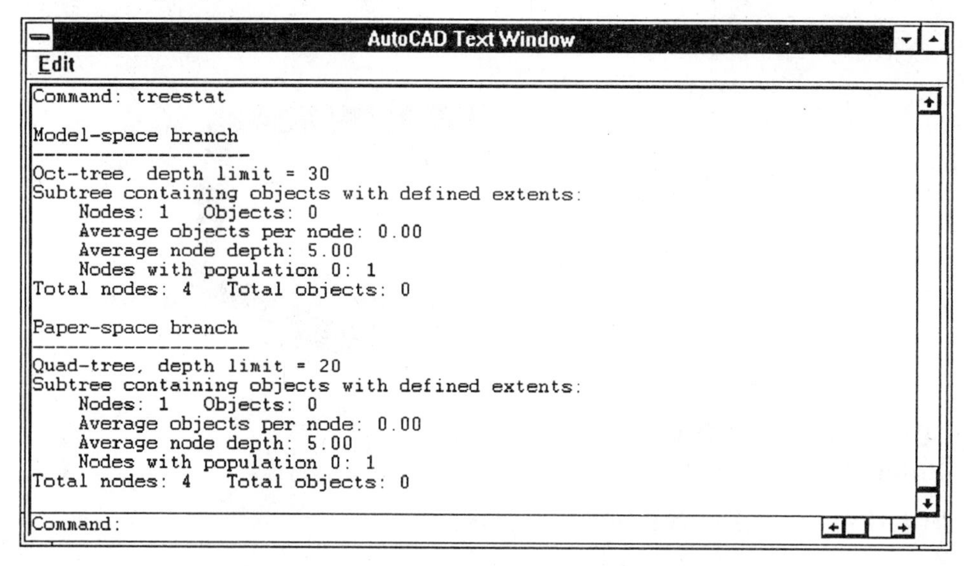

COMMAND OPTIONS

None

RELATED SYSTEM VARIABLES

- **TreeDepth** Size of the tree-structured spatial index in *xxyy* format:
  - **xx** Number of model space nodes (*default = 30*).
  - **yy** Number of paper space nodes (*default = 20*).
  - **-xx** 2D drawing.
  - **+xx** 3D drawing (*default*).
  - **3020** Default value of **TreeDepth**.
- **TreeMax** Maximum number of nodes (*default = 10,000,000*).

TIPS

- Better performance occurs with fewer objects per oct-tree node.

- When redraws and object selection seem slow, increase the value of system variable **TreeDepth**.

- Each node consumes 80 bytes of memory.

DEFINITIONS

Oct tree

■ The model space branch of the spatial index.
■ Objects are either 2D or 3D.
■ "Oct" comes from the eight volumes in x,y,z-coordinate system of 3D space.

Quad tree

■ The paper space branch of the spatial index.
■ All objects are two-dimensional.
■ "Quad" comes from the four areas in the x,y-coordinate system of 2D space.

Spatial index

■ Objects indexed by oct-region to record their position in 3D space.
■ Has a tree structure with two primary branches: oct tree and quad tree.
■ Objects are attached to "nodes"; each node is a branch on the "tree."

Trim

Trims lines, arcs, circles, and 2D polylines back to a real or projected cutting line or view.

Command	Alias	Side Menu	Menu Bar	Tablet
trim	...	[MODIFY] [Trim:]	[Modify] [Trim]	X 18

```
Command: trim
Select cutting edges: (Projmode=UCS, Edgemode=No extend)
Select objects: [pick]
Select objects: [Enter]
<Select object to trim>/Project/Edge/Undo: [pick]
```

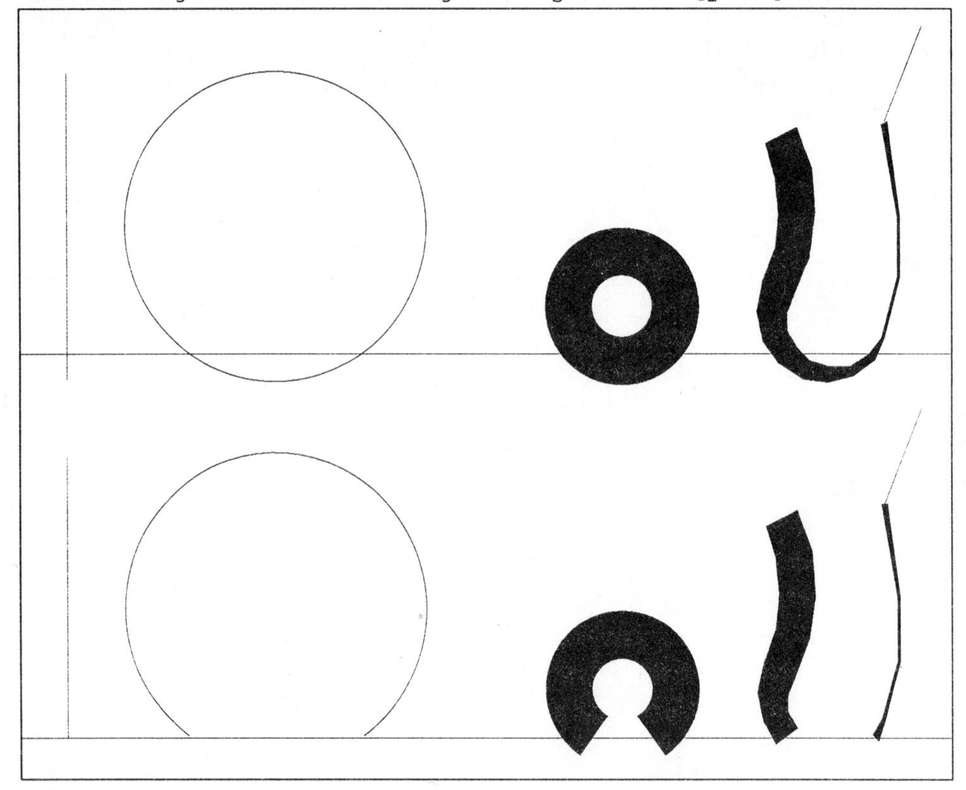

COMMAND OPTIONS

Select object to trim
Picks the objects at the trim end.

Edge Selects type of trim (*new to Release 13*):

Extend Extends cutting edge to trim object.

No extend Only trims at actual cutting edge (*Release 12 compatible*).

Project Selects trim projection mode (*new to Release 13*):

- **None** Uses only objects as cutting edge (*Release 12 compatible*).
- **Ucs** Trims at x,y-plane of current UCS.
- **View** Trims at current view plane.

Undo Untrims the last trim action.

RELATED AUTOCAD COMMANDS

- ■ **Change** Changes the size of lines, arcss and circles.
- ■ **Extend** Lengthens lines, arcss and polylines.
- ■ **Lengthen** Lengthens open objects.
- ■ **PEdit** Changes polylines.
- ■ **Stretch** Lengthens or shortens lines, arcs, and polylines.

U

Ver. 2.5

Undoes the most recent AutoCAD command (*short for Undo*).

Command	Ctrl+	Side Menu	Menu Bar	Tablet
`u`	`Z`	`[ASSIST]`	`[Assist]`	`U 7-8`
		`[U:]`	`[Undo]`	

```
Command: u
```

COMMAND OPTIONS

None

RELATED AUTOCAD COMMANDS

- **Oops** Unerases the most recently erased object.
- **Quit** Exits the drawing, undoing all changes.
- **Redo** Undoes the most recent undo.
- **Undo** Allows more sophisticated control over undo.

RELATED SYSTEM VARIABLE

- **UndoCtl** Determines the state of undo control.

TIPS

- The **U** command is convenient for stepping back through the design process, undoing one command at a time.

- The **U** command is the same as the **Undo 1** command; for greater control over the undo process, use the **Undo** command.

- The **Redo** command undoes the effect of only the most recent undo.

- The **Quit** command restores the drawing to its original state.

- Since the undo mechanism creates a mirror drawing file on disk, disable the **Undo** command with system variable **UndoCtl** (*set it to 0*) when your computer is low on disk space.

- Commands that involve writing to file, plotting, and system variables are not undone.

Ucs

Rel. 10

Defines a new coordinate plane (*short for User Coordinate System*).

Command	Alt+	Side Menu	Menu Bar	Tablet
UCS	V,S	[VIEW:]	[View]	J 4-5
		[UCS:]	[Set UCS]	

```
Command: ucs
Origin/ZAxis/3point/OBject/View/X/Y/Z/Prev/Restore/Save/Del/
   ?/<World>:
```

COMMAND OPTIONS

Del	Deletes the name of a saved UCS.
OBject	Aligns UCS with a picked object.
Origin	Moves the UCS to a new origin point.
Prev	Restores the previous UCS orientation.
Restore	Restores a named UCS.
Save	Saves the current UCS by name.
View	Aligns the UCS with the current view.
<World>	Aligns the UCS with the WCS.
X	Rotates the UCS about the x-axis.
Y	Rotates the UCS about the y-axis.
Z	Rotates the UCS about the z-axis.
ZAxis	Aligns the UCS with a new origin and z-axis.
3point	Aligns the UCS with a point on the positive x-axis and positive x,y-plane.
?	Lists the names of saved UCS orientations.

RELATED AUTOCAD COMMANDS

- **DdUcs** Modifies the UCS via a dialogue box.
- **UcsIcon** Controls the visibility of the UCS icon.
- **Plan** Changes the view to the plan view of the current UCS.

RELATED SYSTEM VARIABLES

- **UcsFollow** Automatically shows plan view in new UCS.
- **UcsIcon** Determines visibility and location of UCS icon.
- **UcsOrg** WCS coordinates of UCS icon.
- **UcsXdir** X-direction of current UCS.
- **UcsYdir** Y-direction of current UCS.
- **WorldUcs** Correlation of WCS and UCS.

TIPS

- Use the **UCS** command to draw objects at odd angles in 3D space.

- Although you can create a UCS in paper space, you cannot use 3D viewing commands.

- A UCS can be aligned with these objects:
  - Point, line, trace, 2D polyline, and solid.
  - Arc and circle.
  - Text, shape, dimension, and attribute definition.
  - 3D face and block reference.

- A UCS will not align with these objects:
  - Mline, ray, xline, and 3D polyline.
  - Spline and ellipse.
  - Leader and viewport.
  - 3D solid, 3D mesh, and region.

DEFINITIONS

UCS:

- User-defined 2D coordinate system oriented in 3D space.
- Sets a working plane, orients 2D objects, defines the extrusion direction, and the axis of rotation.

WCS:

- World coordinate system.
- The default 3D x,y,z-coordinate system.

Ucsicon

Rel. 10

Controls the location and display of the UCS icon.

Command	Alt+	Side Menu	Menu Bar	Tablet
ucsicon	O,U,I	[OPTIONS]	[Options]	...
		[UCSicon:]	[UCS]	
			[Icon]	

```
Command: ucsicon
ON/OFF/All/Noorigin/ORigin <ON>:
```

COMMAND OPTIONS

All Makes the **UcsIcon** command's changes apply to all viewports.
Noorigin Always display UCS icon in lower-left corner.
OFF Turns off display of UCS icon.
ON Turns on display of UCS icon.
ORigin Displays UCS icon at the current UCS origin.

RELATED AUTOCAD COMMAND

- **UCS** Creates and controls user-defined coordinate systems.

RELATED SYSTEM VARIABLE

- **UcsIcon** Determines the display and origin of the UCS icon.

Undefine

Rel. 9

Makes an AutoCAD command unavailable.

Command	Alias	Side Menu	Menu Bar	Tablet
undefine	...	...	...	...

```
Command: undefine
Command name:
```

Example usage:

```
Command: undefine
Command name: line
Command: line
Unknown command.  Type ? for list of commands.
Command: .line
From point:
```

COMMAND OPTION

(Period) Precede undefined command with period to temporarily redefine it.

RELATED AUTOCAD COMMAND

■ **Redefine** Redefines an AutoCAD command.

RELATED SYSTEM VARIABLES

■ *None*

TIP

■ In menu macros written with international language versions of AutoCAD, precede command names with an underscore character (_) to automatically translate the name.

Undo

Ver. 2.5

Undoes the effect of one or more previous commands.

Command	Alias	Side Menu	Menu Bar	Tablet
undo	...	[ASSIST] [UNDO] [Undo:]	...	...

```
Command: undo
Auto/Control/BEgin/End/Mark/Back/<number>:
```

COMMAND OPTIONS

Auto Treats a menu macro as a single command.
Back Undo goes back to the marker.
BEgin Groups a sequence of operations (*the **Group** option in Release 12 and earlier*).
Control Limits the options of the **Undo** command.
  All Toggles on full undo.
  None Turns off undo feature.
  One Limits the **Undo** command to a single undo.
End Ends the **BEgin** option.
Mark Sets a marker.
<number> Indicates the number of commands to undo.

RELATED AUTOCAD COMMANDS

- **Oops** Unerases the most-recently erased object.
- **Quit** Leaves the drawing without saving changes.
- **Redo** Undoes the most recent undo.
- **U** Single-step undo.

RELATED SYSTEM VARIABLES

- **UndoCtl** Indicates the state of the **Undo** command:
  - 0 **Undo** is disabled.
  - 1 **Undo** is enabled.
  - 2 Single-command undo.
  - 3 Auto-group mode enabled.
  - 4 Begin-End (*group*) is currently active.
- **UndoMarks** Number of undo marks placed in the **Undo** control stream.

TIP

- Since the undo mechanism creates a mirror drawing file on disk, disable the **Undo** command with system variable **UndoCtl** (*set it to 0*) when your computer is low on disk space.

Union

Rel. 11

Joins to solids and regions together into a single model (*formerly the **SolUnion** command; an external command in Acis.Dll*).

Command	Alias	Side Menu	Menu Bar	Tablet
union	...	[DRAW 2]	[Modify]	Y 13
		[SOLIDS]	[Explode]	
		[Union:]	[Union]	

```
Command: union
Select objects: [pick]
Select objects: [pick]
```

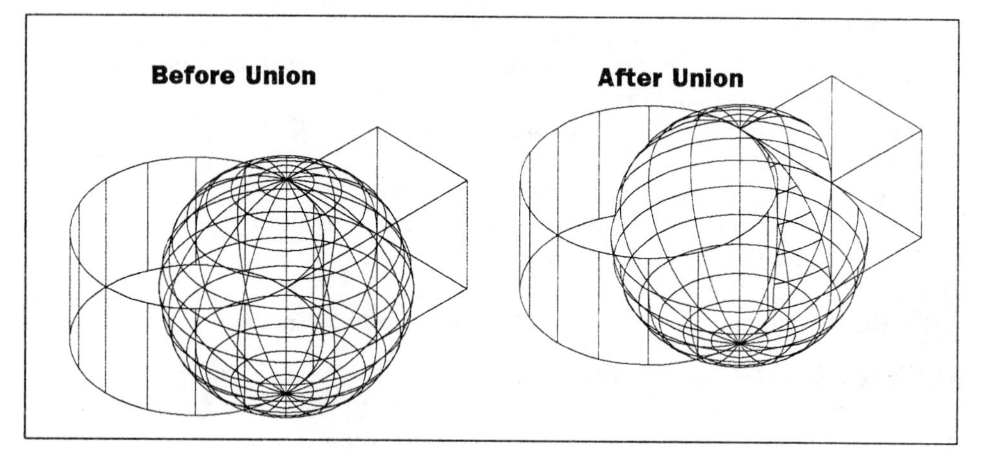

COMMAND OPTIONS

None

RELATED AUTOCAD COMMANDS

- **Intersect** Creates a solid model from the intersection of two objects.
- **Subtract** Creates a solid model by subtracting one from another.

'Units

Ver. 1.4

Controls the display and format of coordinates and angles.

Command	Alias	Side Menu	Menu Bar	Tablet
'units	...	...	...	...

Command: **units**

Switches display to the AutoCAD Text window:

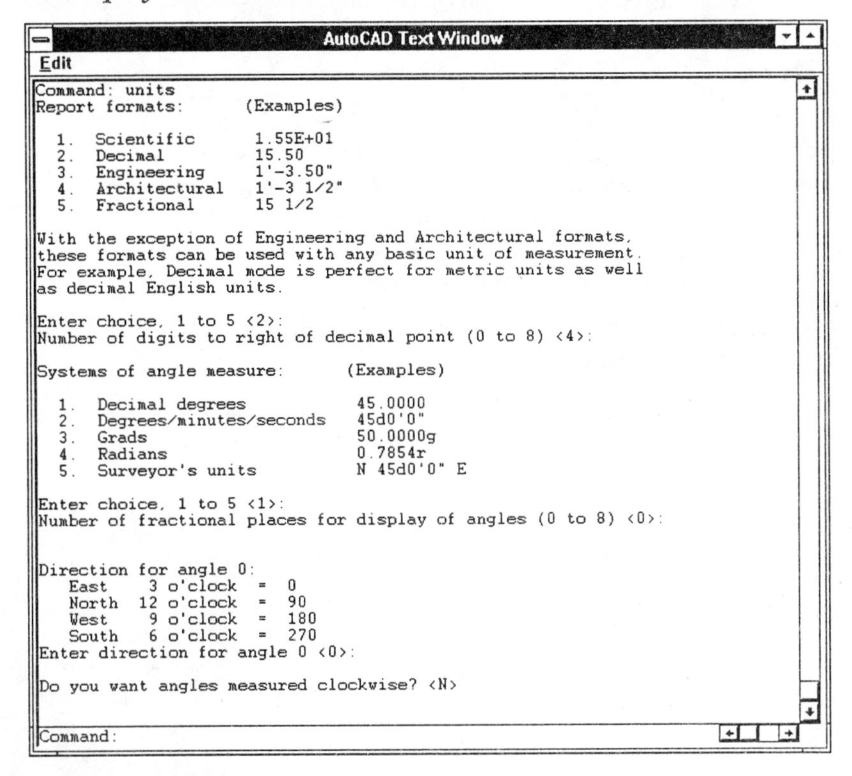

COMMAND OPTION

[F2] Returns to the drawing window.

RELATED AUTOCAD COMMANDS

- **DdUnits** Sets units via a dialogue box.
- **MvSetup** Sets up a drawing with multiple viewports.

RELATED SYSTEM VARIABLES

- **AngBase** Direction of zero degrees.
- **AngDir** Direction of angle measurement.
- **AUnits** Units of angles.

■ **AuPrec** Displayed precision of angles.
■ **LUnits** Units of measurement.
■ **LuPrec** Displayed precision of coordinates.
■ **UnitMode** Toggles type of display of units.

TIPS

■ Since **Units** is a transparent command, you can use it to change units during another command.

■ The 'Direction Angle' prompt lets AutoCAD start angle measurement from any direction.

■ AutoCAD accepts the following notation for angle input:

- ■ < Specify an angle based on current units setting.
- ■ << Bypass angle translation set by **Units** command to use 0-angle-is-east direction and decimal degrees.
- ■ <<< Bypass angle translation; use angle units set by **Units** command, and 0-angle-is-east direction.

■ The system variable **UnitMode** forces AutoCAD to display units in the same manner that you enter them.

■ Do not use a suffix (*such as* **r** *or* **g**) for angles entered as radians or grads; instead, use the **Units** command to set angle measurement to radians and grads.

'View

Ver. 2.0

Saves and displays the view in the current viewport by name.

Command	Alias	Side Menu	Menu Bar	Tablet
'view	...	...	...	...

```
Command: view
?/Delete/Restore/Save/Window: s
View name to save:
```

COMMAND OPTIONS

Delete Deletes a named view.
Restore Restores a named view.
Save Saves the current view with a name.
Window Saves a windowed view with a name.
? Lists the names of views saved in the current drawing.

RELATED AUTOCAD COMMANDS

- **DdView** Creates and displays view names via a dialogue box.
- **Rename** Changes the names of views.

RELATED SYSTEM VARIABLES

- **ViewCtr** The coordinates of the center of the view.
- **ViewSize** The height of the view.

TIPS

- Name views in your drawing to quickly move from one detail to another.

- View names are up to 31 characters long and may not contain spaces.

- The **Plot** command plots named views of a drawing.

- Objects outside of the window created by the **Window** option may be displayed but are not plotted.

- You create separate views in model and paper space.

- When listing named views (*with the ? option*), AutoCAD places an "M" or "P" next to the view name to indicate model or paper space.

ViewRes

Ver. 2.5

Controls the roundness of curved objects; determines whether zooms and pans are performed as redraws or regens (*short for VIEW RESolution*).

Command	Alias	Side Menu	Menu Bar	Tablet
viewres	...	...	...	...

```
Command: viewres
Do you want fast zooms? <Y> [Enter]
Enter circle zoom percent (1-20000) <100>: 1000
Regenerating drawing.
```

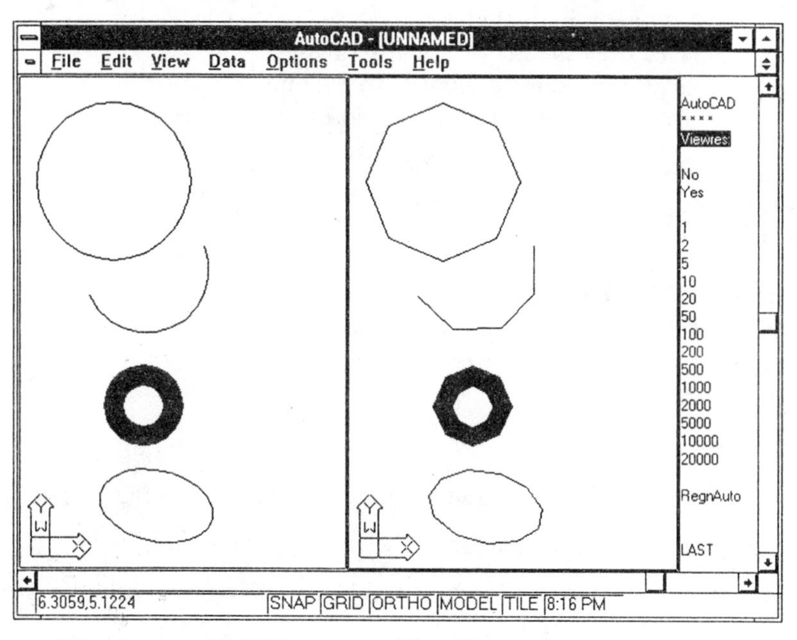

ViewRes = 20,000 **ViewRes = 1**

COMMAND OPTIONS

Do you want fast zooms?

Yes AutoCAD tries to make every zoom and pan a redraw (*faster*).

No Every zoom and pan causes a regeneration (*slower*).

Enter circle zoom percent (1-20000)

Smaller values display faster but make circles look less round (*see figure*).

RELATED AUTOCAD COMMAND

- **RegenAuto** Determines whether AutoCAD uses redraws or regens.

VlConv

Rel. 13

Converts Visual Link data for use by AutoVision (*short for Visual Link CONVert; an external command in Render.Arx*).

Command	Alias	Side Menu	Menu Bar	Tablet
vlconv	...	...	...	...

```
Command: vlconv
```

Displays dialogue box.

COMMAND OPTION

Overwrite	**Off** (*default*)	AutoVision material assignments are preserved.
	On	Visual Link material assignments overwrite AutoVision's.

TIPS

- Visual Link data consists of trees, paths, and material assignments.
- Visual Link data is still available after conversion.

VpLayer

Ver. 2.1

Controls the visibility of layers in viewports when **TileMode** is turned off (*short for ViewPort LAYER*).

Command	Alt+	Side Menu	Menu Bar	Tablet
vplayer	D,V	[DATA]	[Data]	M5 - 05
		[VPlayer:]	[Viewport Layer Controls]	

■

```
Command: vplayer
?/Freeze/Thaw/Reset/Newfrz/Vpvisdflt:
Select a viewport: [pick]
```

COMMAND OPTIONS

Freeze Indicates the names of layers to freeze in this viewport.
Newfrz Creates new layers which will be frozen in newly-created viewports (*short for NEW FReeZe*).
Reset Resets the state of layers based on the Vpvisdflt settings.
Thaw Indicates the names of layers to thaw in this viewport.
Vpvisdflt Determines which layers will be frozen in a newly-created viewport (*short for ViewPort VISibility DeFauLT*).
? Lists the layers frozen in the current viewport.

RELATED AUTOCAD COMMANDS

- **DdLModes** Toggles the visibility of layers in viewports via a dialogue box.
- **Layer** Creates and controls layers in all viewports.
- **MView** Creates and joins viewports when tilemode is off.
- **MvSetup** Sets up a drawing with paper space.

RELATED SYSTEM VARIABLE

- **TileMode** Controls whether viewports are tiled or overlapping.

VPoint

Rel. 10

Changes the viewpoint of a 3D drawing (*short for ViewPOINT*).

Command	Alt+	Side Menu	Menu Bar	Tablet
vpoint	V,E,T	[VIEW] [Vpoint:]	[View] [3D Viewpoint] [Tripod]	K 1

```
Command: vpoint
Rotate/<View point> <0.0000,0.0000,1.0000>:
```

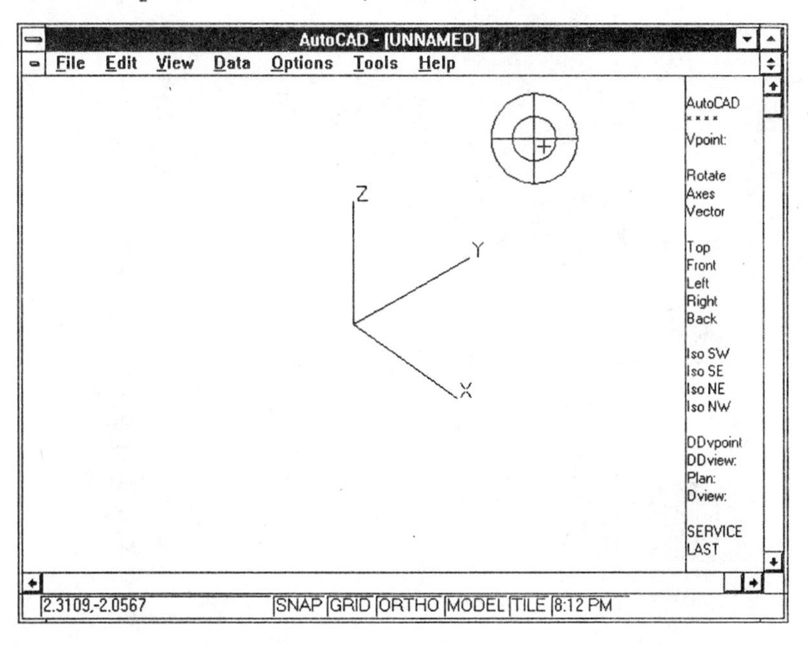

COMMAND OPTIONS

[Enter] Bring up visual guides (*see figure*).

Rotate Indicate the new 3D viewpoint by angle.

<View point> Indicate the new 3D viewpoint by coordinates.

RELATED AUTOCAD COMMANDS

- **DdVpoint** Adjust viewpoint via dialogue box.
- **DView** Changes the viewpoint of 3D objects, plus allows perspective mode.

RELATED SYSTEM VARIABLES

- **VpointX, VpointY, VpointZ** X-, y-, z-coordinates of current 3D view.
- **WorldView** Determines whether **VPoint** coordinates are in WCS or UCS.

ViewPorts *or* VPorts

Ver. 2.0

Creates tiled viewports (*or windows*) of the current drawing.

Command	Alt+	Side Menu	Menu Bar	Tablet
`viewports`	`V,D`	`[VIEW]` `[Vpoint:]`	`[View]` `[Tiled Viewports]`	`R 5`
`vports`				`V 22`

```
Command: vports
Save/Restore/Delete/Join/SIngle/?/2/<3>/4:
```

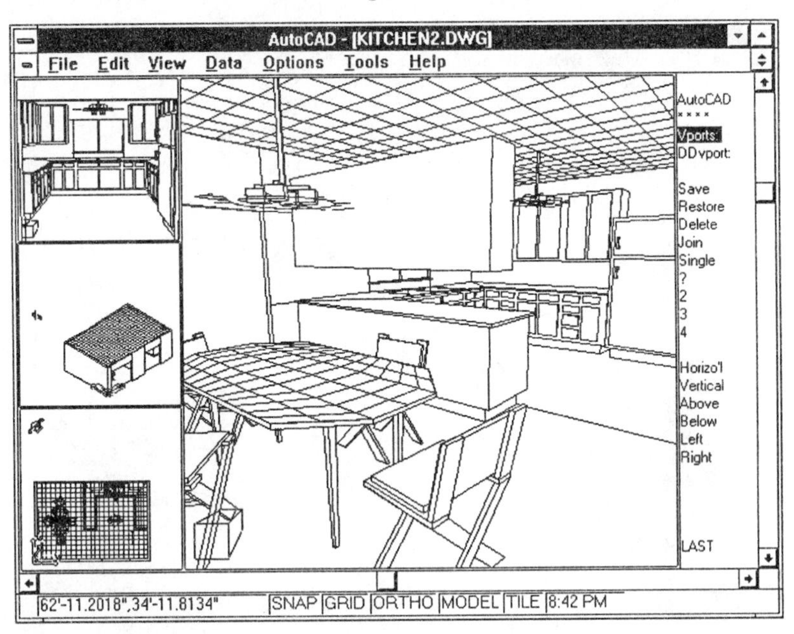

COMMAND OPTIONS

Delete Deletes a viewport definition.
Join Joins two viewports together as one.
Restore Restores a viewport definition.
Save Saves the settings of a viewport by name.
SIngle Joins all viewports into a single viewport.
2 Splits the current viewport into two:
  Horizontal Creates one viewport over another.
  <Vertical> Creates one viewport beside another.

<3> Splits the current viewport into three:
  Horizontal Creates three viewports over each other.
  Vertical Creates three viewports beside each other.

Above	Creates two viewports over one viewport.
Below	Creates two viewports below one viewport.
Left	Creates two viewports left of one viewport.
<Right>	Creates two viewports right of one viewport.

4	Splits the current viewport into four.
?	Lists the names of saved viewport configurations.

RELATED AUTOCAD COMMANDS

- **MView** Creates viewports in paper space.
- **RedrawAll** Redraws all viewports.
- **RegenAll** Regenerates all viewports.
- **[Ctrl]+V** Moves focus to the next viewport.

RELATED SYSTEM VARIABLES

- **CvPort** The current viewport.
- **MaxActVp** The maximum number of active viewports.
- **TileMode** Controls whether viewports can be overlapping or tiled.

VSlide

Ver. 2.0

Displays an SLD-format slide file in the current viewport (*short for View SLIDE*).

Command	Alt+	Side Menu	Menu Bar	Tablet
vslide	T,D,V	[TOOLS]	[Tools]	...
		[VSlide:]	[Slide]	
			[View]	

```
Command: vslide
Slide file <>:
```

COMMAND OPTIONS

~ (*Tilde*) Forces display of the file dialogue box.

RELATED AUTOCAD COMMANDS

- **MSlide** Creates an SLD-format slide file of the current viewport.
- **Redraw** Erases the slide from the screen.

RELATED AUTODESK PROGRAM

- **SlideLib.Exe** Creates an SLB-format library file of a group of slide files.

RELATED SYSTEM VARIABLES

- *None*

TIP

- For faster viewing of a series of slides, an asterisk preceeding **VSlide** preloads the slide file, as in:

```
Command: *vslide filename
```

WBlock

Ver. 1.4

Writes a block, part of the drawing, or all of the drawing to disk (*short for Write BLOCK*).

Command	Alt+	Side Menu	Menu Bar	Tablet
wblock	F,E	[FILE]	[File]	W 8
		[EXPORT]	[Export]	
		[Wblock:]	[DWG]	

```
Command: wblock
File name:
Block name:
```

COMMAND OPTIONS

=	(*Equals*) Block is written to disk using block's name as filename.
*	(*Asterisk*) Entire drawing is written to disk.
[Enter]	Creates a block on disk of selected objects.
[Space]	Moves selected objects to the specified drawing.

RELATED AUTOCAD COMMAND

- **Block** Creates a block of a group of objects.
- **Export** Dialogue-box frontend for the **WBlock** command.

TIPS

The **WBlock** command is the command that exports blocks from the drawing to individual DWG files on disk.

The **WBlock** command is an alternative to the **Purge** command.

Wedge

Rel. 11

Draws a 3D wedge as a solid model *(formerly the SolWedge command; an external command in Acis.Dll)*.

Command	Alias	Side Menu	Menu Bar	Tablet
wedge	...	[DRAW 2]	[Solids]	N 7
		[SOLIDS]	[Wedge]	
		[Wedge:]		

```
Command: wedge
Center/<Corner of wedge> <0,0,0>: [pick]
Cube/Length/<other corner>: [pick]
Height:
```

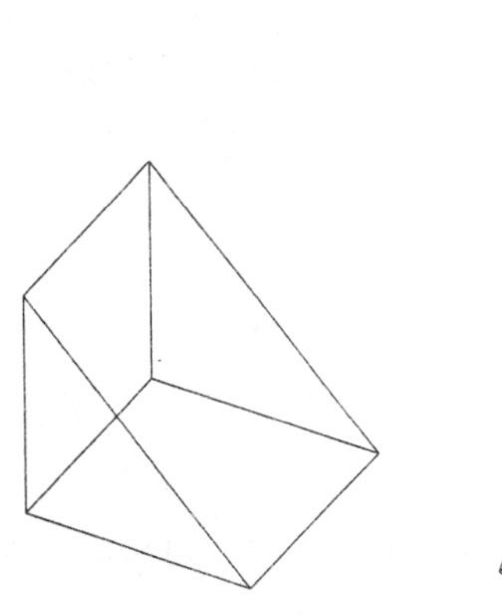

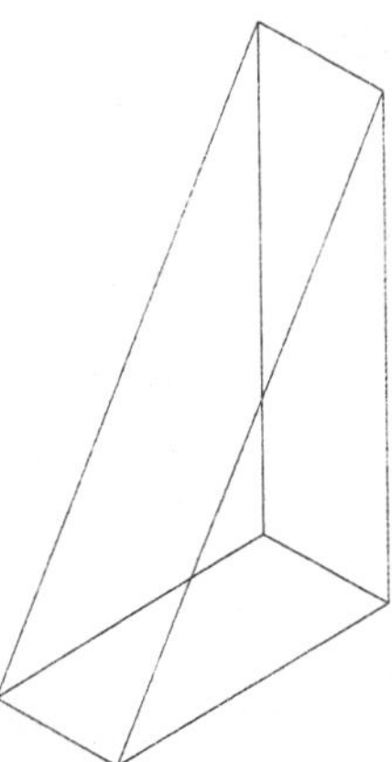

COMMAND OPTIONS

Center Draws wedge base about a center point.

Corner Draws wedge base between two pick points.

Cube Draws a cubic wedge.

Length Specifies length, width, and height of wedge.

RELATED AUTOCAD COMMANDS

- **Ai_Wedge** Draws wedge as a 3D surface model.
- **Box** Draws solid boxes.
- **Cone** Draws solid cones.
- **Cylinder** Draws solid cylinders.
- **Sphere** Draws solid spheres.
- **Torus** Draws solid tori.

NON-EXISTANT COMMAND: WhatsNew

The **WhatsNew** command does not exist in the Windows version of Release 13. Instead, select **Help | What's New** from the menu bar.

Wmfin

Rel. 12

Imports a WMF vector file (*short for Windows MetaFile IN*).

Command	Alt+	Side Menu	Menu Bar	Tablet
wmfin	F,I	...	[File] [Import] [WMF]	...

```
Command: wmfin
```

Displays dialogue box.

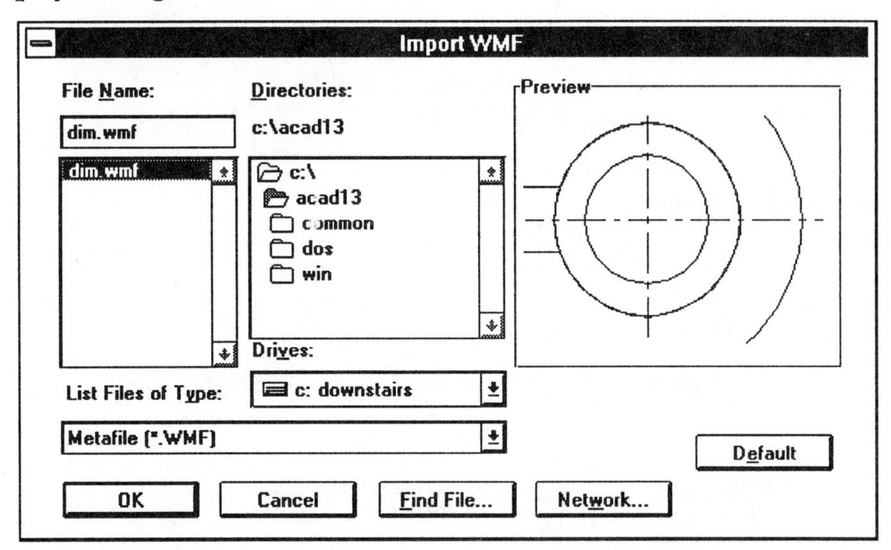


```
Insertion point:    [pick]
X scale factor <1>/Corner/XYZ: [Enter]
```

COMMAND OPTIONS

None

RELATED AUTOCAD COMMANDS

- **WmfOpts** Controls the importation of WMF files.
- **WmfOut** Exports selected objects in WMF format.

TIPS

- The WMF is placed as a block with the name WMF0; subsequent placements increment the digit: WMF1, WMF2, etc.

- Exploding the WMF*n* block results in polylines; circles, arcs, and text are converted to polylines; solid-filled areas are exploded into solid triangles.

WmfOpts

Rel. 12

Controls the importation of WMF files.

Command	Alt+	Side Menu	Menu Bar	Tablet
`wmfopts`	`F,T,T`	...	`[File]` `[Options]` `[WMF Options]`	...

Command: **wmfopts**

Displays dialogue box.

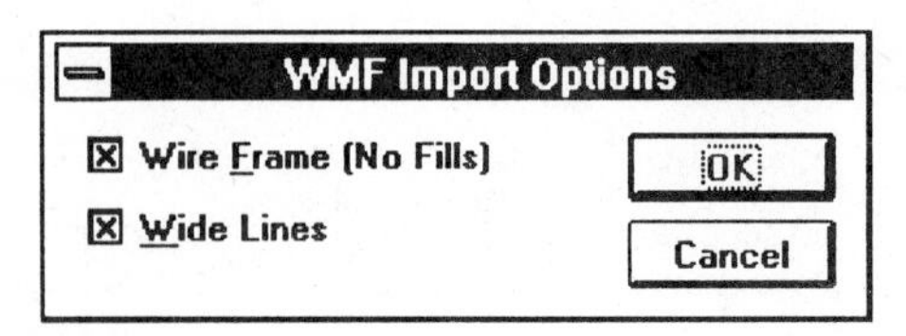

COMMAND OPTIONS

Wireframe	**On**	No area fills (*default*).
	Off	Filled areas are imported as solids.
Wide lines	**On**	Lines retain their width (*default*).
	Off	Lines are collapsed to zero width.

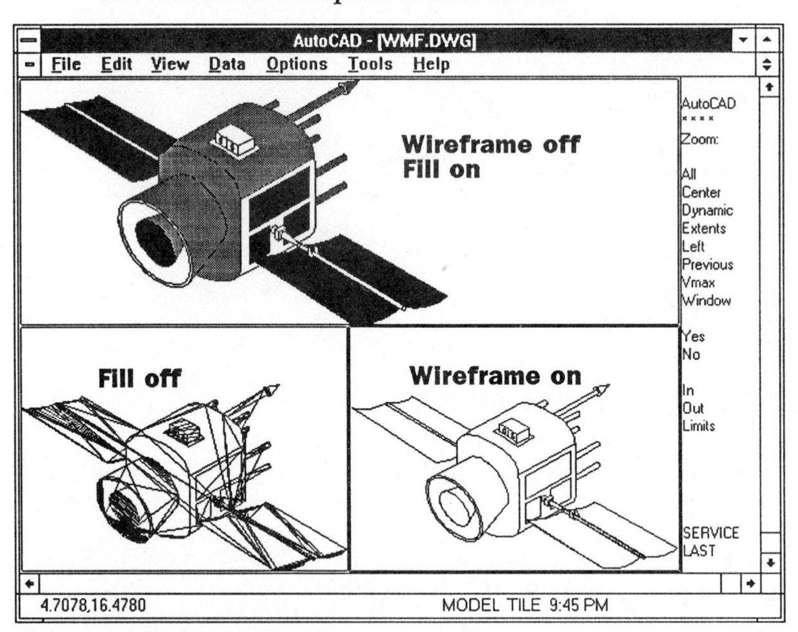

RELATED AUTOCAD COMMANDS

- **WmfIn** Imports WMF files.
- **WmfOut** Exports selected objects in WMF format.

WmfOut

Rel. 12

Exports the drawing in WMF vector file format (*short for Windows MetaFile OUTput*).

Command	Alt+	Side Menu	Menu Bar	Tablet
wmfout	F,E	...	[File] [Export] [WMF]	...

```
Command: wmfout
```

Displays dialogue box.

```
Select objects: [pick]
Select objects: [Enter]
```

COMMAND OPTIONS

None

RELATED AUTOCAD COMMANDS

- **WmfOpts** Controls the importation of WMF files.
- **WmfIn** Imports a WMF format file into AutoCAD.

TIPS

- WMF files created by AutoCAD are resolution-dependent; small circles and arcs lose their roundness.

- The **All** selection does not select all objects in drawing; instead, **WmfOut** selects all objects visible in the current viewport.

XBind

Rel. 11

Binds portions of an externally-referenced drawing to the current drawing (*short for eXternal BINDing*).

Command	Alias	Side Menu	Toolbar	Tablet
xbind	...	[FILE] [Xbind:]	[External Reference] [Bind]	...

```
Command: xbind
Block/Dimstyle/LAyer/LType/Style: b
Dependent Block name(s):
```

COMMAND OPTIONS

Block Binds blocks to current drawing.

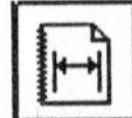

Dimstyle Binds dimension styles to current drawing.

LAyer Binds layer names to current drawing.

LType Binds linetype definitions to current drawing.

Style Binds text styles to current drawing.

RELATED AUTOCAD COMMANDS

- **XRef** Attaches another drawing the current drawing.
- **XrefClip** Inserts an externally-referenced block.

RELATED SYSTEM VARIABLES

- *None*

XLine

Places an infinitely-long construction line.

Command	Alias	Side Menu	Menu Bar	Tablet
xline	...	[DRAW 1] [Xline:]	[Draw] [Line] [XLine]	L 10

```
Command: xline
Hor/Ver/Ang/Bisect/Offset/<From point>: [pick]
Through point: [pick]
```

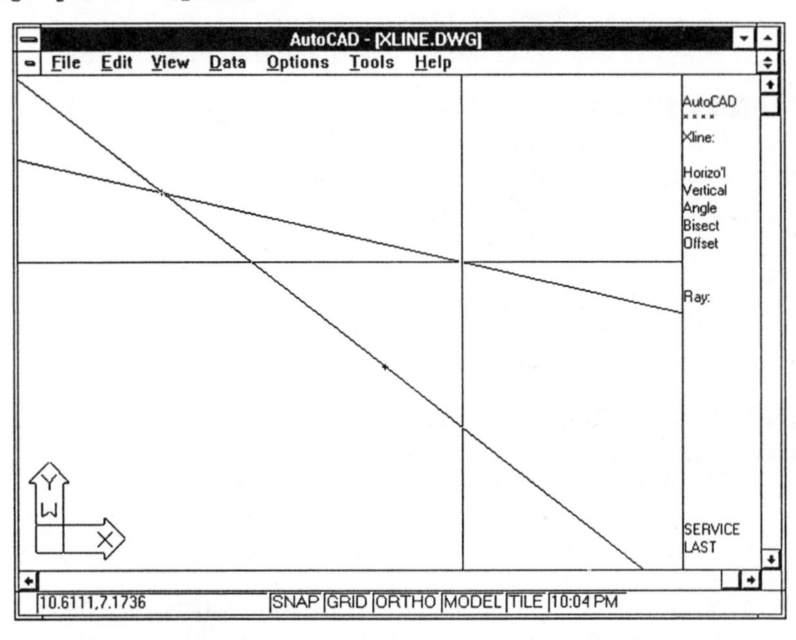

COMMAND OPTIONS

Ang Places the construction line at an angle.
Bisect Bisects an angle with the construction line.
<From point> Places the construction line through a point.
Hor Places a horizontal construction line.
Offset Places the construction line parallel to another object.
Ver Places a vertical construction line.

RELATED AUTOCAD COMMANDS

- **DdModify** Modifies characteristics of the xline and ray objects.
- **Ray** Places a semi-infinite construction line.

TIP

- The ray and xline construction lines do not plot.

'Xplode

Rel. 12

Reduces complex objects to their primitive constituent parts; provides greater user control than the **Explode** command (*short for eXplode; an external command in Xplode.Lsp*).

Command	Alias	Side Menu	Menu Bar	Tablet
'xplode	xp	...	...	W 20

```
Command: xplode
Select objects: [pick]
Select objects: [Enter]
Xplode Individually/<Globally>: [Enter]
All/Color/LAyer/LType/Inherit from parent block/<Explode>:
```

COMMAND OPTIONS

Individually Explodes one object at a time.
Globally Applies explode to all selected objects.
All Prompts for changes *after* exploding.
Color Specifies color after explosion.
LAyer Specifies layer.
LType Specifies lineytpe.
Inherit Sets color, linetype, and layer to that of the original block.
<Explode> Mimics the **Explode** command.

RELATED AUTOCAD COMMANDS

- **Explode** Original explode command.
- **U** Undoes the effects of the **Xplode** command.

TIPS

- **Color** option:
  - BYLAYER Exploded objects inherit color from the original object's layer.
  - BYBLOCK Exploded objects inherit color from the original object.
- **LAyer** option:
  - By default, exploded objects inherit the current layer, not the original object's layer.
- **LType** option:
  - BYLAYER Exploded objects inherit linetype from the original object's layer.
  - BYBLOCK Exploded objects inherit linetype from the original object.

Xref

Rel. 11

Attaches a drawing to the current drawing (*short for eXternal REFerence*).

Command	Alias	Side Menu	Toolbar	Tablet
xref	...	[FILE:] [Xref:]	[External Reference] [Xref]	P 8-10

```
Command: xref
?/Bind/Detach/Path/Reload/Overlay/<Attach>:
```

COMMAND OPTIONS

<Attach> Attaches another drawing to the current drawing.

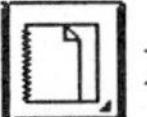

Bind Binds the xref drawing to the current drawing.

Detach Removes the externally-referenced drawing.

Overlay Overlays the externally-referenced drawing (*new to Release 13*).

Path Respecifies the path to the externally-referenced drawing.

Reload Updates the externally-referenced drawing.

? Lists the names of externally-referenced drawings.

RELATED AUTOCAD COMMANDS

- **Insert** Adds another drawing to the current drawing.
- **XBind** Binds parts of the externally-referenced drawing to the current drawing.
- **XrefClip** Lets you clip an area of an externally-referenced drawing to attach to the current drawing.

RELATED SYSTEM VARIABLE

- **XRefCtl** Controls whether XLG external reference log files are written:

TIPS

- The **XRef** command lets you view other drawings at the same time as the currently-loaded drawing; however, you cannot edit the externally-referenced drawing.

- If you are working with AutoCAD on a network, no other user can access the externally referenced drawing while you are loading it; this is called "soft file locking."

XrefClip

Rel. 12

Clips and inserts a portion of an externally-referenced drawing (*short for eXternal REFerence CLIP; an external command in XRefClip.Lsp*).

Command	Alias	Side Menu	Menu Bar	Tablet
xrefclip	...	...	[External Reference] [Clip]	...

```
Command: xrefclip
Enable paper space? <Y> [Enter]
Xref name:
Clip onto what layer?
First corner of clip box: [pick]
Other corner: [pick]
Number of paper space units <1.0>:  [Enter]
Number of model space units <1.0>: [Enter]
Insertion point for clip: [pick]
```

COMMAND OPTIONS

None

RELATED AUTOCAD COMMANDS

- **Insert** Adds another drawing to the current drawing.
- **XBind** Binds parts of the externally-referenced drawing to the current drawing.
- **Xref** Displays an externally-referenced drawing in the current drawing.

RELATED SYSTEM VARIABLE

- **TileMode** Must be set to 0 for **XRefClip** to work.

TIPS

- **TileMode** must be set to 0 before starting the **XRefClip** command.
- During the **XRefClip** command, layers, and viewports are turned off.
- The layer name must exist before you start the **XRefClip** command.
- The 'clip box' becomes a paper space viewport.
- You cannot have an irregularly-clipped xref.

'Zoom

Ver. 1.0

Displays a drawing larger or smaller in the current viewport.

Command	Alias	Side Menu	Menu Bar	Tablet
'zoom	z	[VIEW]	[View]	Q 7-9
		[ZOOM:]	[Zoom]	R 7-10

```
Command: zoom
All/Center/Dynamic/Extents/Left/Previous/Vmax/Window/
   <Scale(X/XP)>:
```

Zoom Dynamic screen:

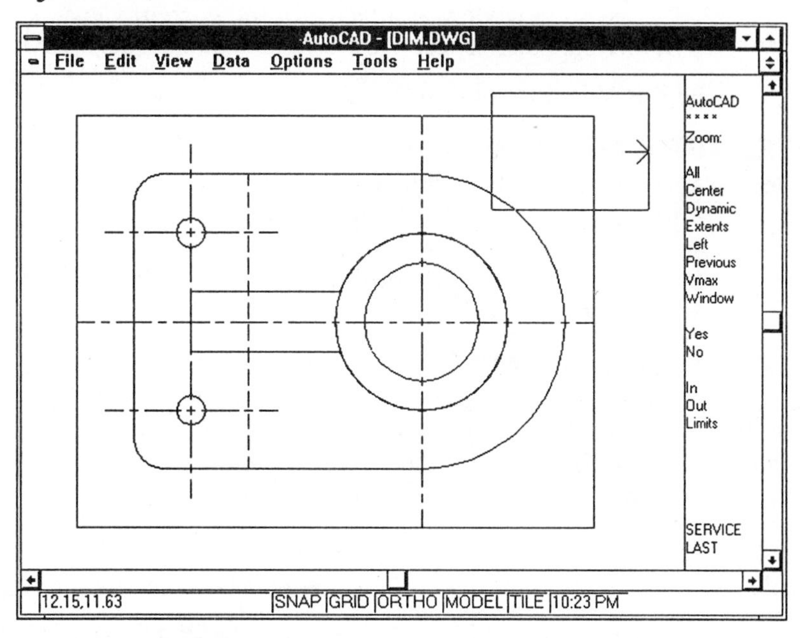

COMMAND OPTIONS

All Displays the drawing limits or extents, whichever is greater.

Center Zooms in about a center point:

Center point
Indicates the center point of the new view.

Magnification or Height
Indicates a magnification value or height of view.

Dynamic Brings up the dynamic zoom view.

Extents Displays the current drawing extents.

Left Displays a view with a new lower-left corner:

Lower left corner point

Indicates the lower-left corner of the new view.

Magnification or Height

Indicates a magnification value or height of view.

Previous Displays the previous view generated by **Pan**, **View**, or **Zoom**.

Vmax Displays the current virtual screen limits (*short for Virtual MAXimum*).

Window Indicate the two corners of the new view.

<Scale(X/XP)> Displays a new view as a factor of the drawing limits:

- **X** Displays a new view as a factor of the current view.
- **XP** Displays a paper space view as a factor of model space.

[pick] Begins **Window** option.

RELATED AUTOCAD COMMANDS

- **DsViewer** Aerial View, available when AutoCAD is configured with the Windows Accelerated Display Driver.
- **Limits** Specifies the limits of the drawing.
- **Pan** Moves the view to a different location.
- **View** Saves views by name.

RELATED SYSTEM VARIABLES

- **ViewCtr** Coordinates of the current view's center point.
- **ViewSize** Height of the current view.
- **VsMax** Upper-right corner of the virtual screen.
- **VsMin** Lower-left corner of the virtual screen.

TIPS

- A scale factor of 1.0 displays the entire drawing as defined by the limits.

- A zoom factor of 2 enlarges objects (*zooms in*), while 0.5 makes objects smaller (*zooms out*).

3D

Rel. 11

Draws 3D primitives with poly meshes (*an external command in 3D.Lsp*).

Command	Alias	Side Menu	Toolbar	Tablet
3d	...	[DRAW 2] [SURFACES]	[Surfaces]	...

```
Command: 3d
Box/Cone/DIsh/DOme/Mesh/Pyramid/Sphere/Torus/Wedge:
```

*See **Ai_Box** command for more details.*

COMMAND OPTIONS

Box Draws a 3D box or cube.
Cone Draws cone shapes.
Dish Draws a dish (*bottom-half of a sphere*).
Dome Draws a dome (*top-half of a sphere*).
Mesh Draws a 3D mesh.
Pyramid Draws pyramid shapes.
Sphere Draws a sphere.
Torus Draw torus (*3D donut*) shapes.
Wedge Draws wedge shapes.

RELATED AUTOCAD COMMANDS

- **Ai_Box** Draws a 3D surface box or cube.
- **Ai_Cone** Draws 3D surface cone shapes.
- **Ai_Dish** Draws a 3D surface dish.
- **Ai_Dome** Draws a 3D surface dome.
- **Ai_Mesh** Draws a 3D mesh.
- **Ai_Pyramid** Draws a 3D surface pyramid.
- **Ai_Sphere** Draws a 3D surface sphere.
- **Ai_Torus** Draws a 3D surface torus.
- **Ai_Wedge** Draws 3D surface wedge.
- **Box** Draws a 3D solid box or cube.
- **Cone** Draws a 3D solid cone.
- **Cylinder** Draws a 3D solid cylinder.
- **Sphere** Draws a 3D solid sphere.
- **Torus** Draws a 3D solid torus.
- **Wedge** Draws a 3D solid wedge.

TIP

- The **3D** command creates three-dimensional objects made of 3D meshes and not 3D ACIS solids.

3dArray

Rel. 11

Creates three-dimensional rectangular and polar arrays (*an external command in 3dAray.Lsp*).

Command	Alias	Side Menu	Toolbar	Tablet
3darray	...	[CONSTRCT] [3Darray:]	[Modify] [Copy] [3dArray]	...

```
Command: 3darray
Select objects: [pick]
Select objects: [Enter]
Rectangular or Polar array (R/P):
```

Command prompts for a rectangular array:

```
Number of rows (---) <1>:
Number of columns (|||) <1>:
Number of levels (...) <1>:
Distance between rows (---) <1>:
Distance between columns (|||) <1>:
Distance between levels (...) <1>:
```

Command prompts for a polar array:

```
Number of items:
Angle to fill <360>:
Rotate objects as they are copied? <Y>:
Center point of array:
Second point on axis of rotation:
```

COMMAND OPTIONS

R	Creates rectangular 3D array.
P	Creates polar 3D array.
[Esc]	Interrupts drawing of array.

RELATED AUTOCAD COMMANDS

- **Array** Creates rectangular or polar array in 2D space.
- **Copy** Creates one or more copies of the selected object.
- **MInsert** Creates a rectangular block array of blocks.

3dFace

Ver. 2.6

Draws 3D faces with three or four corners.

Command	Alias	Side Menu	Toolbar	Tablet
3dface	...	[DRAW 2] [SURFACES] [3Dface:]	[Surfaces] [3dFace]	P 9

```
Command: 3dface
First point: [pick]
Second point: [pick]
Third point: [pick]
Fourth point: [pick]
```

COMMAND OPTION

i Prefix for corner coordinate to make edge invisible.

RELATED AUTOCAD COMMANDS

- **3D** Draws 3D objects: box, cone, dome, dish, pyramid, sphere, torus, and wedge.
- **Edge** Changes the visibility of the edges of 3D faces.
- **EdgeSurf** Draws 3D surfaces made of 3D meshes.
- **PEdit** Edits 3D meshes.
- **PFace** Draws generalized 3D meshes.

RELATED SYSTEM VARIABLE

- **SplFrame** Controls the visibility of edges.

TIPS

■ A 3D face is the same as a 2D solid, except that each corner can have a different z-coordinate.

■ Unlike the **Solid** command, corner coordinates are entered in natural order.

■ The **i** (*short for invisible*) suffix must be entered before object snap modes, point filters, and corner coordinates.

■ Invisible 3D faces – where all four edges are invisible – do not appear in wireframe views; however, they hide objects behind them in hidden-line mode and are rendered in shaded views.

■ 3D faces cannot be extruded.

3dMesh

Rel. 10

Draws open three-dimensional rectangular meshes made of 3D faces.

Command	Alias	Side Menu	Toolbar	Tablet
3dmesh	...	[DRAW 2] [SURFACES] [3Dmesh:]	[Surfaces] [3dMesh]	...

```
Command: 3dmesh
Mesh M size:
Mesh N size:
Vertex (0, 0):
Vertex (0, 1):
```
...etc.

COMMAND OPTIONS

None

RELATED AUTOCAD COMMANDS

- **3D** Draws a variety of 3D objects.
- **Explode** Explodes a 3D mesh into individual 3D faces.
- **PEdit** Edits a 3D mesh.
- **PFace** Draws a generalized 3D face.
- **Xplode** Explodes a group of 3D meshes.

RELATED SYSTEM VARIABLES

- **SurfU** Surface density in m-direction.
- **SurfV** Surface density in n-direction.

TIPS

- It is more convenient to use the **EdgeSurf**, **RevSurf**, **RuleSurf**, and **TabSurf** commands than the **3dMesh** command.

- The range of values for the m- and n-mesh size is 2 to 256.

3dPoly

Rel. 10

Draws three-dimensional polylines (*short for 3D POLYline*).

Command	Alias	Side Menu	Toolbar	Tablet
3dpoly	...	[DRAW 1]	[Draw]	P 10
		[3Dpoly:]	[Polyline]	
		[3D Surfs]	[3dPoly]	

```
Command: 3dpoly
From point: [pick]
Close/Undo/<Endpoint of line>: [pick]
```

COMMAND OPTIONS

Close Joins the last endpoint with the start point.

<Endpoint of line>
Indicates the endpoint of the current segment.

Undo Erases the last-drawn segment.

RELATED AUTOCAD COMMANDS

- **Explode** Reduces a 3D polyline into lines and arcs.
- **PEdit** Edits 3D polylines.
- **PLine** Draws 2D polylines.
- **Xplode** Explodes a group of 3D polylines.

RELATED SYSTEM VARIABLES

None

TIPS

- Since 3D polylines are made of straight lines, use the **PEdit** command to spline the polyline as a curve.
- 3D polylines do not support linetypes or widths.

3dsIn

Imports a 3DS file created by 3D Studio (*an external command in Render.Arx*).

Command	Alt+	Side Menu	Menu Bar	Tablet
3dsin	F,I	[FILE]	[File]	...
		[IMPORT]	[Import]	
		[3DSin:]	[3DS]	

Command: **3dsin**

Displays dialogue box.

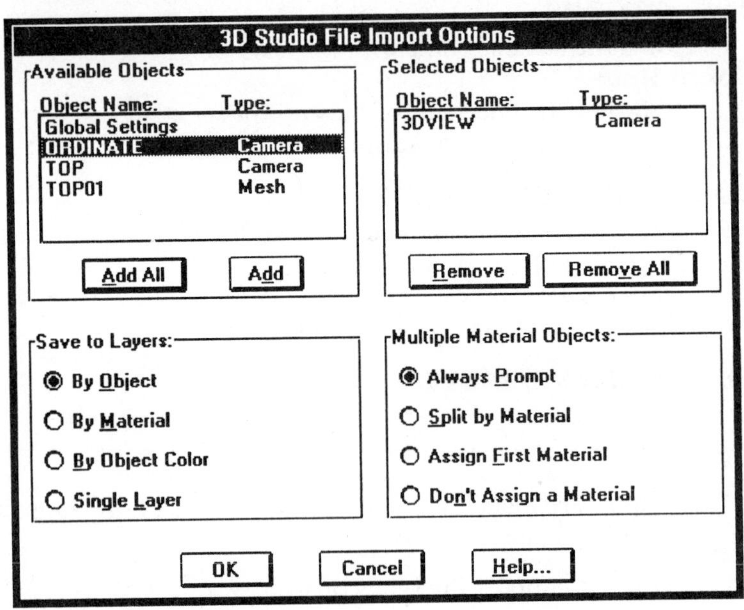

COMMAND OPTIONS

Available Objects

Names and types of objects in a 3D Studio drawing.

Object Name	Name of object.
Type	Type of object.
Add	Adds object to **Selected Objects** list.
Add All	Adds all objects to **Selected Objects** list.
Remove	Removes object from **Selected Objects** list.
Remove All	Removes all objects from **Selected Objects** list.

Save to Layers

Allows you to control the assignment of 3D Studio objects to layers.

By Object Each object placed on its own layer.

By Material Places objects on layers named after materials.

By Object Color Places objects on layers named "Color*nn*."

Single Layer Places all objects on layer "AvLayer."

Multiple Material Objects

Allows you to control how materials are assigned:

Always Prompt

Prompts you for each material.

Split by Material

Splits objects with more than one material into multiple objects, each with one material.

Assign First Material

Assigns first material to entire object.

Don't Assign to a Material

Loses all 3D Studio material definitions.

RELATED AUTOCAD COMMAND

- **3dsOut** Exports the drawing as a 3DS file.

RELATED FILES

- ***.3DS** 3D studio files.
- ***.TGA** Converted bitmap and animation files.

TIPS

- Objects:
  - You are limited to selecting a maximum of 70 3D Studio objects.
  - Conflicting object names are truncated and given a sequence number.
  - The **By Object** option gives the AutoCAD layer the name of the object.
  - The **By Object Color** option places all objects on layer "ColorNone" when no colors are defined in 3DS file.
  - 3D Studio assigns materials to faces, elements, and objects; AutoCAD only assigns materials to object, color, and layer.
- Bitmap and animation conversion:
  - 3D Studio bitmaps are converted to TGA (*Targa format*) bitmaps.
  - Only the first frame of an animation file (*CEL, CLI, FLC, or IFL files*) is converted to a Targa bitmap file.
  - Converted TGA files are saved to the 3DS file's subdirectory.
- Light conversion:
  - 3D Studio ambient lights lose their color.
  - 3D Studio 'omni lights' become point lights in AutoCAD.
  - 3D Studio cameras become a named view in AutoCAD.

3dsOut

Exports the AutoCAD drawing as a 3DS file for 3D Studio (*an external command in Render.Arx*).

Command	Alias	Side Menu	Menu Bar	Tablet
3dsout	...	[FILE]	[File]	...
		[EXPORT]	[Export]	
		[3DSout:]	[3D Studio]	

Command: **3dsout**

Displays dialogue box.

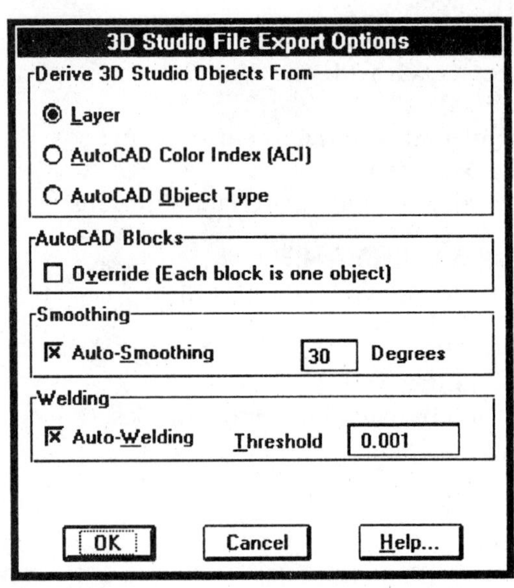

COMMAND OPTIONS

Layer All objects on an AutoCAD layer become a single 3D Studio object.

ACI All objects of an ACI color become a single 3D Studio object.

Object Type All objects on an AutoCAD object type become a single 3D Studio object.

Override Each AutoCAD block becomes a single 3D Studio object; overrides above three options.

Autosmoothing
Creates a 3D Studio smoothing group.

Autowelding Creates a 3D Studio welded vertex.

RELATED AUTOCAD COMMAND

- **3dsIn** Imports 3DS file to the drawing.

RELATED FILE

- ***.3DS** 3D studio files.

TIPS

- Exported objects:
  - AutoCAD objects with 0 thickness are not exported, with the exception of circles, polygon, and polyface meshes.
  - Solids and 3D faces must have at least 3 vertices.
  - 3D solids and bodies are converted to meshes.
  - AutoSurf and AME objects must be converted to meshes with the **SolMesh** command in AutoCAD Release 12.
  - AutoCAD blocks are exploded unless **Override** is turned on.
- The weld threshold distance:
  - **Minimum** 0.00 000 001
  - **Default** 0.001
  - **Maximum** 99,999,999
- Camera and light conversion:
  - AutoCAD named views become 3D Studio cameras.
  - AutoCAD point lights become 3D Studio 'omni lights.'

System Variables

APPENDIX A

AutoCAD stores information about the current state of itself, the drawing and the operating system in *system variables*. The variables help programmers – who often work with menu macros and AutoLISP – determine the state of the AutoCAD system.

TIPS

■ You get a list of system variables at the Command: prompt with the ? options of the **SetVar** command:

```
Command: setvar
Variable name or ?: ?
```

■ The **SetVar** command lets you can change the value of most variables.

■ *Italicized system variables* are not listed by the **SetVar ?** command nor in AutoCAD's *Command Reference*.

■ Some system variables have the same name as a command, such as Area; other variables do not work at the Command: prompt. These are prefixed by the ⌨ character.

■ **Default Value:** The table lists all known system variables, along with the default values as set in the Acad.Dwg prototype drawing.

■ **Ro:** Some system variables cannot be changed by the user or by programming; these are labeled "R/o" (*short for "read only"*) in the table below.

■ **Loc:** System variables are stored in a variety of places:

- **Acad** — AutoCAD executable (*hard-coded*)
- **Cfg** — Acad.Cfg or Acad.XmX files
- **Dwg** — Current drawing
- **...** — Not saved

Variable	Default	Ro	Loc	Meaning
_PKSER	*117-999999*	*R/o*	*Acad*	*Software package serial number*
_SERVER	*0*	*R/o*	*Cfg*	*Network authorization code*
A				
ACADPREFIX	"d:\ACAD13\"	R/o	...	Path spec'd by ACAD environment var
ACADVER	"13"	R/o	...	AutoCAD version number
AFLAGS	0	...	...	Attribute display code: 0 No mode specified 1 Invisible 2 Constant 4 Verify 8 Preset
ANGBASE	0	...	Dwg	Direction of zero degrees relative to UCS

Variable	Default	Ro	Loc	Meaning
ANGDIR	0	...	Dwg	Rotation of angles: 0 Clockwise 1 Counterclockwise
APERTURE	10	...	Cfg	Object snap aperture in pixels: 1 Minimum size 10 Default size 50 Maximum size
AREA	0.0000	R/o	...	Area measured by Area, List or Dblist
ATTDIA	0	...	Dwg	Attribute entry interface: 0 Command-line prompts 1 Dialogue box
ATTMODE	1	...	Dwg	Display of attributes: 0 Off 1 Normal 2 On
ATTREQ	1	...	Dwg	Attribute values during insertion are: 0 Default values 1 Prompt for values
AUDITCTL	0	...	Cfg	Determines creation of ADT audit log file: 0 File not created 1 ADT file created
AUNITS	0	...	Dwg	Mode of angular units: 0 Decimal degrees 1 Degrees-minutes-seconds 2 Grads 3 Radians 4 Surveyor's units
AUPREC	0	...	*Dwg*	*Dec*imals places displayed by angles
AUXSTAT	*0*	...	*Dwg*	*-32768 Minimum value* *32767 Maximum value*
AXISMODE	*0*	...	*Dwg*	*Obsolete system variable*
AXISUNIT	*0.0000*	...	*Dwg*	*Obsolete system variable*

B

Variable	Default	Ro	Loc	Meaning
BACKZ	0.0000	R/o	Dwg	Back clipping plane offset
BLIPMODE	1	...	Dwg	Display of blip marks: 0 Off 1 On

C

Variable	Default	Ro	Loc	Meaning
CDATE	19950105.15560660	R/o	...	Current date and time in YyyyMmDd.HhMmSsDd format
CECOLOR	"BYLAYER"	...	Dwg	Current entity color
CELTSCALE	1.0000	...	Dwg	Global linetype scale
CELTYPE	"BYLAYER"	...	Dwg	Current layer color
CHAMFERA	0.0000	...	Dwg	First chamfer distance
CHAMFERB	0.0000	...	Dwg	Second chamfer distance

Variable	Default	Ro	Loc	Meaning
CHAMFERC	0.0000	...	Dwg	Chamfer length
CHAMFERD	0	...	Dwg	Chamfer angle
CHAMMODE	0	...	..	Chamfer input mode: 0 Chamfer by two lengths 1 Chamfer by length and angle.
CIRCLERAD	0.0000	...	...	Most-recent circle radius: 0 No default
CLAYER	"0"	...	Dwg	Current layer name
CMDACTIVE	1	R/o	...	Type of current command: 1 Regular command 2 Transparent command 4 Script file 8 Dialogue box
CMDDIA	1	...	Cfg	Plot command interface: 0 Command line prompts 1 Dialogue box
CMDECHO	1	...	...	AutoLISP command display: 0 No command echoing 1 Command echoing
CMDNAMES	"SETVAR"	R/o	...	Current command
CMLJUST	0	...	Cfg	Multiline justification mode: 0 Top 1 Middle 2 Bottom
CMLSCALE	1.0000	...	Cfg	Scales width of multiline: -1 Flips offsets of multiline 0 Collapses to single line 1 Default 2 Doubles multiline width
CMLSTYLE	"STANDARD"	Cfg	...	Current multiline style name
COORDS	1	...	Dwg	Coordinate display style: 0 Updated by screen picks 1 Continuous display 2 Polar display upon request
CVPORT	2	...	Dwg	Current viewport number 2 Minimum (default)

D

Variable	Default	Ro	Loc	Meaning
DATE	2448860.54043252	R/o	...	Current date and fraction in Julian format
DBGLISTALL	*0*	...	...	*Toggle:*
DBMOD	0	R/o	...	Drawing modified in these areas: 0 No modification made 1 Entity database 2 Symbol table 4 Database variable 8 Window 16 View

Variable	Default	Ro	Loc	Meaning
DCTCUST	""	...	Cfg	Name of custom spelling dicitonary
DCTMAIN	"enu"	...	Cfg	Code for spelling dictionary: ca Catalan cs Czech da Danish de German - sharp 's' ded German - double 's' ena English - Australian ens English - British: 'ise' enu English - American enz English - British: 'ize' es Spanish - unaccented capitals esa Spanish - accented captitals fi Finish fr French - unaccented capitals fra French - accented captials it Italian nl Dutch - primary nls Dutch - secondary no Norwegian - Bokmal non Norwegian - Nynorsk pt Portuguese - Iberian ptb Portuguese - Brazilian ru Russian - infrequent 'io' rui Russian - frequent 'io' sv Swedish
DELOBJ	1	...	Dwg	Toggle source objects deletion: 0 Objects deleted 1 Objects retained
DIASTAT	1	R/o	...	User exited dialogue box by clicking on: 0 Cancel button 1 OK button

Dimension Variables

Variable	Default	Ro	Loc	Meaning
DIMALT	0	...	Dwg	Alternate units selected
DIMALTD	2	...	Dwg	Alternate unit decimal places
DIMALTF	25.4000	...	Dwg	Alternate unit scale factor
DIMALTTD	2	...	Dwg	Tolerance alternate unit decimal places
DIMALTTZ	0	...	Dwg	Alternate tolerance units zeros: 0 Zeros not suppressed 1 Zeros suppressed
DIMALTU	2	...	Dwg	Alternate units: 1 Scientific 2 Decimal 3 Engineering 4 Architectural 5 Fractional

Variable	Default	Ro	Loc	Meaning
DIMALTZ	0		Dwg	Zero suppresion for alternate units: 0 Zeros not suppressed 1 Zeros suppressed
DIMAPOST	“”	...	Dwg	Suffix for alternate text
DIMASO	1	...	Dwg	Create associative dimensions
DIMASZ	0.1800	...	Dwg	Arrow size
DIMAUNIT	0	...	Dwg	Angular dimension format: 0 Decimal degrees 1 Degrees.Minutes.Seconds 2 Grad 3 Radian 4 Surveyor units
DIMBLK	“”	R/o	Dwg	Arrow block name
DIMBLK1	“”	R/o	Dwg	First arrow block name
DIMBLK2	“”	R/o	Dwg	Second arrow block name
DIMCEN	0.0900	...	Dwg	Center mark size
DIMCLRD	0	...	Dwg	Dimension line color
DIMCLRE	0	...	Dwg	Extension line & leader color
DIMCLRT	0	...	Dwg	Dimension text color
DIMDEC	4	...	Dwg	Primary tolerance decimal places.
DIMDLE	0.0000	...	Dwg	Dimension line extension
DIMDLI	0.3800	...	Dwg	Dimension line continuation increment
DIMEXE	0.1800	...	Dwg	Extension above dimension line
DIMEXO	0.0625	...	Dwg	Extension line origin offset
DIMFIT	3	...	Dwg	Placement of text and arrowheads: 0 Between extension lines if possible 1 Text has priority over arrowheads 2 Whichever fits between ext lines 3 Whatever fits 4 Place text at end of leader line
DIMGAP	0.0900	...	Dwg	Gap from dimension line to text
DIMJUST	0	...	Dwg	Horizontal text positioning: 0 Center justify 1 Next to first extension line 2 Next to second extension line 3 Above first extension line 4 Above second extension line
DIMLFAC	1.0000	...	Dwg	Linear unit scale factor
DIMLIM	0	...	Dwg	Generate dimension limits
DIMPOST	“”	...	Dwg	Default suffix for dimension text
DIMRND	0.0000	...	Dwq	Rounding value
DIMSAH	0	...	Dwg	Separate arrow blocks
DIMSCALE	1.0000	...	Dwg	Overall scale factor
DIMSD1	Off	...	Dwg	Suppress first dimension line
DIMSD2	Off	...	Dwg	Suppress second dimension line
DIMSE1	0	...	Dwg	Suppress the first extension line
DIMSE2	0	...	Dwg	Suppress the second extension line
DIMSHO	1	...	Dwg	Update dimensions while dragging

Variable	Default	Ro	Loc	Meaning
DIMSOXD	0	...	Dwg	Suppress outside extension dimension
DIMSTYLE	"STANDARD"	R/o	Dwg	Current dimension style (read-only)
DIMTAD	0	...	Dwg	Place text above the dimension line
DIMTDEC	4	...	Dwg	Primary tolerance decimal places
DIMTFAC	1.0000	...	Dwg	Tolerance text height scaling factor
DIMTIH	1	...	Dwg	Text inside extensions is horizontal
DIMTIX	0	...	Dwg	Place text inside extensions
DIMTM	0.0000	...	Dwg	Minus tolerance
DIMTOFL	0	...	Dwg	Force line inside extension lines
DIMTOH	1	...	Dwg	Text outside extensions is horizontal
DIMTOL	0	...	Dwg	Generate dimension tolerances
DIMTOLJ	1	...	Dwg	Tolerance vertical justification: 0 Bottom 1 Middle 2 Top
DIMTP	0.0000	...	Dwg	Plus tolerance
DIMTSZ	0.0000	...	Dwg	Tick size
DIMTVP	0.0000	...	Dwg	Text vertical position
DIMTXSTY	"STANDARD"	...	Dwg	Dimension text style
DIMTXT	0.1800	...	Dwg	Text height
DIMTZIN	0	...	Dwg	Tolerance zero suppression
DIMUNIT	2	...	Dwg	Dimension unit format 1 Scientific 2 Decimal 3 Engineering 4 Architectural 5 Fractional
DIMUPT	Off	...	Dwg	User-positioned text: 0 Cursor positions dimension line 1 Cursor also positions text
DIMZIN	0	...	Dwg	Suppression of zero in feet-inches units: 0 Suppress 0 feet and 0 inches 1 Include 0 feet and 0 inches 2 Include 0 feet; suppress 0 inches 3 Suppress 0 feet; include 0 inches
DISPSILH	0	...	Dwg	Silhouette display of 3D solids: 0 Off 1 On
DISTANCE	0.0000	R/o	...	Distance measured by Dist command
DONUTID	0.5000	...	...	Inside radius of donut
DONUTOD	1.0000	...	...	Outside radius of donut
DRAGMODE	2	...	Dwg	Drag mode: 0 No drag 1 On if requested 2 Automatic
DRAGP1	10	...	Cfg	Regen drag display
DRAGP2	25	...	Cfg	Fast drag display

Variable	Default	Ro	Loc	Meaning
DWGCODEPAGE	"dos850"	Dwg	...	Drawing code page
DWGNAME	"UNNAMED"	R/o	...	Current drawing filename
DWGPREFIX	"d:\"	R/o	...	Drawing's drive and subdirectory
DWGTITLED	0	R/o	...	Drawing has filename: 0 "Untitled.Dwg" 1 User-assigned name
DWGWRITE	1	...	...	Drawing read-write status: 0 Read-only 1 Read-write

E

Variable	Default	Ro	Loc	Meaning
EDGEMODE	0	...	...	Toggle edge mode for Trim & Extend: 0 No extension 1 Extends cutting edge.
ELEVATION	0.0000	...	Dwg	Current elevation relative to current UCS
ENTMODS	*193*	*R/o*	...	
ERRNO	*0*	...	...	*Error number from AutoLISP,ADS,Arx*
EXPERT	0	...	...	Controls prompts: 0 Normal prompts 1 Supress these messages: "About to regen, proceed?" "Really want to turn the current layer off?" 2 Also suppress: "Block already defined. Redefine it?" "A block with this name already exists. Overwrite it?" 3 Also suppress messages related to the Linetype command. 4 Also suppress messages related to the UCS Save and VPorts Save commands. 5 Also suppress messages related to the DimStyle Save and DimOverride commands.
EXPLMODE	1	...	Dwg	Toggle whether Explode and Xplode commands explode non-uniformly scaled blocks: 0 Does not explode. 1 Does explode
EXTMAX	11.3706,10.0130,0,000	R/o	Dwg	Upper right coordinate of drawing extents
EXTMIN	1.0158,5.6333,0.000	R/o	Dwg	Lower left coordinate of drawing extents

Variable	Default	Ro	Loc	Meaning
F				
FACETRES	0.5	...	Dwg	Adjusts smoothness of shaded and hidden-line objects: 0.01 Minimum value 0.05 Default value 10.0 Maximum value
FFLIMIT	0	...	Cfg	Maximum number of PostScript and TrueType fonts loaded into memory: 0 No limit 1 One font 100 Maximum value
FILEDIA	1	...	Cfg	User interface: 0 Command-line prompts 1 Dialogue boxes (when available)
FILLETRAD	0.0000	...	Dwg	Current fillet radius
FILLMODE	1	...	Dwg	Fill of solid objects: 0 Off 1 On
FLATLAND	*0*	*R/o*	...	*Obsolete system variable*
FONTALT	"txt"	...	Cfg	Name for substituted font.
FONTMAP	""	...	Cfg	Name of font mapping file.
FORCE_PAGING	*0*	...	...	*0 Minimum (default)* *1,410,065,408 Maximum*
FRONTZ	0.0000	R/o	Dwg	Front clipping plane offset
G				
GLOBCHECK	*0*	...	...	*Reports statistics on dialogue boxes:* *0 Turn off* *1 Warns if larger than 640x400* *2 Also reports size in pixels* *3 Additional info*
GRIDMODE	0	...	Dwg	Display of grid: 0 Off 1 On
GRIDUNIT	0.0000,0.0000	...	Dwg	X,y-spacing of grid
GRIPBLOCK	0	...	...	Display of grips in blocks: 0 At insertion point 1 At all entities within block
GRIPCOLOR	5	...	Cfg	Color of unselected grips 1 Minimum color number 5 Default color: blue 255 Maximum color number
GRIPHOT	1	...	Cfg	Color of selected grips 1 Default: red 255 Maximum color number

Variable	Default	Ro	Loc	Meaning
GRIPS	1	...	Cfg	Display of grips: 0 Off 1 On
GRIPSIZE	3	...	Cfg	Size of grip box, in pixels 1 Minimum size 3 Default size 255 Maximum size

H

Variable	Default	Ro	Loc	Meaning
HANDLES	1	R/o	...	Obsolete system variable
HIGHLIGHT	1	...	...	Object selection highlighting: 0 Disabled 1 Enabled
HPANG	0	...	...	Current hatch pattern angle
HPBOUND	1	...	Dwg	Object created by BHatch and Boundary commands: 0 Polyline 1 Region
HPDOUBLE	0	...	...	Double hatching: 0 Disabled 1 Enabled
HPNAME	"ANSI31"	...	...	Current hatch pattern name "" No default . Set no default
HPSCALE	1.0000	...	...	Current hatch pattern scale factor
HPSPACE	1.0000	...	...	Current spacing of user-defined hatching

I

Variable	Default	Ro	Loc	Meaning
INSBASE	0.0000,0.0000,0.0000	...	Dwg	Insertion base point relative to current UCS
INSNAME	""	...	...	Current block name . Set to no default "" No default
ISOLINES	4	...	Dwg	Isolines on 3D solids: 0 Minimum 4 Default 16 Good-looking 2,047 Maximum

L

Variable	Default	Ro	Loc	Meaning
LASTANGLE	0	R/o	...	Ending angle of last-drawn arc
LASTPOINT	0.0000,0.0000,0.0000	...	Dwg	Last-entered point
LAZYLOAD	*0*	...	...	*Toggle 0 or 1*
LENSLENGTH	50.0000	R/o	Dwg	Perspective view lens length, in mm
LIMCHECK	0	...	Dwg	Drawing limits checking: 0 Disabled 1 Enabled

Variable	Default	Ro	Loc	Meaning
LIMMAX	12.0000,9.0000	...	Dwg	Upper right drawing limits
LIMMIN	0.0000,0.0000	...	Dwg	Lower left drawing limits
LOCALE	"en"	R/o		ISO language code
LOGINNAME	"??"	R/o	Dwg	User's login name
LTSCALE	1.0000	...	Dwg	Current linetype scale factor
LUNITS	2	...	...	Linear units mode: 1 Scientific 2 Decimal 3 Engineering 4 Architectural 5 Fractional
LUPREC	4	...	Dwg	Decimal places of linear units

M

Variable	Default	Ro	Loc	Meaning
MACROTRACE	*0*	...	...	*Diesel debug mode:* *0 Off* *1 On*
MAXACTVP	16	...	...	Maximum viewports to regenerate: 0 Minimum 16 Default 32767 Maximum
MAXSORT	200	...	Cfg	Maximum filenames to sort alphabetically: 0 Minimum 16 Default 32767 Maximum
MAXOBJMEM	2,147,483,647	...	...	Maximum number of objects in memory
MENUCTL	1	...	...	Submenu display: 0 Only with menu picks 1 Also with keyboard entry
MENUECHO	0	...		Menu and prompt echoing: 0 All prompts displayed 1 Suppress menu echoing 2 Suppress system prompts 4 Disable ^P toggle 8 Display all input-output strings
MENUNAME	"acad"	R/o	...	Current menu filename
MIRRTEXT	1	...	Dwg	Text handling during Mirror command: 0 Mirror text 1 Retain text orientation
MODEMACRO	""	...	...	Invoke Diesel programming language
MTEXTED	""	...	Cfg	External mtext editor

Variable	Default	Ro	Loc	Meaning
N				
NODENAME	*"AC$"*	*R/o*	*Cfg*	*Name of network node (1 to 3 chars)*
O				
OFFSETDIST	-1.0000	...	...	Current offset distance; Through mode, if negative
ORTHOMODE	0	...	Dwg	Orthographic mode: 0 Off 1 On
OSMODE	0	...	Dwg	Current object snap mode: 0 NONe 1 ENDpoint 2 MIDpoint 4 CENter 8 NODe 16 QUAdrant 32 INTersection 64 INSertion 128 PERpendicular 256 TANgent 512 NEARest 1024 QUIck 2048 APPint
P				
PDMODE	0	...	Dwg	Point display mode: 0 Dot 1 No display 2 +-symbol 3 x-symbol 4 Short line 32 Circle 64 Square
PDSIZE	0.0000	...	Dwg	Point display size, in pixels -1 Absolute size 0 5% of drawing area height +1 Percentage of viewport size
PELLIPSE	0	...	Dwg	Toggle Ellipse creation: 0 True ellipse 1 Polyline
PERIMETER	0.0000	R/o	...	Perimeter calculated by Area command
PFACEVMAX	4	R/o	...	Maximum vertices per 3D face
PHANDLE	*0*	...	...	*2,803,348,672 Maximum*
PICKADD	1	...	Cfg	Effect of [Shift] key on selection set: 0 Adds to selection set 1 Removes from selection set

Variable	Default	Ro	Loc	Meaning
PICKAUTO	1	...	Cfg	Selection set mode: 0 Single pick mode 1 Automatic windowing and crossing
PICKBOX	3	...	Cfg	Object selection pickbox size, in pixels
PICKDRAG	0	...	Cfg	Selection window mode: 0 Pick two corners 1 Pick 1 corner; drag to 2nd corner
PICKFIRST	1	...	Cfg	Command-selection mode: 0 Enter command first 1 Select objects first
PICKSTYLE	1	...	Dwg	Included groups and associative hatches in selection: 0 Neither included 1 Include groups 2 Include associative hatches 3 Include both
PLATFORM	"386 DOS Extender"	R/o	Acad	AutoCAD platform name: "386 DOS Extender" "DECstation" "Microsoft Windows" "Silicon Graphics Iris Indego" "Sun4/SPARCstation"
PLINEGEN	0	...	Dwg	Polyline linetype generation: 0 From vertex to vertex 1 From end to end
PLINEWID	0.0000	...	Dwg	Current polyline width
PLOTID	""	...	Cfg	Current plotter
PLOTROTMODE	1	...	Dwg	Orientation of plots: 0 Lowerleft = 0 1 Lowerleft plotter area = lowerleft of media
PLOTTER	1	...	Cfg	Current plotter configuration number: 0 No plotter configured 29 Maximum configurations
POLYSIDES	4	...	...	Current number of polygon sides: 3 Minimum sides 4 Default 1024 Maximum sides
POPUPS	1	R/o	...	Display driver support of AUI: 0 Not available 1 Available
PROJMODE	1	...	Cfg	Projection mode for Trim & Extend: 0 No projection 1 Project to x,y-plane of current UCS 2 Project to view plane
PSLTSCALE	1	...	Dwg	Paper space linetype scaling: 0 Use model space scale factor 1 Use viewport scale factor

Variable	Default	Ro	Loc	Meaning
PSPROLOG	“”	...	Cfg	PostScript prologue filename
PSQUALITY	75	...	Dwg	Resolution of PostScript display, in pixels:
				-n Display as outlines; no fill
				0 No display
				+n Display filled

Q

Variable	Default	Ro	Loc	Meaning
QAFLAGS	*1*	...	...	*Quality assurance flags*
QTEXTMODE	0	...	Dwg	Quick text mode:
				0 Off
				1 On

R

Variable	Default	Ro	Loc	Meaning
RASTERPREVIEW	0	...	Dwg	Preview image:
				0 BMP format
				1 BMP and WMF formats
				2 WMF format
				3 None saved.
REGENMODE	1	...	Dwg	Regeneration mode:
				0 Regen with each new view
				1 Regen only when required
RE-INIT		...	...	Reinitialize I/O devices:
				1 Digitizer port
				2 Plotter port
				4 Digitizer
				8 Plotter
				16 Reload PGP file
RIASPECT	1.0000	...	Rasterin	Raster image aspect ration
RIBACKG	0	...	Rasterin	Raster image background color
				0 Black; default
				7 White
				255 Maximum value
RIEDGE	0	...	Rasterin	Raster image edge detection mode:
				0 Off
				1 On
				255 Maximum value
RIGAMUT	256	...	Rasterin	Raster image gamut of colors
				8 Minimum value
				256 Maximum (default)
RIGREY	0	...	Rasterin	Raster image gray scale conversion:
				0 Off
				1 On
RITHRESH	0	...	Rasterin	Raster image brightness threshold
				0 Off
				255 Maximum threshold value

Variable	Default	Ro	Loc	Meaning

S

Variable	Default	Ro	Loc	Meaning
SAVEFILE	"AUTO.SV$"	R/o	Cfg	Automatic save filename
SAVENAME	""	R/o	...	Drawing save-as filename
SAVETIME	120	...	Cfg	Automatic save interval, in minutes 0 Disable auto save 120 Default
SCREENBOXES	26	R/o	Cfg	Maximum number of menu items 0 Screen menu turned off
SCREENMODE	0	R/o	Cfg	State of AutoCAD display screen: 0 Text screen 1 Graphics screen 2 Dual-screen display
SCREENSIZE	575.0000,423.0000	R/o	...	Current viewport size, in pixels
SHADEDGE	3	...	Dwg	Shade style: 0 Shade faces (256-color shading) 1 Shade faces; edges background color 2 Simulate hidden-line removal 3 16-color shading
SHADEDIF	70	...	Dwg	Percent of diffuse to ambient light 0 Minimum 70 Default 100 Maximum
SHPNAME	""	...	...	Current shape name Set to no default "" No default
SKETCHINC	0.1000	...	Dwg	Sketch command's recording increment
SKPOLY	0	...	Dwg	Sketch line mode: 0 Record as lines 1 Record as polylines
SNAPANG	0	...	Dwg	Current rotation angle for snap and gird
SNAPBASE	0.0000,0.0000	...	Dwg	Current origin for snap and grid
SNAPISOPAIR	0	...	Dwg	Current isometric drawing plane: 0 Left isoplane 1 Top isoplane 2 Right isoplane
SNAPMODE	0	...	Dwg	Snap mode: 0 Off 1 On
SNAPSTYL	0	...	Dwg	Snap style: 0 Normal 1 Isometric
SNAPUNIT	1.0000,1.0000	...	Dwg	X,y-spacing for snap

Variable	Default	Ro	Loc	Meaning
SORTENTS	96	...	Cfg	Entity display sort order: 0 Off 1 Object selection 2 Object snap 4 Redraw 8 Slide generation 16 Regeneration 32 Plots 64 PostScript output
SPLFRAME	0	...	Dwg	Polyline and mesh display: 0 Polyline control frame not displayed; Display polygon fit mesh; 3D faces invisible edges not displayed 1 Polyline control frame displayed; display polygon defining mesh; 3D faces invisible edges displayed
SPLINESEGS	8	...	Dwg	Number of line segments that define a splined polyline
SPLINETYPE	6	...	Dwg	Spline curve type: 5 Quadratic bezier spline 6 Cubic bezier spline
SURFTAB1	6	...	Dwg	Density of surfaces and meshes: 2 Minimum 6 Default 32766 Maximum
SURFTAB2	6	...	Dwg	Density of surfaces and meshes 2 Minimum 6 Default 32766 Maximum
SURFTYPE	6	...	Dwg	Pedit surface smoothing: 5 Quadratic bezier spline 6 Cubic bezier spline 8 Bezier surface
SURFU	6	...	Dwg	Surface density in m-direction 2 Minimum 6 Default 200 Maximum
SURFV	6	...	Dwg	Surface density in n-direction 2 Minimum 6 Default 200 Maximum
SYSCODEPAGE	"dos850"	R/o	Dwg	System code page

Variable	Default	Ro	Loc	Meaning
T				
TABMODE	0	...	...	Tablet mode: 0 Off 1 On
TARGET	0.0000,0.0000,0.0000	R/o	Dwg	Target in current viewport
TDCREATE	2448860.54014699	R/o	Dwg	Time and date drawing created
TDINDWG	0.00040625	R/o	Dwg	Duration drawing loaded
TDUPDATE	2448860.54014699	R/o	Dwg	Time and date of last update
TDUSRTIMER	0.00040694	R/o	Dwg	Time elapsed by user-timer
TEMPPREFIX	“”	R/o	...	Path for temporary files
TEXTEVAL	0	...	...	Interpretation of text input: 0 Literal text 1 Read (and ! as AutoLISP code
TEXTFILL	0	...	Dwg	Toggle fill of PostScript &TrueType fonts: 0 Outline text 1 Filled text
TEXTQLTY	50	...	Dwg	Resolution of PostScript &TrueType fonts: 0 Minimum resolution 50 Default 100 Maximum resolution
TEXTSIZE	0.2000	...	Dwg	Current height of text
TEXTSTYLE	“STANDARD”	...	Dwg	Current name of text style
THICKNESS	0.0000	...	Dwg	Current entity thickness
TILEMODE	1	...	Dwg	Viewport mode: 0 Display tiled viewports 1 Display overlapping viewports
TOOLTIPS	1	...	Cfg	Display tooltips (works in Windows only): 0 Off 1 On
TRACEWID	0.0500	...	Dwg	Current width of traces
TREEDEPTH	3020	...	Dwg	Maximum branch depth in xxyy format: *xx* Model-space nodes *yy* Paper-space nodes +*n* 3D drawing -*n* 2D drawing
TREEMAX	10000000	...	Cfg	Limits memory consumption during drawing regeneration
TRIMMODE	1			Trim toggle for Chamfer & Fillet: 0 Leave selected edges in place 1 Trim selected edges

Variable	Default	Ro	Loc	Meaning
U				
UCSFOLLOW	0	...	Dwg	New UCS views: 0 No change 1 Automatic display of plan view
UCSICON	1	...	Dwg	Display of UCS icon: 0 Off 1 On 2 Display at UCS origin, if possible
UCSNAME	""	R/o	Dwg	Name of current UCS view "" Current UCS is unnamed
UCSORG	0.0000,0.0000,0.0000	R/o	Dwg	Origin of current UCS relative to WCS
UCSXDIR	1.0000,0.0000,0.0000	R/o	Dwg	X-direction of current UCS relative to WCS
UCSYDIR	0.0000,1.0000,0.0000	R/o	Dwg	Y-direction of current UCS relative to WCS
UNDOCTL	5	R/o	...	State of undo: 0 Undo disabled 1 Undo enabled 2 Undo limited to one command 4 Auto-group mode 8 Group currently active
UNDOMARKS	0	R/o	...	Current number of undo marks
UNITMODE	0	...	Dwg	Units display: 0 As set by Units command 1 As entered by user
USERI1–I5	*0*	...	...	*Five user-definable integer variables*
USERR1–R5	*0.0000*	...	...	*Five user-definable real variables*
USERS1–S5	*""*	...	...	*Five user-definable string variables*
V				
VIEWCTR	6.2433,4.5000,0.0000	R/o	Dwg	X,y-coordinate of center of current view
VIEWDIR	0.0000,0.0000,1.0000	R/o	Dwg	Current view direction relative to UCS
VIEWMODE	0	R/o	Dwg	Current view mode: 0 Normal view 1 Perspective mode on 2 Front clipping on 4 Back clipping on 8 UCS-follow on 16 Front clip not at eye
VIEWSIZE	9.0000	R/o	Dwg	Height of current view
VIEWTWIST	0	R/o	Dwg	Twist angle of current view

Variable	Default	Ro	Loc	Meaning
VISRETAIN	0	...	Dwg	Determination of xref drawing's layers: 0 Current drawing 1 Xref drawing
VSMAX	37.4600,27.00,0.00	R/o	Dwg	Upper right corner of virtual screen
VSMIN	-24.9734,-18.00,0.00	R/o	Dwg	Lower left corner of virtual screen

W

Variable	Default	Ro	Loc	Meaning
WORLDUCS	1	R/o	...	Matching of WCS with UCS: 0 Current UCS is not WCS 1 UCS is WCS
WORLDVIEW	1	...	Dwg	Display during Dview &Vpoint commands: 0 Display UCS 1 Display WCS

X

Variable	Default	Ro	Loc	Meaning
XREFCTL	0	...	Cfg	Determines creation of XLG xref log files: 0 File no written 1 XLG file written

Obsolete Commands

The following commands have been removed from AutoCAD DOS over the years:

Command	Introduced	Removed	Replacement	Reaction
3dLine	R9	R11	Line	"Line"
AmeLite	R11	R12	Region	"Unknown command"
AscText	R11	R13	MText	"Unknown command"
Ase...	R12	R12	ASE...	"Unknown command"
	(Most R12 ASE commands were combined into fewer ASE commands.)			
Axis	v1.4	R12	*none*	"Discontinued command."
DL, DLine	R11	R13	MLine	"Unknown command"
EndRep	v1.0	v2.5	MInsert	"Discontinued command."
EndSv	v2.0	v2.5	End	"End"
Filmroll	v2.6	R13	*none*	"Unknown command"
Flatland	R10	R11	*none*	"Cannot set Flatland to that value."
IgesIn,IgesOut	v2.5	R13	*none*	"Discontinued command"
PrPlot	v2.1	R12	Plot	"Discontinued command."
QPlot	v1.1	v2.0	SaveImg	*no reaction*
Repeat	v1.0	v2.5	MInsert	"Discontinued command."
Snapshot	v2.0	v2.1	SaveImg	"Unknown command."
Sol...	R11, R12	R13	*ACIS-based solid modelling commands.*	
	(AME commands lost their SOL-prefix.)			

TIPS

■ In addition to the commands listed above, the following features (found in Release 12) were dropped from Release 13:

- **SolChP**, the ability to edit primitives in a solid model.
- **SolMat**, to assign a material property to a solid model, including density.
- Export of AutoShade's RND (*ReNDer*) and import of AutoShade's FLM (*FiLmRoll*) file formats.
- The "Release 11" hidden-line removal algorithm.
- Freeplot, starting AutoCAD with -p.
- **AcsText** command, for ASCII text file import.
- Some device drivers, including a basic VGA display driver.
- **Handles** toggle is now always turned on.

Topical Index

A

C

D

M

O

P

Q

Tools Menu *(continued)*

V

View Menu:

W

Windows Elements:

Dear Reader,

I would like to take this opportunity to extend to you an invitation to subscribe to the *CAD++ Newsletter*. *CAD++* is an exciting new concept that delivers:

- CAD industry news, and interviews with CAD vendor representatives.
- Analysis (and predictions) of CAD industry trends.
- CAD software programming tips.
- Issues in CAD data exchange.
- Tips on starting and running your own CAD business.

Every month, subscribers in 14 countries and at seven of the top CAD companies read articles by Cliff Jennings (Fitting Solutions), Scott Taylor (Tailor Made Software), Jake Richter (Panacea), Dietmar Rudolph (CR/LF GmbH), and other leaders of the CAD industry.

I look forward to having you join our readership!

Ralph Grabowski
Author, *The Illustrated AutoCAD Quick Reference*
Editor, *CAD++ Newsletter*

Special Subscription Offer

Save 20% Today!
This discount is available to readers of *The Illustrated AutoCAD Quick Reference:*

- ☐ $240.00 Corporations (*more than 20 employees*) and libraries.
  The corporate rate is higher because the pass-along readership rate is higher. Think of this as a site licence.
- ☐ $ 48.00 Individual subscribers (*includes the 20% discount*).
- ☐ $ 5.00 Back issues, each (*many are still available*).
- ☐ $ 18.00 Additional postage for addresses outside of North America.

Payment Instructions
Please make out checks, money orders, and purchase orders to XYZ Publishing, Ltd..
No Visa, Master Card, or AmEx credit card orders, please! Mail your order to:

XYZ Publishing, PO Box 3053, Sumas WA 98295-3053, USA

Corporations can fax their order (*with purchase order number*) to +1 (604) 859-9597, 24 hours a day, 7 days a week.

International Orders
Canadians addresses remit in Canadian funds; please add 7% GST (*R138245949*).
Outside North America: please remit in US funds drawn on a bank with an American address. Most international banks have a US branch (or have an arrangement with an American bank) that allows you to have the check drawn on the US bank address.